MW00844817

Paul A. Tres

Designing Plastic Parts for Assembly

Paul A. Tres

Designing Plastic Parts for Assembly

8th Edition updated

HANSER

Hanser Publishers, Munich

Hanser Publications, Cincinnati

The Author:

Paul A. Tres, ETS Inc.
www.ets-corp.com
P.O. Box 7747
Bloomfield Hills MI 48302-7747

Distributed in North and South America by:
Hanser Publications
6915 Valley Avenue, Cincinnati, Ohio 45244-3029, USA
Fax: (513) 527-8801
Phone: (513) 527-8977
www.hanserpublications.com

Distributed in all other countries by
Carl Hanser Verlag
Postfach 86 04 20, 81631 München, Germany
Fax: +49 (89) 98 48 09
www.hanser-fachbuch.de

The use of general descriptive names, trademarks, etc., in this publication, even if the former are not especially identified, is not to be taken as a sign that such names, as understood by the Trade Marks and Merchandise Marks Act, may accordingly be used freely by anyone. While the advice and information in this book are believed to be true and accurate at the date of going to press, neither the authors nor the editors nor the publisher can accept any legal responsibility for any errors or omissions that may be made. The publisher makes no warranty, express or implied, with respect to the material contained herein.

Author's Disclaimer:
The data and information presented in this book have been collected by the authors and publisher from many sources that are believed to be reliable. However, the authors and publisher make no warranty, expressed or implied, to this book's accuracy or completeness. No responsibility or liability is assumed by the authors and publisher for any loss or damage suffered through reliance on any information presented in this book.
The authors do not purport to give any toxicity or safety information.

The final determination of the suitability of any information for the use contemplated for a given application remains the sole responsibility of the user.

Cataloging-in-Publication Data is on file with the Library of Congress

Bibliografische Information Der Deutschen Bibliothek
Die Deutsche Bibliothek verzeichnet diese Publikation in der Deutschen Nationalbibliografie;
detaillierte bibliografische Daten sind im Internet über <http://dnb.d-nb.de> abrufbar.

ISBN 978-1-56990-668-2
E-Book ISBN 978-1-56990-669-9

All rights reserved. No part of this book may be reproduced or transmitted in any form or by any means, electronic or mechanical, including photocopying or by any information storage and retrieval system, without permission in writing from the publisher.

The illustration for the cover is provided by Daimler Trucks North America (DTNA) of Portland, Oregon, United States of America. It shows the new Freightliner Cascadia Truck undergoing aerodynamic testing in DTNA's proprietary full-scale wind tunnel.

© Carl Hanser Verlag, Munich 2017
Editor: Mark Smith
Production Management: Jörg Strohbach
Coverconcept: Marc Müller-Bremer, www.rebranding.de, München, Germany
Coverdesign: Stephan Rönigk
Typesetting: Kösel Media GmbH, Krugzell, Germany
Printed and bound by Hubert & Co GmbH, Göttingen, Germany
Printed in Germany

Foreword to the Eighth Edition

In this 8th edition of *Designing Plastic Parts for Assembly*, the addition of the "*Fasteners*" chapter provides a great update to an already indispensable resource for design engineers. Paul Tres starts with an introduction to plastics, covering many topics from the basics of resin to what makes up a plastic, to a variety of plastic properties. This is fundamental for anyone working with plastics and Paul presents the information in a concise, organized manner. By providing the basis of plastic materials, he is able to lay the foundation for solid design. Of particular interest to DENSO as a global automotive supplier, Paul covers both the principle and detail of safety factors, essential in today's automotive environment as well as within other industries. Additionally, an overview of plastic material strength is provided, which serves as both a refresher for some, and a detailed calculation guide for all.

The main focus of the book, of course, is assembly techniques and the needed design considerations for plastic parts. Clear explanations for numerous types of joining techniques and the desired design features to ensure appropriate product quality can be easily found by the reader, while the design engineer is also provided with a general understanding of the manufacturing process. Paul covers in detail the design of press fits, snap fits, and living hinges. In particular, Paul's approach to the case studies based on a "right way" and a "wrong way" allows for ease of understanding from the novice engineer through to management. The book does a great job in illustrating real-life examples and case studies to make the key points clear for nearly every topic that Paul presents. The easy-to-follow text provides a standardized approach and step-by-step calculation methods and examples for a wide variety of assembly techniques.

As a global manufacturer of numerous automotive parts, DENSO utilizes many of the topics presented in this book, which has been a go-to resource, along with Paul's local training class. In addition to the book, Paul's supplemental training course has been of benefit to numerous DENSO engineers. While Paul's enthusiasm during the

Automotive Plastic Part Design class is difficult to put to paper, the book *Designing Plastic Parts for Assembly* incorporates the essence of Paul's enthusiasm for this subject. In all, Paul's explanation and expertise in this area is highly recommended for those involved in plastic product development.

Southfield, Michigan *Douglas E. Patton*

2017 President–Society of Automotive Engineers (SAE)

Executive Vice President/Chief Technical Officer DENSO International America, Inc.

Preface to the Eighth Edition

The world of plastic products has been growing at an average rate of over five percent per year in the last decades, which makes plastics by far the fastest growing material class on the globe. This growth is related to the versatile properties of polymers and plastics. Perhaps some of the most important aspects contributing to this spectacular growth are design and assembly. Making sure that plastic parts are designed optimally for assembly has become an intrinsic and fundamental part of engineering of new products. Paul Tres has done a tremendous job in bringing together in one concise book all the technical aspects of this process.

The book is a "must" for all those engineers, developers, and designers who are working on tomorrow's products. The detailed presentations provide many examples, clearly illustrated with succinct yet very effective drawings. This makes the book the ultimate outline of the perfect syllabus for all engineering classes in plastics and product development programs at any university.

At SPE, we know that we have to educate the OEMs and their product developers on the great potential and opportunities there are in designing and combining plastic parts with each other or with other materials. Paul Tres' *Designing Plastic Parts for Assembly* should be mandatory literature for all of them.

I truly believe that this book can open more doors and add opportunities in our plastics industry. The book is very well conceived and therefore experienced plastics engineers, as well as product developers (who might have only limited knowledge of the properties and behavior of plastics), can use it as a perfect, extremely helpful synopsis.

Ronse, East Flanders, Belgium

Wim De Vos
Chief Executive Officer
Society of Plastics Engineers (SPE)

Foreword to the First Edition

Knowing well the work and many special talents of Paul A. Tres, I take delight in the opportunity to introduce his new book, Designing Plastic Parts for Assembly, and recommend it to a broad range of readers. Material engineers, design and manufacturing engineers, graduate and under-graduate students, and all others with an interest in design for assembly or plastic components development now have a clearly written, method-oriented resource.

This practical book is an outgrowth of the like-named University of Wisconsin–Madison course which is being offered nationally and internationally. Just as his lectures in the course provide a detailed yet simplified discussion of material selection, manufacturing techniques, and assembly procedures, this book will make his unique expertise and effective teaching method available to a much larger audience.

Mr. Tres' highly successful instructional approach is evident throughout the book. Combining fundamental facts with practical techniques and a down-to-earth philosophy, he discusses in detail joint design and joint purpose, the geometry and nature of the component parts, the type of loads involved, and other vital information crucial to success in this dynamic field. Treatment of this material is at all times practice-oriented and focuses on everyday problems and situations.

In addition to plastics, Mr. Tres has expert knowledge in computer software, having directed the development of DuPont's design software. The course at the University of Wisconsin–Madison is indirectly an outgrowth of the software he designed for living hinges and snap fits at DuPont.

Mr. Tres holds numerous patents in the plastics field. He is known worldwide for his expertise in computer programming, manufacturing processes, material selection and project management on both a national and international scale.

Most recently, Mr. Tres' accomplishments have earned him the DuPont Automotive Marketing Excellence Award as well as recognition in the 1994–1995 edition of Who's Who Worldwide.

Whether you are just entering the field, or are a seasoned plastic parts designer, Designing Plastic Parts for Assembly is an excellent tool that will facilitate cost-effective design decisions, and help to ensure that the plastic parts and products you design stand up under use.

Madison, WI

Dr. Donald E. Baxa

University of Wisconsin-Madison

Preface to the First Edition

It gives me great pleasure to write this preface for such an important contribution to engineering design. It is rather sad fact that while the creative use of plastics has changed the very structure of consumer products over the past decade, many engineering students graduate with very little knowledge of polymer engineering or plastic design principles. This book written by a recognized expert and practitioner in the field of plastic component design is both a valuable text for engineering courses and a resource for practicing design engineers.

The full potential for the use of plastics in consumer products became recognized in the mid 1980s through the pioneering development of the IBM ProPrinter. The ProPrinter destroyed the myth, prevalent amongst product engineers at that time, that such design elements as plastic springs, plastic bearings, plastic securing elements, etc., lacked the structural integrity of their more common metal counterparts. In the ProPrinter, not only were these plastic design features shown to have the required reliability in regular use and abuse, they were combined into single parts to produce a new level of design elegance. For example, the injection molded side-frames of the ProPrinter, which support the rollers and lead screw, incorporated bearings for all of these rotating members, springs to maintain the required paper pressure, and cantilever securing elements to allow the frames to be snap fitted into the base. The result of such innovative design details produced a desktop printer which could be assembled in only 32 final assembly steps compared to the 185 steps required to assemble its main competitor in the marketplace.

Since the emergence of the ProPrinter, smart plastic design has become an essential tool in the competitive battle to produce products which have simpler structures with smaller numbers of discrete parts. Part count reduction, in particular, has been shown, through numerous case studies published over the past five years, to have a ripple effect on product manufacture which improves the efficiency of the entire organization. Fewer parts mean fewer manufacturing and assembly steps, and fewer joints and interfaces, all of which have a positive effect on quality and reliability. Moreover, a reduction in the number of the parts results in a direct attack on the hidden or overhead cost of an organization. Thus, fewer parts also mean fewer vendors for purchasing to deal with, less documentation, smaller inventory levels, less inspection, simpler production scheduling and so on.

Designing Plastic Parts for Assembly tackles all of the important issues to be faced in designing multi-feature complex plastic parts. The book is thus much more than its title suggests. It deals with essential fundamentals for the development of competitive consumer products.

Providence, Rhode Island

Dr. Peter Dewhurst
Department of Industrial and Manufacturing Engineering
University of Rhode Island

Acknowledgments

A special thank you to Tiffany and David Karrow, parents of seven-month-old baby girl Abigail, who passed away due to the product failure of a plastic component.

The author thanks Jennifer Edwards, Manager, On-Highway Marketing & Communications, with *Daimler Trucks North America of Fort Mill, South Carolina,* Katherine Tucker, Chassis Engineering Team Leader, David Meyer, Senior Design Engineer, Cab Engineering, Molded Vehicle Panels, and Kevin Neale, Manager, Cab Engineering, Molded Vehicle Panels Team with Daimler Trucks North America of Portland, Oregon, who made it possible to use the new Cascadia truck for the 8th edition book cover.

A special thank you to Plante Renald, director, and Martin Ethier, product manager with BRP (Bombardier Recreational Products) of Valcourt, Quebec, Canada, who made it possible to use the Can-Am Roadster Spider RT Limited with its well-known Ace 1330 high-torque Rotax engine on the cover of the 7th edition.

The author thanks Giles & Lucas Dillingham with Brighton Technologies Group in the United States, Jörg Neukum with DILAS Diodenlaser GmbH in Germany, Hervé Lamy of Scholle Packaging in France, Louis Lundberg with 3M Company in the United States, and Astrid Hermann with Herrmann Ultraschalltechnik GmbH & Co. KG in Germany for their contributions to the 7th edition of the book.

The author also gratefully thanks Tom Matano, excutive director with the School of Industrial Design at the Academy of Art University in San Francisco, California, who made it possible to use the futuristic car designs on the cover for the 6th edition, which were created by his students John Liu and John Lazorack. A special thank you to Horia Blendea with Schukra North America, a division of Leggett & Platt in Canada, and Dr. Andrew D'Souza with 3M Company in the United States for their contributions to the 6th edition of the book.

A special thank you to Dr. Jessica Schroeder, Lisa Stanick, and Mike Meyerand, all three with General Motors LLC, who made it possible to use Chevrolet Corvette C5 see-through image on the cover of the 5th edition.

I also thank BASF Corporation, Borg-Warner Automotive, Bronson Ultrasonic (a business of Emerson Electric Company), CAMPUS Consortium, Fiat Chrysler Automotive Group, Dukane Corporation, ETS Inc., Forward Technology Industries Inc., Hewlett-Packard Company, Corvalis Division, Leister Corporation, Miniature Precision Components, Inc., Molding Technology, Inc., Sonics and Materials, Inc., Solvay Automotive, and TWI, who supplied illustrations for this book.

Contents

Foreword to the Eighth Edition V

Preface to the Eighth Edition VII

Foreword to the First Edition IX

Preface to the First Edition XI

Acknowledgments XIII

1 Understanding Plastic Materials 1

1.1 Basic Resins 1
1.1.1 Thermoplastics 1
1.1.2 Thermosets 2
1.2 Basic Structures 2
1.2.1 Crystalline 2
1.2.2 Amorphous 3
1.2.3 Liquid Crystal Polymer (LCP) 4
1.2.4 New Polymer Technologies 4
1.2.4.1 Inherently Conductive Polymers (ICP) 4
1.2.4.2 Electro-Optic Polymers (EOP) 5
1.2.4.3 Biopolymers 6
1.3 Homopolymer vs. Copolymer 7
1.4 Reinforcements 7
1.5 Fillers 8
1.5.1 Glass Spheres 8
1.5.1.1 Microsphere Properties 10
1.5.1.2 Compounding 10
1.5.1.3 Injection Molding 11
1.5.1.4 Mechanical Properties in Injection-Molded Thermoplastic Applications 11
1.6 Additives 13
1.7 Physical Properties 14
1.7.1 Density and Specific Gravity 14
1.7.2 Elasticity 15
1.7.2.1 Case History: Elasticity and Denier 16

1.7.3 Plasticity 18
1.7.4 Ductility 18
1.7.5 Toughness 19
1.7.6 Brittleness 19
1.7.7 Notch Sensitivity 20
1.7.8 Isotropy 24
1.7.9 Anisotropy 24
1.7.10 Water Absorption 24
1.7.11 Mold Shrinkage 25
1.8 Mechanical Properties 27
1.8.1 Normal Stress 27
1.8.2 Normal Strain 27
1.8.3 Stress-Strain Curve 28
1.9 Creep 30
1.9.1 Introduction 30
1.9.2 Creep Experiments 30
1.9.3 Creep Curves 31
1.9.4 Stress-Relaxation 33
1.10 Impact Properties 33
1.11 Thermal Properties 34
1.11.1 Melting Point 35
1.11.2 Glass Transition Temperature 35
1.11.3 Heat Deflection Temperature 35
1.11.4 Coefficient of Thermal Expansion 35
1.11.5 Thermal Conductivity 38
1.11.6 Thermal Influence on Mechanical Properties 38
1.11.7 Case History: Planetary Gear Life Durability 39

2 Understanding Safety Factors 45
2.1 What Is a Safety Factor 45
2.2 Using the Safety Factors 46
2.2.1 Design Safety Factors 46
2.2.1.1 Design Static Safety Factor 46
2.2.1.2 Design Dynamic Safety Factor 46
2.2.1.3 Design Time-Related Safety Factor 46
2.2.2 Material Properties Safety Factor 47
2.2.3 Processing Safety Factors 48
2.2.4 Operating Condition Safety Factor 48

3 Strength of Material for Plastics 49
3.1 Tensile Strength 49
3.1.1 Proportional Limit 50
3.1.2 Elastic Stress Limit 50
3.1.3 Yield Stress 51
3.1.4 Ultimate Stress 51
3.2 Compressive Stress 52

3.3 Shear Stress ... 53
3.4 Torsion Stress ... 54
3.5 Elongations ... 55
3.5.1 Tensile Strain ... 55
3.5.2 Compressive Strain ... 56
3.5.3 Shear Strain ... 56
3.6 True Stress and Strain vs. Engineering Stress and Strain ... 57
3.7 Poisson's Ratio ... 58
3.8 Modulus of Elasticity ... 60
3.8.1 Young's Modulus ... 60
3.8.2 Tangent Modulus ... 60
3.8.3 Secant Modulus ... 61
3.8.4 Creep (Apparent) Modulus ... 62
3.8.5 Shear Modulus ... 62
3.8.6 Flexural Modulus ... 63
3.8.7 The Use of Various Moduli ... 64
3.9 Stress Relations ... 64
3.9.1 Introduction ... 64
3.9.2 Experiment ... 65
3.9.3 Equivalent Stress ... 65
3.9.4 Maximum Normal Stress ... 65
3.9.5 Maximum Normal Strain ... 66
3.9.6 Maximum Shear Stress ... 66
3.9.7 Maximum Deformation Energy ... 67
3.10 ABCs of Plastic Part Design ... 68
3.10.1 Constant Wall ... 68
3.10.2 Fillets ... 70
3.10.3 Boss Design ... 72
3.10.4 Rib Design ... 73
3.10.5 Case History: Ribs ... 75
3.11 Conclusions ... 78

4 Nonlinear Considerations ... 79
4.1 Material Considerations ... 79
4.1.1 Linear Material ... 79
4.1.2 Nonlinear Materials ... 79
4.2 Geometry ... 80
4.2.1 Linear Geometry ... 80
4.2.2 Nonlinear Geometry ... 81
4.3 Finite Element Analysis (FEA) ... 81
4.3.1 FEA Method Application ... 81
4.3.2 Using FEA Method ... 82
4.3.3 Most Common FEA Codes ... 82
4.4 Conclusions ... 83

5 Welding Techniques for Plastics ... 85
5.1 Ultrasonic Welding ... 85
5.1.1 Ultrasonic Equipment ... 85
5.1.2 Horn Design ... 89
5.1.3 Ultrasonic Welding Techniques ... 91
5.1.4 Control Methods ... 94
5.1.4.1 Common Issues with Welding ... 98
5.1.4.2 Joint Design ... 101
5.1.4.3 Butt Joint Design ... 102
5.1.4.4 Shear Joint Design ... 103
5.1.4.5 Torsional Ultrasonic Welding ... 106
5.1.4.6 Case History: Welding Dissimilar Polymers ... 108
5.2 Ultrasonic (Heat) Staking ... 112
5.2.1 Standard Stake Design ... 113
5.2.2 Flush Stake Design ... 114
5.2.3 Spherical Stake Design ... 115
5.2.4 Hollow (Boss) Stake Design ... 115
5.2.5 Knurled Stake Design ... 116
5.3 Ultrasonic Spot Welding ... 118
5.4 Ultrasonic Swaging ... 118
5.5 Ultrasonic Stud Welding ... 119
5.6 Spin Welding ... 119
5.6.1 Process ... 120
5.6.2 Equipment ... 123
5.6.3 Welding Parameters ... 123
5.6.4 Joint Design ... 125
5.7 Hot Plate Welding ... 128
5.7.1 Process ... 130
5.7.2 Joint Design ... 131
5.8 Vibration Welding ... 134
5.8.1 Process ... 136
5.8.2 Equipment ... 138
5.8.3 Joint Design ... 139
5.8.4 Common Issues with Vibration Welding ... 142
5.9 Electromagnetic Welding ... 144
5.9.1 Equipment ... 145
5.9.2 Process ... 145
5.9.3 Joint Design ... 146
5.10 Radio Frequency (RF) Welding ... 148
5.10.1 Equipment ... 149
5.10.2 Process ... 149
5.11 Laser Welding ... 151
5.11.1 Equipment ... 152
5.11.2 Process ... 153
5.11.3 Noncontact Welding ... 154

5.11.4 Transmission Welding ... 155
5.11.5 Intermediate Film & ClearWeld™ Welding ... 160
5.11.6 Polymers ... 162
5.11.7 Applications ... 162
5.12 Conclusion ... 167

6 Press Fitting ... 169
6.1 Introduction ... 169
6.2 Definitions and Notations ... 170
6.3 Geometric Definitions ... 170
6.4 Safety Factors ... 171
6.5 Creep ... 171
6.6 Loads ... 172
6.7 Press Fit Theory ... 173
6.8 Design Algorithm ... 175
6.9 Case History: Plastic Shaft and Plastic Hub ... 176
6.9.1 Shaft and Hub Made of Different Polymers ... 176
6.9.2 Safety Factor Selection ... 176
6.9.3 Material Properties ... 177
6.9.4 Shaft Material Properties at 23°C ... 179
6.9.4.1 Shaft Material Properties at 93°C ... 181
6.9.4.2 Creep Curves at 23°C ... 181
6.9.4.3 Creep at 93°C ... 183
6.9.4.4 Pulley at 23°C ... 184
6.9.4.5 Pulley at 93°C ... 187
6.9.4.6 Creep, Pulley at 23°C ... 188
6.9.4.7 Creep, Pulley at 93°C ... 189
6.10 Solutions: Plastic Shaft, Plastic Hub ... 190
6.10.1 Case A ... 190
6.10.2 Case B ... 192
6.10.3 Case C ... 193
6.10.4 Case D ... 194
6.11 Case History: Metal Ball Bearing and Plastic Hub ... 195
6.11.1 Fusible Core Injection Molding ... 195
6.11.2 Upper Intake Manifold Background ... 197
6.11.3 Design Algorithm ... 200
6.11.4 Material Properties ... 201
6.11.4.1 CAMPUS ... 202
6.11.5 Solution ... 204
6.11.5.1 Necessary IF at Ambient Temperature ... 209
6.11.5.2 IF Available at 118°C ... 210
6.11.5.3 IF Verification at -40°C ... 210
6.11.5.4 Verification of Stress Level at -40°C, Time = 0 ... 211
6.11.5.5 Stress Level at -40°C, Time = 5,000 h ... 211

6.11.5.6 Stress Level at 23°C, Time = 5,000 h 212
6.11.5.7 Stress Level at 118°C, Time = 5,000 h 212
6.12 Successful Press Fits 213
6.13 Conclusion 217

7 Living Hinges 219
7.1 Introduction 219
7.2 Classic Design for PP and PE 220
7.3 Common Living Hinge Design 221
7.4 Basic Design for Engineering Plastics 222
7.5 Living Hinge Design Analysis 222
7.5.1 Elastic Strain Due to Bending 223
7.5.1.1 Assumptions 223
7.5.1.2 Geometric Conditions 224
7.5.1.3 Strain Due to Bending 224
7.5.1.4 Stress Due to Bending 224
7.5.1.5 Closing Angle of the Hinge 225
7.5.1.6 Bending Radius of the Hinge 225
7.5.2 Plastic Strain Due to Pure Bending 226
7.5.2.1 Assumptions 226
7.5.2.2 Strain Due to Bending 226
7.5.3 Plastic Strain Due to a Mixture of Bending and Tension 227
7.5.3.1 Tension Strain 228
7.5.3.2 Bending Strain 230
7.5.3.3 Neutral Axis Position 231
7.5.3.4 Hinge Length 231
7.5.3.5 Elastic Portion of the Hinge Thickness 234
7.6 Computer Flow Chart 235
7.6.1 Computer Notations 235
7.7 Computer Flow Chart Equations 237
7.8 Example: Case History 239
7.8.1 World-Class Connector 239
7.8.1.1 Calculations for the “Right Way” Assembly 240
7.8.1.2 Calculations for the “Wrong Way” Assembly 242
7.8.2 Comparison Material 243
7.8.2.1 “Right Way” Assembly 244
7.8.2.2 “Wrong Way” Assembly 245
7.8.3 Ignition Cable Bracket 245
7.8.3.1 Initial Design 246
7.8.3.2 Improved Design 247
7.9 Processing Errors for Living Hinges 248
7.10 Coined Hinges 250
7.11 Oil-Can Designs 253
7.12 Conclusion 255
7.13 Exercise 255

8 **Snap Fitting** ... 261
8.1 Introduction ... 261
8.2 Material Considerations ... 262
8.3 Design Considerations ... 265
8.3.1 Safety Factors ... 267
8.4 Snap Fit Theory ... 268
8.4.1 Notations ... 268
8.4.2 Geometric Conditions ... 270
8.4.3 Stress/Strain Curve and Formulae ... 271
8.4.4 Instantaneous Moment of Inertia ... 272
8.4.5 Angle of Deflection ... 273
8.4.6 Integral Solution ... 273
8.4.7 Equation of Deflection ... 275
8.4.8 Integral Solution ... 275
8.4.9 Maximum Deflection ... 276
8.4.10 Self-Locking Angle ... 279
8.5 Case History: One-Way Continuous Beam with Rectangular Cross Section ... 279
8.5.1 Geometrical Model ... 281
8.6 Annular Snap Fits ... 284
8.6.1 Case History: Annular Snap Fit, Rigid Beam with Soft Mating Part ... 285
8.6.2 Notations ... 285
8.6.3 Geometric Definitions ... 286
8.6.4 Material Selections and Properties ... 287
8.6.5 Basic Formulas ... 287
8.6.6 Angle of Assembly ... 289
8.6.7 Case History: Digital Wristwatch ... 289
8.7 Torsional Snap Fits ... 295
8.7.1 Notations ... 295
8.7.2 Basic Formulae ... 297
8.7.3 Material Properties ... 298
8.7.4 Solution ... 298
8.8 Case History: Injection Blow Molded Bottle Assembly ... 300
8.9 Tooling ... 302
8.10 Case History: Snap Fits That Kill ... 303
8.11 Assembly Procedures ... 307
8.12 Issues with Snap Fitting ... 310
8.13 Serviceability ... 311
8.14 Exercise ... 311
8.14.1 Solution ... 313
8.15 Conclusions ... 317

9 **Bonding** ... 319
9.1 Failure Theories ... 319
9.2 Surface Energy ... 320

9.3 Surface Treatment ... 324
9.4 Types of Adhesives ... 327
9.5 Advantages and Limitations of Adhesives ... 329
9.6 Stress Cracking in Bonded Joints of Adhesives ... 330
9.7 Joint Design ... 331
9.8 Conclusion ... 334

10 In-Mold Assembly ... 335
10.1 Overmolding ... 336
10.2 In-Mold Assembly ... 337
10.3 Joint Design ... 338
10.4 Tool Design ... 341
10.5 Case Histories: Automotive IMA ... 347
10.6 Conclusion ... 349

11 Fasteners ... 351
11.1 Thread Forming ... 352
11.2 Case History: Automotive Undercarriage Splash Shield ... 362
11.3 Thread Cutting ... 368
11.4 Conclusion ... 369

Appendix A: Enforced Displacement ... 371

Appendix B: Point Force ... 379

Appendix C: Molding Process Data Record ... 387

Appendix D: Tool Repair & Inspection Record ... 389

References ... 391

World Wide Web References Related to Plastic Part Design ... 401

About the Author Paul A. Tres ... 409

Index ... 411

1

Understanding Plastic Materials

1.1 Basic Resins

Polymers are divided into two major groups: *thermoplastics* and *thermosets.*

Thermoplastic resins are formed from individual molecular chains, which have a linear structure and exhibit no chemical linkage between the individual molecules.

Thermoset resins have molecules that are chemically linked together by crosslinks and form a sort of network structure.

1.1.1 Thermoplastics

A major characteristic of *thermoplastic* polymers is that they repeatedly soften when heated and harden when cooled. The molecules are held together by intermolecular forces, such as van der Waals. During the molding process, when heat and pressure are applied to the thermoplastic resin, the intermolecular joints break and the molecules move, changing positions in relation to one another. In the holding phase of the molding cycle, the molecules are allowed to cool, doing so in their new locations. The intermolecular bonds are restored in the new shape.

Thermoplastic polymers are ideal for recycling purposes because of their ability to rebond many times over (Fig. 1.1).

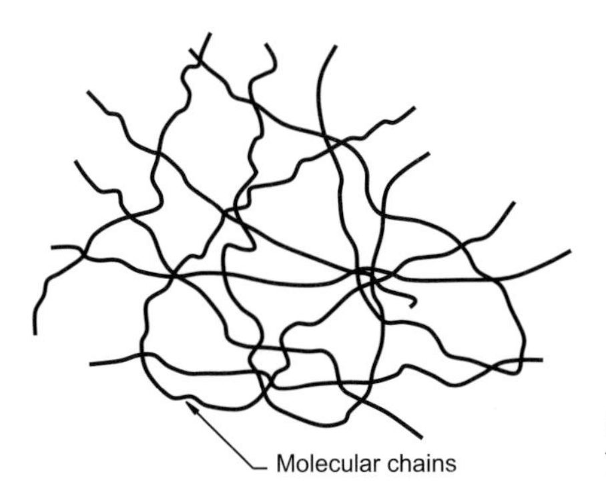

Figure 1.1
Thermoplastics: molecular chains

When heated, the individual chains slip, causing plastics to flow. When cooled, the molecular chains are strongly held together once again. There are practical limitations to the number of times the material can be heated and cooled, depending upon the thermoplastic being used.

Polycarbonate, nylon, acetal, acrylic, thermoplastic elastomers (TPEs), and polyethylene are examples of thermoplastics.

1.1.2 Thermosets

Thermoset polymers undergo chemical change during processing.

During the molding process, when thermoset resins are heated and cured, crosslinks form between molecular chains (Fig. 1.2). This reaction is also called a *polymerization reaction*. When reheated, these cross bonds prevent individual chains from slipping. Chemical degradation occurs if more heat is added after the cross bonding is complete. Therefore, applying heat and pressure cannot remelt thermosets, and they cannot be recycled.

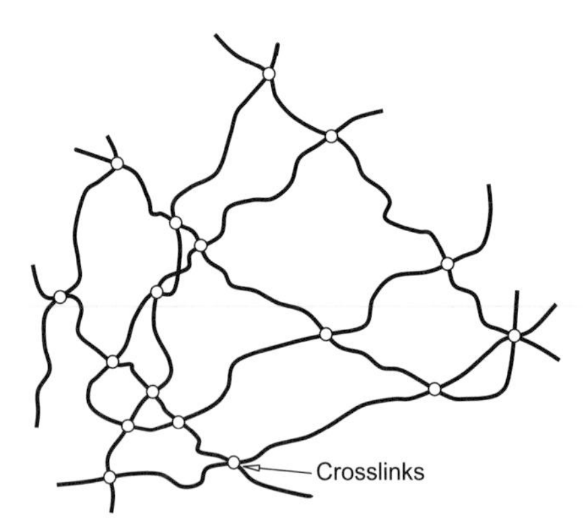

Figure 1.2
Thermosets: crosslinks

1.2 Basic Structures

1.2.1 Crystalline

Crystalline polymers are orderly, densely packed arrangements of molecular chains (see Fig. 1.3). These molecular chains have the appearance of a shoestring when magnified many times under a microscope. The highly organized regions show the behavioral characteristics of crystals.

It should be noted that complete crystallinity is seldom achieved during polymer processing. There will almost always be some amorphous areas left in the part. During processing, parts cool from the outside in, so the skin of the part is the area most likely to lack the necessary crystallinity. Many crystalline polymers achieve a degree of crystallinity of only 35 to 40%, even under ideal processing conditions. That means that slightly more than one-third of the component structure becomes self-organized.

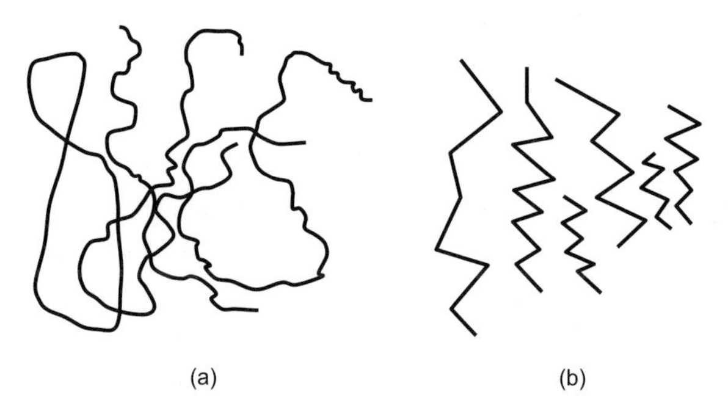

Figure 1.3
Molecular chains for crystalline polymers: (a) melt; (b) solid

Some typical examples of crystalline polymers include acetal, polyamide (nylon), polyethylene (PE), polypropylene (PP), polyester (PET, PBT), and polyphenylene sulfide (PPS).

1.2.2 Amorphous

Polymers having *amorphous* structures represent a disordered or random mass of molecules (Fig. 1.4). A typical noncrystalline or amorphous structure tends to give the resin a higher elongation and flexibility. It will also exhibit higher impact strength than would a crystalline structure.

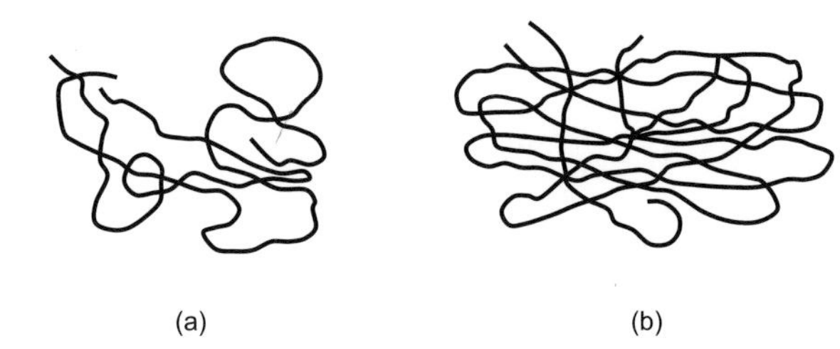

Figure 1.4
Molecular chains for amorphous polymers: (a) melt; (b) solid

Some examples of amorphous polymers are acrylonitrile-butadiene-styrene (ABS), styrene-acrylonitrile copolymer (SAN), polyvinyl chloride (PVC), polycarbonate (PC), and polystyrene (PS). Table 1.1 highlights various properties of semicrystalline polymers compared to amorphous polymers.

Table 1.1 Comparison of Typical Properties between Amorphous and Semicrystalline Polymers

Properties	Amorphous	Semicrystalline
Chemical resistance	Poor	Very good
Creep capabilities	Very good	Good
Elongation at yield	Average 0.4–0.8%	Average 0.5–0.8%
Fatigue strength	Poor	Very good
Mechanical properties	Good	Very good
Temperature	Softening range	Defined point
Notch sensitivity	Poor	Good
Service temperature	Good	Very good
Shrinkage	Very good	Poor

1.2.3 Liquid Crystal Polymer (LCP)

Liquid crystal polymers (LCP) are generally considered a separate and unique class of polymers. Their molecules are stiff, rod-like structures that are organized in large parallel arrays in both the molten and solid state (see Fig. 1.5). This parallel organization of molecules gives LCP characteristics similar to both crystalline and amorphous materials.

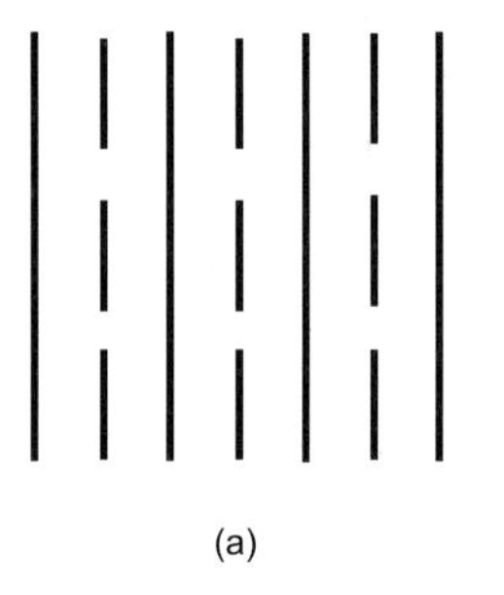
(a)

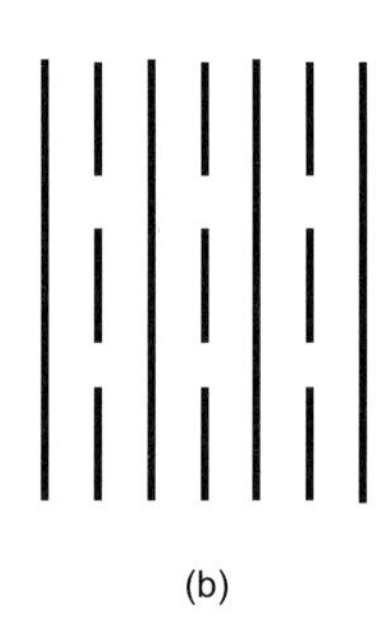
(b)

Figure 1.5 Molecular chains for liquid crystal polymers: (a) melt; (b) solid

1.2.4 New Polymer Technologies

1.2.4.1 Inherently Conductive Polymers (ICP)

Polymers, since their inception, have been known as insulators both electrically and thermally. In the last few decades a number of suppliers have tried to make polymers conductive by using metal fillers or reinforcements. The result of improving polymer conductivity has been marginal.

The discovery of inherently conductive polymers (ICP) has changed all that. Alan Heeger, Alan MacDiarmid, and Hideki Shirakawa were awarded the Nobel Prize for Chemistry in 2000 for their pioneering work related to ICPs.

The above scientists, as well as many others, discovered that adding or subtracting atoms to or from polymers allows the plastic to become electrically or thermally conductive. This process, known as doping, typically removes or adds conducting electrons, leaving a polymer with some positive or negative charges. Dopants are chemical substances that either supply additional electrons to conduct a charge or, alternatively, take electrons away to create what are known as holes or places in a molecule that conduct the charge by accepting electrons. Having fewer electrons that remain in the polymer, they will be capable of moving more openly, allowing conduction.

The most promising ICPs are polyacetylene, polyaniline (PAni), and polypyrrole (PPY). They can be added to known polymers, such as acrylics (known as poly methyl methacrylate or PMMA), poly vinyl chloride (PVC), polyproylene (PP), and others, to make them conductive.

1.2.4.2 Electro-Optic Polymers (EOP)

Electro-optic polymers (EOP) are distinct resins, but they do have some overlapping characteristics with ICPs (inherently conductive polymers). Upon the application of an electric field, EOPs exhibit optical characteristics: they glow. The effect is due to the large molecules making up the polymers. When voltage is applied it raises the molecules' electrons to higher energy levels, after which they drop back to their original energy levels, emitting light in the process (electroluminescence). Richard Friend and Jeremy Burroughes from Cambridge University, England, developed the first electroluminescent polymer, PPV (polyphenylene vinylene). They showed that sandwiching a thin layer of resin between a pair of electrodes–one of which was transparent–made the polymer glow.

There are a number of polymers emitting various colors. For example, green is emitted by PPV, red by PT (polythiopene), and blue by PF (polyfluorene). When an electric charge (typically low voltage of 3 to 5 volts) is applied, the benzene electrons of each polymer are excited. Then the benzene electrons, returning to their original state of energy levels, emit light in a color specific to their resin, which is vibrant and soft.

The simplified manufacturing process consists of an electrical conductor laid down on a carrier foil, glass, or plastic. Then a thin LEP (light-emitting polymer) layer is applied, which is less than 1 μm (0.00004 in.) thick. Finally, another electrode is deposited on top and the display is realized (see Fig. 1.6).

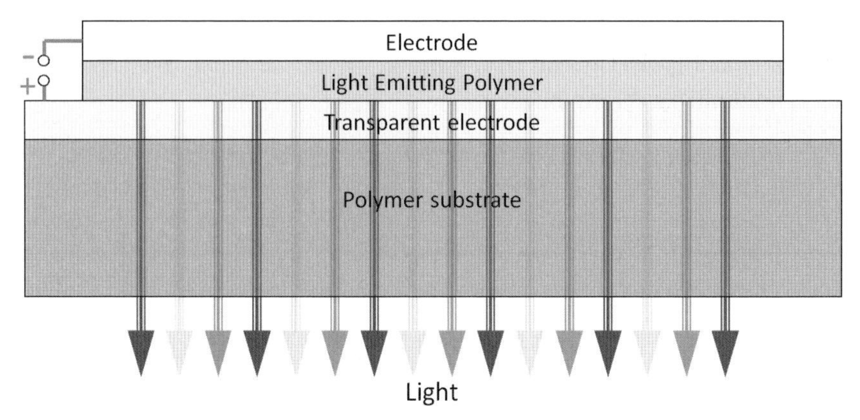

Figure 1.6 Cross section through a light-emitting display (LED) containing a thin layer of light-emitting polymer (LEP)

It should be noted that display stability and performance is greatly improved when sealing techniques are used to prevent air and moisture from reaching the LEP sandwiched between the two electrodes. Certain LEDs (light-emitting displays) have reached continuous working lifetimes exceeding 50,000 hours.

1.2.4.3 Biopolymers

Polylactic (PLA) polymers are one type of biopolymers that consists of rather long chains of lactic acid. The lactic acid is manufactured from fermented and polymerized sugars extracted from starch-producing plants, such as corn and potatoes.

Table 1.2 Physical and Mechanical Property Comparison between PLA, PET, and PS Polymers

Property	ASTM Method	PLA (Polylactide)	PET (Polyethylene Terephthalate)	PS (Polystyrene)
Specific gravity	D792	1.21	1.37	1.05
Melt index (g/min @ 190°C)	D1238	10–30	1–10	1–25
Clarity	15	Transparent	Opaque	Transparent
Tensile yield strength (psi)	D638	7,000	9,000	6,000
Tensile strain (%)	D638	2.5	3	1.5
Notched Izod impact (ft-lb/in.)	D790	0.3	0.7	0.01
Flexural strength (psi)	D790	12,000	16,000	11,500
Flexural modulus (psi)	D790	555,000	500,000	430,000

The PLAs have mechanical properties, such as tensile strength and elastic modulus, comparable to PET (polyethylene terephthalate), while their appearance is comparable to polystyrene clarity and gloss.

1.3 Homopolymer vs. Copolymer

Homopolymers have a single type of repeating unit throughout the molecular chain. All monomer links in the chain are of the same type (Fig. 1.7).

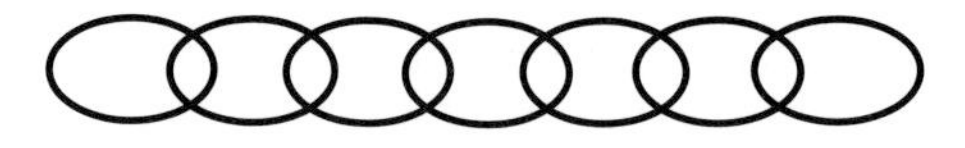

Figure 1.7
Molecular chain for homopolymer resins

A *copolymer* is a polymer with links of more than one kind, which are randomly placed in the chain. Copolymers have different properties from homopolymers because they have different repeating units. When the base unit is altered, its physical and mechanical properties change in different areas (Fig. 1.8).

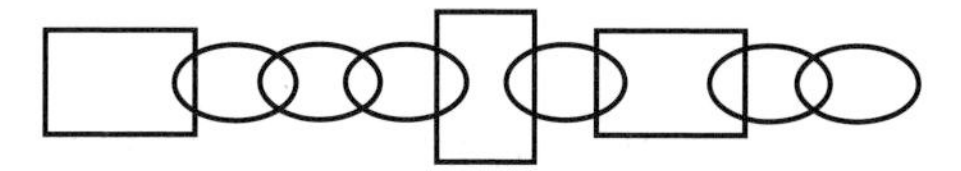

Figure 1.8
Molecular chain for copolymer resins

1.4 Reinforcements

Reinforcing fibers significantly improve most mechanical and thermal properties of thermoplastic and thermoset materials. Glass fibers, carbon fibers, and aramid fibers are some typical reinforcements for plastic materials.

Glass fiber reinforcements are strands of thin filaments drawn from glass furnaces. They are often coated with polymeric films that hold glass strands together. Typical fiber diameters are between 0.002 and 0.02 mm (0.0001 to 0.001 in.). The coated fibers with polymeric films are used as intermediate products or are directly processed into reinforced resins.

The glass fibers have a tensile strength that varies between 3,000 and 5,000 MPa (435,000 to 725,000 psi), have a Young's modulus (see Section 3.8) in the range of 70,000 to 90,000 MPa (10 to 13 million psi), and have an ultimate strain between 4 and 5%.

Reinforcements are available in different forms, such as continuous strand, woven, roving, or chopped fiber. Continuous strands are mostly used in processes like sheet molding compounding (SMC). Chopped fibers are used to reinforce a variety of resins used in processes such as injection molding and compression molding. These fibers usually range from 3 to 12 mm (0.125 to 0.5 in.) or more.

The carbon fibers have a tensile strength of 340 to 5,500 MPa (50,000 psi to 800,000 psi) and a Young's modulus of 35,000 MPa to 700,000 MPa (5 to 100 million psi). The aramid fibers have a tensile strength of 3,500 to 5,000 MPa (508,000 to 725,000 psi) and a Young's modulus of 80,000 to 175,000 MPa (11.6 to 25.4 million psi).

1.5 Fillers

Fillers affect the physical nature of the material without significantly improving the mechanical properties. Some examples of materials used as fillers are talc, wollastonite, mica, glass spheres, silica, and calcium carbonate.

There are two basic types of micas: phlogopite and muscovite. Their size varies between 325 and 40 mesh in particle size (MPS).

Glass spheres can be solid or hollow. Their size varies between 0.005 and 5 mm (0.0002 to 0.2 in. or 5 μm to 5,000 μm). Both solid and hollow glass spheres can be covered with special coatings to improve the bonding between the filler and the matrix.

Nanocomposites are a new technology. Clay additive particles or nanofillers that are added to the polymer matrix are extremely small, having a thickness of about one-millionth of a millimeter (one twenty-five millionth of an inch). In comparison, the filler in conventional thermoplastic polymers–polyolefins, for example–is usually about 1,000 or more times thicker. A polyolefin polymer having as little as 2.5% nanoclay additive filler can be as stiff and at the same time much lighter than parts with 10 times the amount of conventional talc filler. The weight savings can reach 20%, depending on the part and the material that is being replaced by the nanocomposite polymer.

1.5.1 Glass Spheres

Glass bubbles are hollow, spherical particles typically in the range of 10 μm to 200 μm, which can be used to reduce the weight of components made of thermoplastic polymers [170]. They are known by a variety of names, including glass microballoons and hollow glass microspheres. On a per-pound basis, glass spheres are more expensive than conventional fillers such as talc and calcium carbonate. However, most glass bubbles are sold on a volumetric basis with loading levels of 3–5%, making cost less of a concern. The most common glass bubbles for thermoplastic applications are made by a high-temperature melt process and are a borosilicate glass composition. Figures 1.9 and 1.10 show some scanning electron microscopy (SEM) micrographs of glass bubbles.

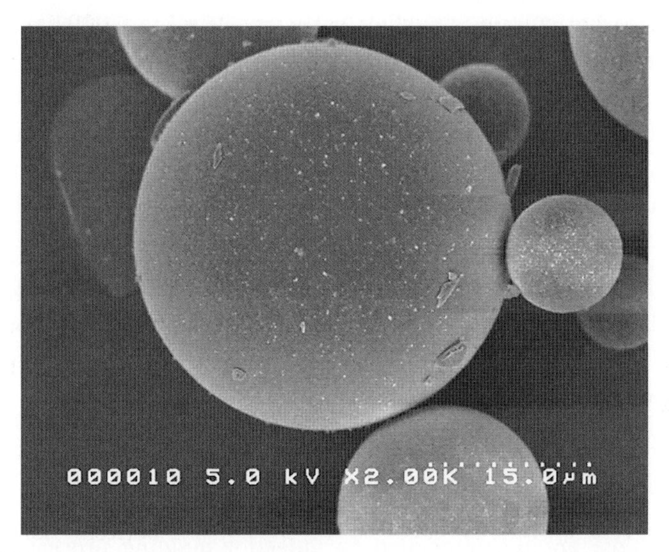

Figure 1.9 SEM micrograph of hollow glass microsphere, grade S60HS *(Courtesy of 3M Company)*

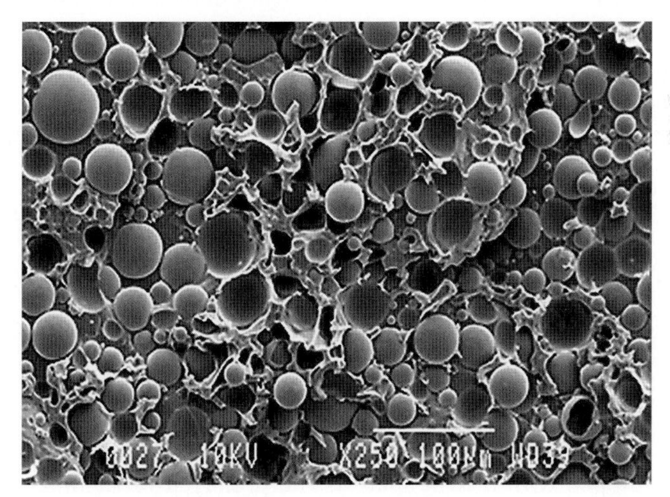

Figure 1.10 SEM micrograph of the fracture surface of S60HS in polypropylene at 40% by volume loading *(Courtesy of 3M Company)*

The microspheres are available in a wide range of densities, ranging from 0.1 grams per cubic centimeter to over 1.0 g/cc [172]. The higher density microspheres have higher compressive strengths due to their increased wall thickness. This strength-to-density correlation is an important consideration when choosing a hollow microsphere for a particular application, since most end users desire the lowest density microspheres that can survive their processing conditions.

1.5.1.1 Microsphere Properties

The most common use for glass bubbles in thermoplastics is as lightweight filler. In addition to providing the benefit of light weight, glass bubbles also provide attributes similar to those of other traditional fillers, such as reduced mold shrinkage, warpage, and coefficient of linear thermal expansion. Other attributes unique to glass bubbles are their low thermal conductivity and improved cycle times because materials containing hollow glass bubbles have less mass to heat and cool.

The downside to using hollow glass bubbles is that they are more expensive than traditional fillers. Their reduced density compared to solid fillers mitigates some of the cost imbalance. These issues are discussed in more detail in the following sections.

1.5.1.2 Compounding

In order to survive the hydrostatic and shear forces experienced during compounding and injection molding, glass bubbles with strengths greater than about 3,500 psi are recommended for extrusion processes; glass bubbles with strengths greater than 10,000–15,000 psi are recommended for injection-molding processes. The exact grade of bubble that should be used depends on the formulation and processing conditions. Heavily filled systems require stronger bubbles, as do high screw speeds, high extruder outlet pressures, and aggressive mixing elements in the extruder.

To minimize bubble breakage it is imperative to use a twin-screw extruder or a Buss kneader equipped with a downstream feed port for adding the glass bubbles. Extruder screw flights beyond this point should be set up to impart a minimum amount of shear stress. Distributive mixing elements, such as gear mixers, are preferred. Aggressive dispersive mixing elements, such as reverse flight elements and kneading blocks, are not recommended. Single-screw extruders are also not recommended, as they typically do not have a downstream port and often contain "barrier" designs or other narrow tolerance/high shear features. A system designed for adding chopped fiberglass to a polymer would be a suitable starting point when developing a low shear extrusion system for glass bubbles.

Glass bubbles can be fed with a volumetric feeder, although a gravimetric feeder–ideally one equipped with a twin shaft–is the preferred method. An auger-driven side feeder at the downstream port, rather than a simple open hopper, will provide the most consistent feeding behavior. Glass bubbles will become fluidized when aerated (which often occurs when initially filling a hopper), and this may lead to flooding of the hopper. To prevent this possibility, it is recommended that the feeder discharge be covered until the hopper is filled with glass bubbles.

Any pelletizing method suitable for the polymer of choice is suitable for a system containing glass bubbles. For highly bubble-filled polymers, it is advisable to use an underwater strand pelletizer or water-slide pelletizer.

1.5.1.3 Injection Molding

A general-purpose, three-zone screw (feed, compression, and metering) is suitable for processing glass bubbles. As mentioned in the previous section, dispersive mixing screws, such as barrier, vented, or double wave, are not recommended. Distributive mixers will give acceptable results.

A generous nozzle/sprue orifice dimension should be used. Additionally, it is best not to use internally tapered tips or tips without a constant-diameter pathway, as they can cause additional stress on the glass bubbles. For optimum mold filling, gate designs should incorporate full, round runners and sprues that are as short as possible.

1.5.1.4 Mechanical Properties in Injection-Molded Thermoplastic Applications

Although hollow microspheres are not considered to be reinforcing due to their 1:1 aspect ratio, careful control of the formulation can minimize the negative effect of the microspheres on mechanical properties. In 1995, Trexel became sole licensee of MuCell® technology (see Fig. 1.11), which was developed by Massachusetts Institute of Technology. The technology uses nitrogen and sometimes carbon dioxide as the foaming agent. Figure 1.12 shows an engine cover made from thermoplastic polyamide 6,6 using 18% glass fibers and 8% glass microspheres as filler content. The cover is molded using the MuCell® microcellular foam injection-molding process, which employs the controlled use of gas in its supercritical state to create the foamed part.

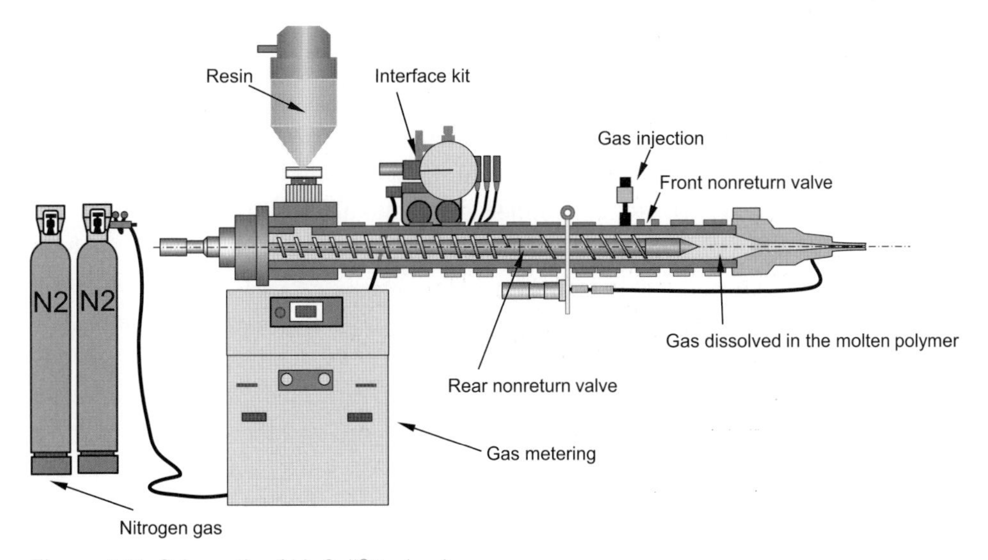

Figure 1.11 Schematic of MuCell® technology

Figure 1.12
Engine cover for 3.5 liter, six-cylinder engine made using 18% glass-reinforced fibers and 8% glass microspheres reinforcedthermoplastic polyamide 6,6 *(Courtesy of ETS Inc.)*

The polymer is melted prior to gas introduction in the injection-molding barrel. Then the gas is injected into the polymer during the screw rotation. The nitrogen or carbon dioxide gas dissolves in the molten polymer. After it is fully dissolved, it is kept under pressure by controlling the screw position and by shutting off the nozzle of the injection-molding press and by employing valve gates when hot runner tools are used. The next step includes nucleation, where a large number of nucleation sites are formed throughout the resin during the injection-molding process. A substantial and rapid pressure drop is necessary to create the large number of uniform sites. Thereafter, cell growth is controlled by processing conditions, including precise control of the molding pressure and temperature.

The process generally offers a 50–75% improvement in key quality measures such as flatness, roundness, and warpage while also eliminating all sink marks. Sink marks result when uniform stress patterns, rather than a nonuniform stress characteristic, are created in the molded part in a solid part that was not foamed. Parts produced in this way tend to comply better with the dimensional specifications as a direct result of the uniform stress and shrinkage that occurs because the pack-and-hold phase of the molding cycle is eliminated. Table 1.3 shows the weight reduction achieved using the MuCell® for an engine cover used in a 3.5 liter V-6 engine.

Table 1.3 Weight Reduction Achieved Using Glass Microspheres in Combination with Glass Fibers for the Polyamide 6,6 Used in the Engine Cover versus 20% Glass Reinforced PA 6,6 *(Courtesy of 3M Company)*

Polymer	20% Glass Reinforced Polyamide 6,6	18% Glass Reinforced +8% Glass Microsphere Polyamide 6,6	Change (%)
Process	Classic injection molding	MuCell® injection molding	
Flexural modulus (MPa)	4,233	3,712	−12

Flexural strength (MPa)	158	128.5	−18
Tensile strength (MPa)	88	66.2	−25
Elongation (%)	4.9	5.7	+16
Notched Izod (J/m^2)	3,962	3,046	−23
Density (g/cm^3)	1.27	1.029	-
Weight reduction			+19

Glass bubbles have the ability to produce lightweight thermoplastic parts with improved dimensional stability.

1.6 Additives

Additives are used to improve specific properties of the plastic material. Flame retardants, thermal stabilizers, and UV stabilizers are some examples (Table 1.4).

Table 1.4 The Influence Additives Have on Polymer Properties

Additive Type	Max Content	Modulus	Impact	Strain	Dimensional Stability	Flammability
Aramid fibers	20	Better	Lower	Lower	Lower	Better
Antistatic agents	5	Lower	Much lower	Much lower	No effect	No effect
Elastomers	15	Lower	Excellent	Much better	Lower	Lower
Glass fibers	60	Excellent	Lower	Much lower	Lower	Better
Inorganic flame retardants	40	Lower	Worst	Much lower	Better	Excellent
Minerals	40	Better	Lower	Lower	Much better	Better
Organic flame retardants	20	Lower	Much lower	Much lower	Better	Excellent
UV stabilizers	1	Lower	Lower	Lower	No effect	No effect

1.7 Physical Properties

Density, specific gravity, elasticity, plasticity, ductility, toughness, brittleness, notch sensitivity, isotropy, anisotropy, water absorption, and mold shrinkage are important physical properties we will be exploring.

1.7.1 Density and Specific Gravity

Table 1.5 Typical Density Values for Polymers

Material	Density (g/cm^3)	Density (lb/in^3)
ABS	1.05	0.0382
ABS GR	1.2	0.0433
Acetal	1.4	0.051
Acetal GR	1.6	0.0582
Acrylic	1.2	0.0433
Cast epoxy	1.8	0.0655
Phenolic	1.85	0.0673
Polyamide (PA)	1.15	0.0415
Polyamide (PA) GR	1.35	0.0487
Polyamide imide	1.55	0.0564
Polycarbonate (PC)	1.2	0.0433
Polycarbonate GR	1.45	0.0523
Polyester (PET, PBT)	1.14	0.0415
Polyester GR	1.63	0.0588
Polyethylene	0.9	0.0325
Polyphenylene oxide (PPO)	1.08	0.0393
Polyphenylene sulfide (PPS)	1.55	0.0564
Polypropylene (PP)	0.9	0.0325
Polypropylene (PP) GR	1.1	0.0397
Polysulfone (PSU)	1.25	0.0451
Polystyrene (PS)	1.05	0.0382

Material	Density (g/cm³)	Density (lb/in³)
Polyvinyl chloride (PVC), rigid	1.35	0.491
Polyvinyl chloride (PVC), flexible	1.25	0.0451
Styrene acrylonitrile (SAN)	1.07	0.0389
Styrene acrylonitrile (SAN) GR	1.28	0.4657

Density is a measure of the mass per unit volume, expressed in pounds per cubic inch or grams per cubic centimeter. Table 1.5 shows densities for various polymers.

Specific gravity is the density of a material divided by the density of water. This is a dimensionless measurement.

Both density and specific gravity are used in determining part weight and cost (Fig. 1.13).

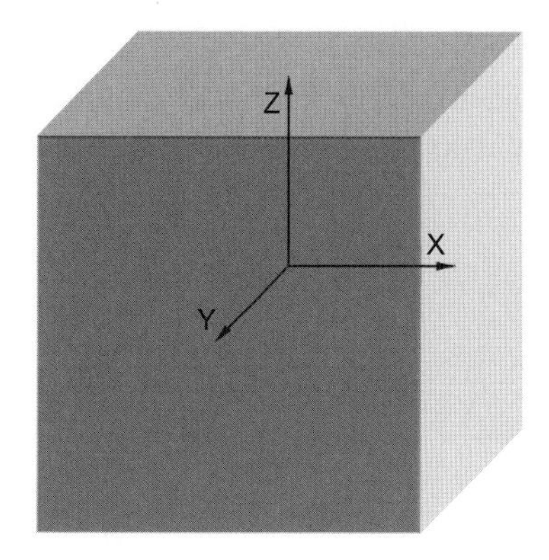

Figure 1.13
Unit volume to calculate density and specific gravity

1.7.2 Elasticity

Elasticity is the ability of a material to return, partially or completely, to its original size and shape after being deformed (Fig. 1.14). Materials that recover fully to their initial size are perfectly elastic. Those that partially recover are partially elastic.

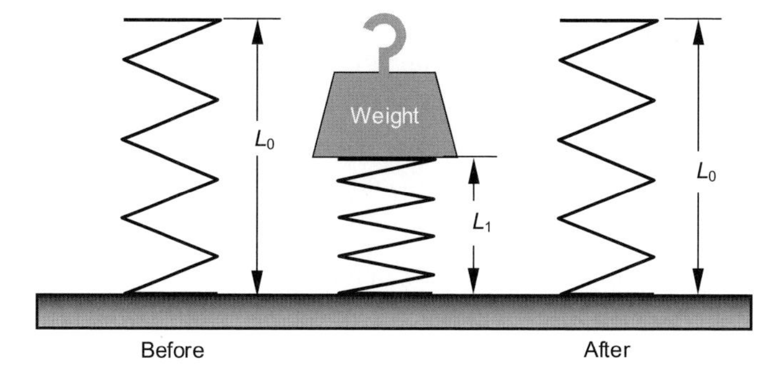

Figure 1.14
Elasticity

The elastic region for thermoplastic or thermoset material is a very important domain for linear analysis. Brittle materials show generally reduced elastic domains. Rubber and thermoplastic elastomers have excellent elasticity.

1.7.2.1 Case History: Elasticity and Denier

Denier is a term derived from the old French word *denier*, which stands for silver or copper coins that were used in France until 1794. It also represents the measurement that is used to identify the *fiber thickness* or diameter of individual filaments used in the creation of textiles. Initially, the term was applied mainly to natural fibers, such as silk and then cotton. Over time, the unit of thickness for synthetic fibers such as polyester and nylon also came to be identified with the same term. The denier, used as textile unit, describes the density of yarn. A 9,000-meter strand of yard that weighs exactly a gram has a density of one denier.

A company from the state of Ohio manufactures playpens for infants (see Fig. 1.15). The playpen incorporates a baby mattress which originally used a 300 denier fabric to cover it. Then later, to save money during the manufacturing process (a few pennies per yard or meter of the finished fabric), it reduced the 70/30 polyester-cotton filament blend used to weave the fabric from 300 to 100 denier, thus making the covering mattress material non-breathable.

Figure 1.15
Playpen

A baby girl, Abigail, was dropped off in 2010 by her grandparents to a day-care center in the morning. Later that morning, the seven-month-old infant was placed into a playpen to rest. About an hour later, the day-care personnel found Abigail dead, face down, in the playpen. Later, during the civil court trial, it was determined that the baby girl suffocated on the mattress (see Fig. 1.16).

The 100 denier fabric uses a very thin diameter yarn, comparable to the yarn used to manufacture cargo parachutes for the U.S. Army, as specified in the Parachute

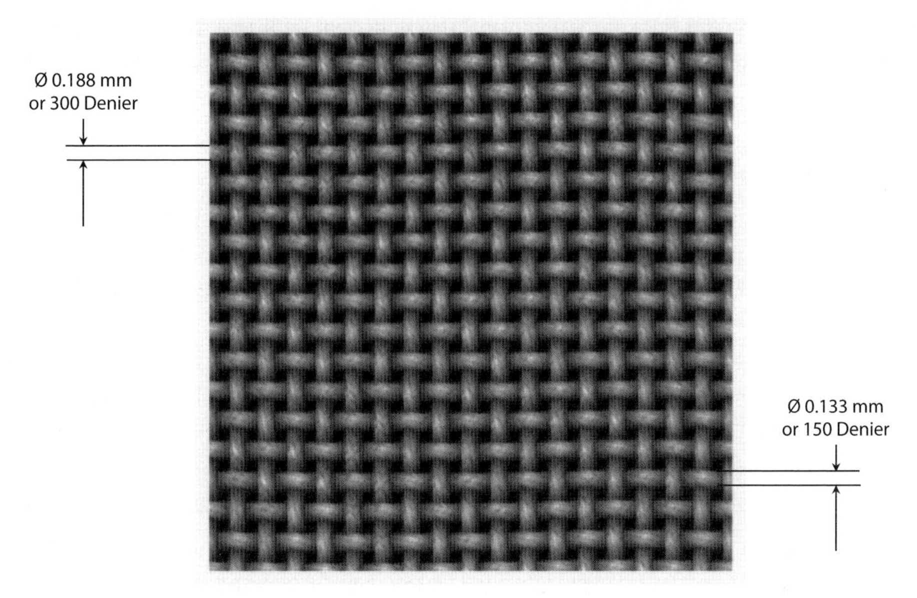

Figure 1.16 Detail of playpen mattress fabric

Industry Association commercial specification 7350 from April 2007. The 100 denier fabric is very elastic because its fibers have small diameter, allowing a much tighter weaving process as compared with 300 denier filaments. However, this fabric quality was decreased because the manufacturing company chose to place an imprint on the material which covered the interfilament spaces between fibers (see Fig. 1.17) [181].

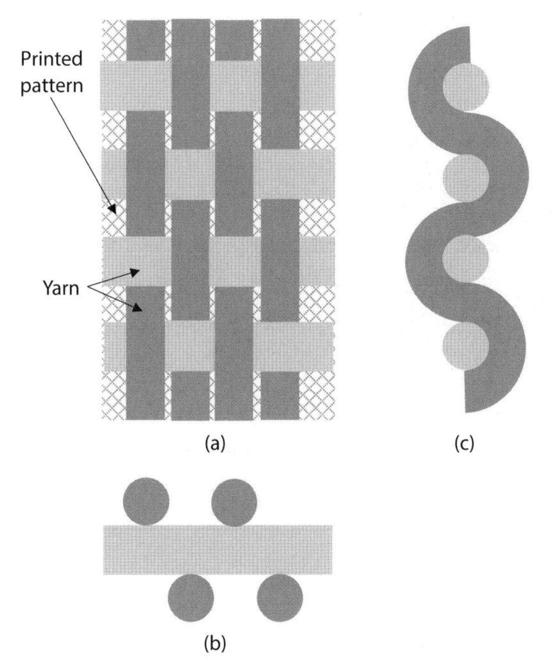

Figure 1.17
Ink printed fabric: (a) top view; (b) perpendicular cross-section; (c) parallel cross-section

At the end of the civil trial the jury found the mattress pad defective and the manufacturer of the playpen guilty for the wrongful death, awarding Abigail's parents a multimillion dollar verdict. Now her parents' goal is that no other baby should have the same fate; they are lobbying to change the law so that all playpen manufacturers use only breathable fabrics for their mattresses.

1.7.3 Plasticity

Plasticity is the property of a material to preserve the shape or size to which it is deformed (Fig. 1.18). Plasticity occurs when the stress goes beyond the yield point on the stress-strain curve for any given material.

This material property can be utilized in cold-forming processes for some plastics. Increases in temperature greatly affect plasticity, especially in thermoplastic resins.

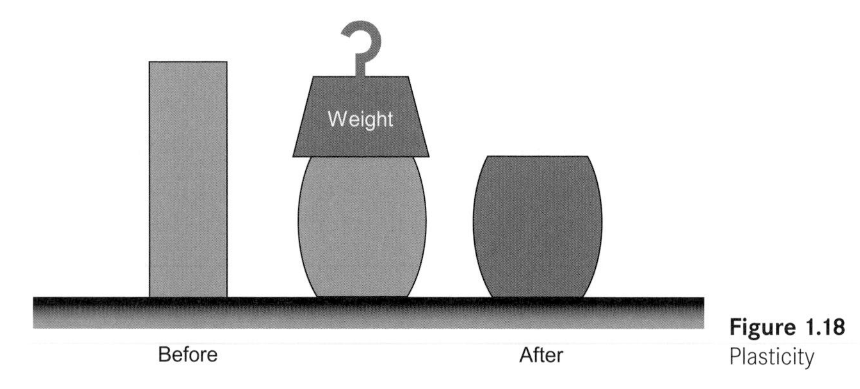

Figure 1.18
Plasticity

1.7.4 Ductility

The material's ability to be stretched, pulled, or rolled into shape without destroying its integrity is called *ductility* (Fig. 1.19). Polymers are categorized as ductile or brittle at a given temperature. Typical failure of ductile polymers occurs when molecules slide along or over each other. This causes large elongation, usually with necking down of the cross-sectional area, and breakage.

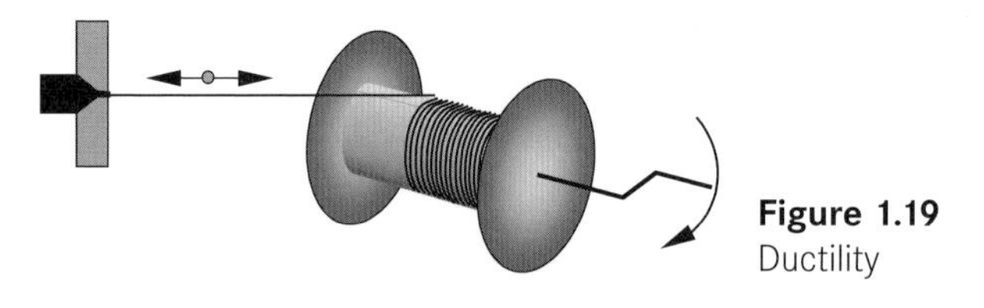

Figure 1.19
Ductility

1.7.5 Toughness

Toughness is the ability of polymeric materials to absorb mechanical energy without fracturing. This is done with either elastic or plastic deformation. Toughness is often measured as the area under the stress-strain curve, as in Fig. 1.20.

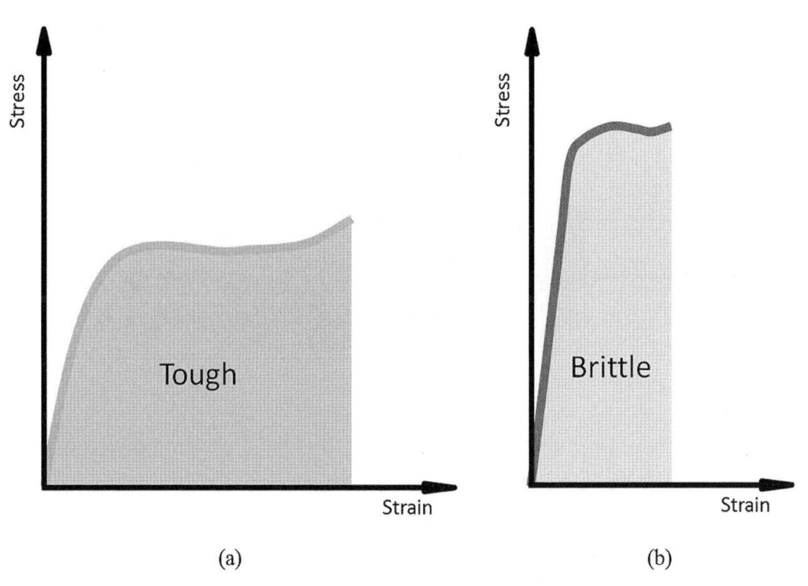

Figure 1.20 Toughness: (a) tough material; (b) brittle material

1.7.6 Brittleness

Brittleness is the property of polymeric materials that fracture easily when absorbing mechanical energy (Fig. 1.21). Many reinforced plastics are brittle and therefore show lower impact and higher stiffness properties.

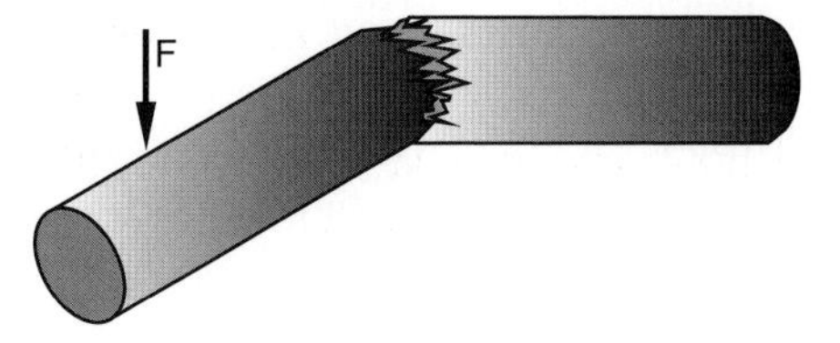

Figure 1.21 Brittleness

1.7.7 Notch Sensitivity

Notch sensitivity is the ease of crack propagation through a material from a preexisting notch, crack, or sharp corner (Fig. 1.22). Excessive stress concentrations can occur as a result of three types of conditions: grooves and holes, changes in the cross-sectional area brought about by shoulders or offsets, and various mounting methods. To account for stress concentration areas, a stress concentration factor *k* has been established.

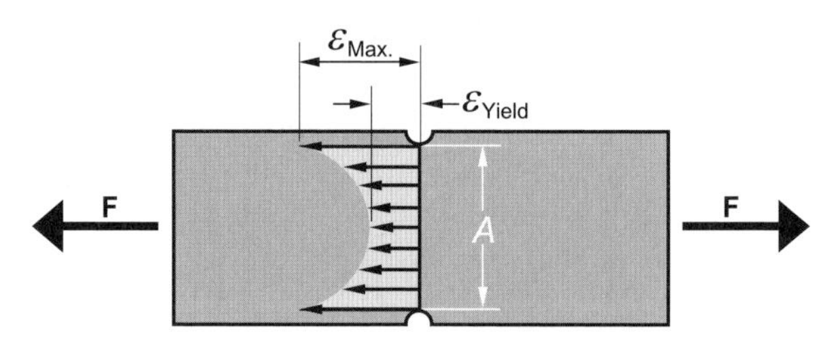

Figure 1.22 Notch sensitivity created by grooves and holes

For example, let's consider a plate as shown in Fig. 1.23. In the course of designing a part, we need to determine the ideal fillet between two right-angle surfaces that will allow us to avoid a sharp corner. In order to determine the transition radius *r* the following steps should be followed.

Let's assume that *D* is the part thickness and *d* is the thickness of the thinner portion. Then *d/D* is the ratio of the two thicknesses. The appropriate curve *d/D* should be selected from the lower graph (Fig. 1.23). If none is available in Fig. 1.23, a *d/D* curve is established by graphic interpolation. Then a fillet radius, *r*, is selected.

Next, the ratio *r/d* is calculated.

The calculated ratio *r/d* is then located on the horizontal axis (Fig. 1.23, lower). A vertical line from that point is drawn to intersect with the corresponding *d/D* curve.

From there, we move horizontally to the left to intersect the vertical axis. This gives us the stress concentration factor (*k*).

The transition radius gives the part a higher stress level because the fillet will ease the stress concentration created by a sharp corner. To calculate the new stress level, multiply *k* by the stress level the part was designed for with the sharp corner design.

In order to lessen the dimensional unpredictability of a plastic part, there should be no sharp corners. A small radius at the junction of the nominal wall and the rib is necessary to achieve good tolerances. Not having such a radius weakens the junction between the rib and the rest of the part. Also, the lack of the radius makes it more difficult for the molten polymer to change directions and flow into a restricted rib cavity; ribs are thinner than the normal wall stock.

The minimum practical width of a rib is limited by mold filling concerns, while the maximum possible width is limited by the shrinkage characteristics of the resin selected for the specific application. While the use of larger radii should theoretically reduce the local stress concentration, the thicker wall section associated with the excessive radii leads to shrinkage stresses and shrinkage voids, which actually adversely affect dimensional tolerances.

Stress concentration factors have been determined for a variety of designs, and many technical books dealing with material strength provide these graphs. Figures 1.23 through 1.27 show some of the most common stress concentration factors as functions of part dimensions when the part is loaded in tension, compression, and bending. For a part whose dimension ratio is not represented by a curve on the graph, one can be generated by graphic interpolation. This will determine the k factor and from that the maximum stress the part can undertake.

It should be noted that k is always greater than 1. After a transition fillet is employed, the maximum stress the same part will be capable of sustaining is

$$\sigma_{\text{Max}} = k\sigma_{\text{Nominal}} \tag{1.1}$$

where k = stress concentration factor and

$$\sigma_{\text{Nominal}} = \frac{\text{Force}}{\text{Area}} = \frac{\text{F}}{A} \tag{1.2}$$

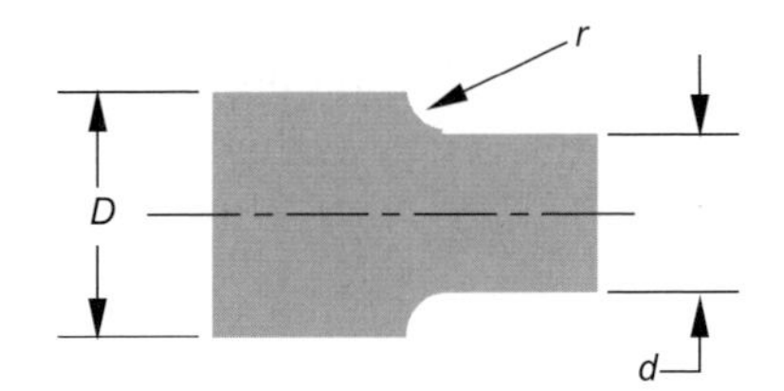

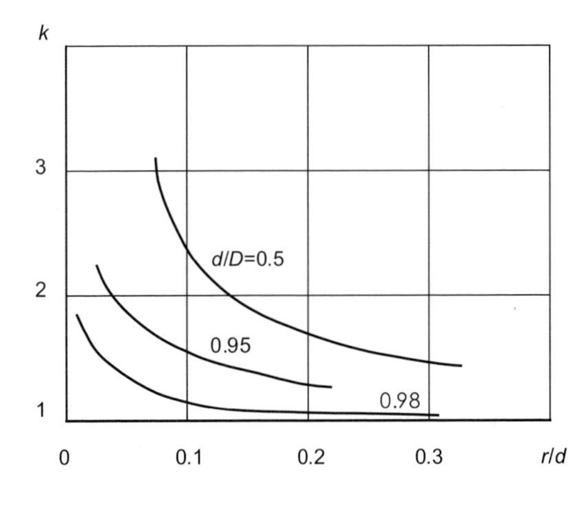

Figure 1.23
Stress concentration factor, k, for flat bars loaded in tension or compression

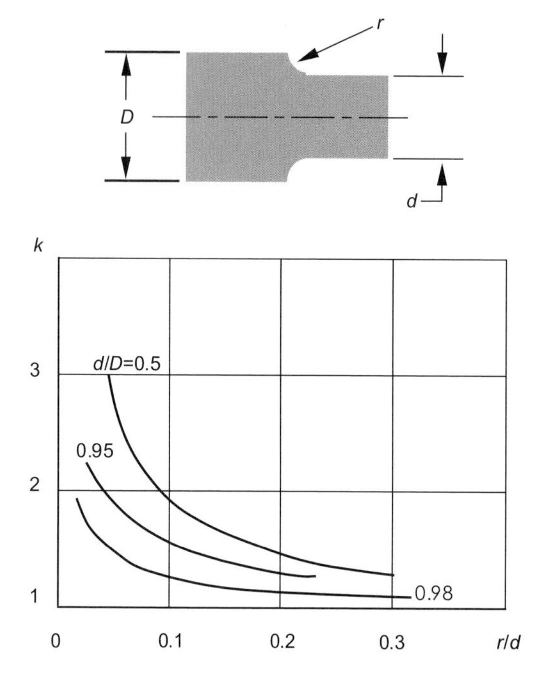

Figure 1.24
Stress concentration factor, *k,* for flat bars loaded in bending

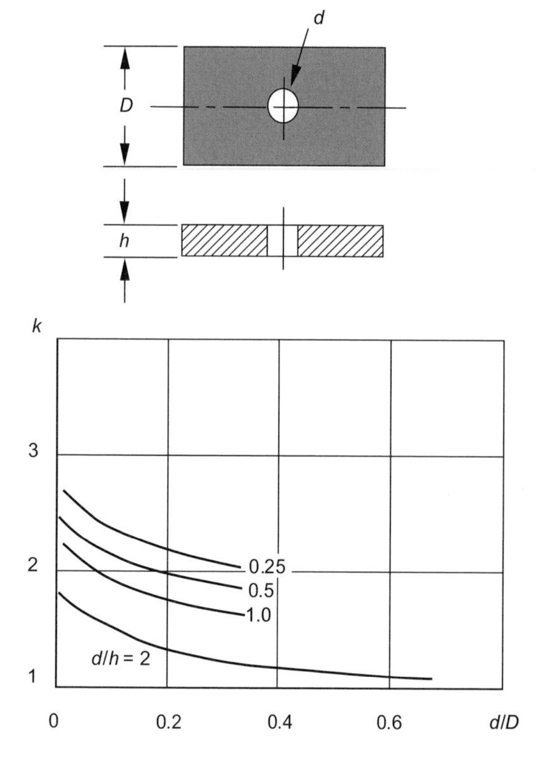

Figure 1.25
Stress concentration factor, *k,* for flat bars loaded in bending

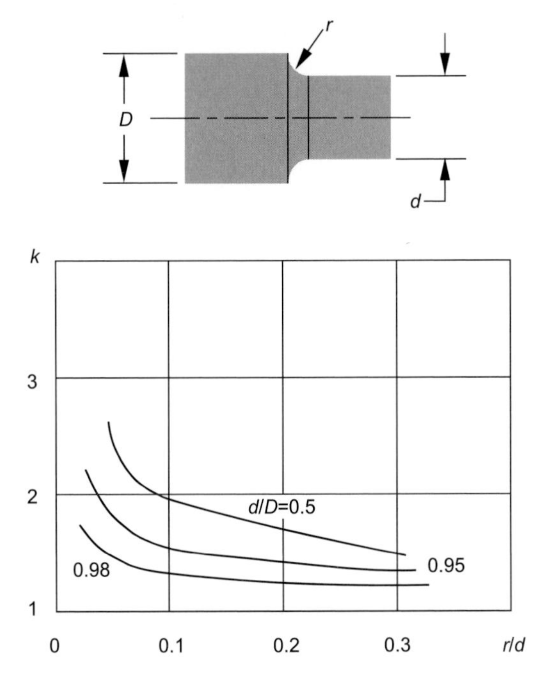

Figure 1.26
Stress concentration factor, *k*, for rods loaded in tension or compression

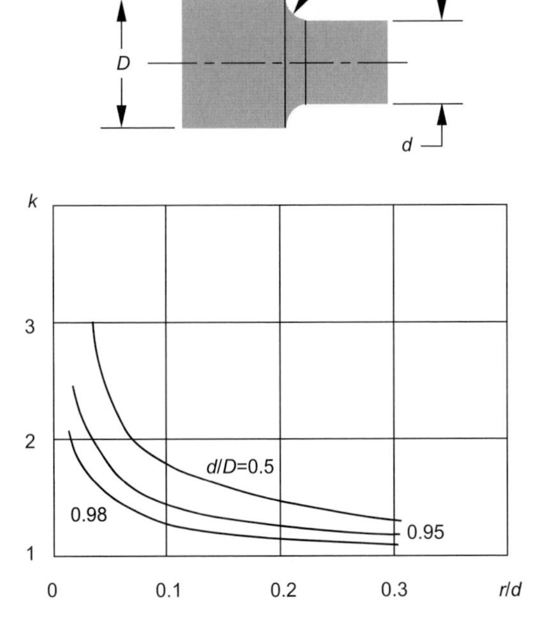

Figure 1.27
Stress concentration factor, *k*, for rods loaded in bending

1.7.8 Isotropy

An *isotropic* material is a thermoplastic or thermoset material that has the same physical properties when measured in any direction (Fig. 1.28).

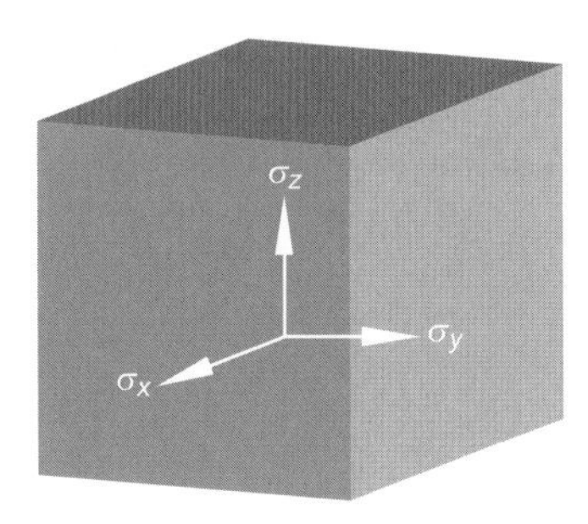

Figure 1.28
Isotropy

1.7.9 Anisotropy

Anisotropic material properties depend on the direction in which they are measured (Fig. 1.29). Reinforced plastics have a high degree of property orientation in the direction of fiber reinforcements.

Extruded and laminated plastics also have different properties in the machine longitudinal and transverse directions. Wood, which has a "grain," is a good example of an anisotropic material, having distinct properties in three directions.

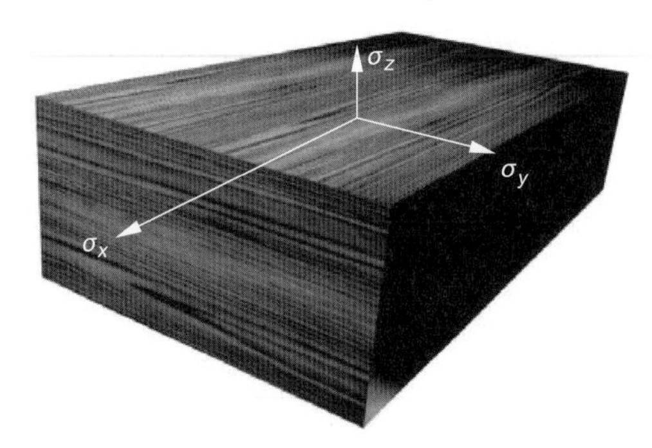

Figure 1.29
Anisotropy

1.7.10 Water Absorption

Water absorption represents the percentage increase in weight of a material due to the absorption of water (Fig. 1.30). Plastic materials can be either absorbent (hygroscopic) or nonabsorbent. Most plastics show hygroscopic tendencies in their "dry" condition. They absorb water by direct exposure or from airborne water vapors at a rate specific to each material. Absorption ends when the saturation level is reached.

Generally, the rate of water absorption is measured when the material is exposed to 50% relative humidity air (50% RH), and saturation is given by the percentage of part dry weight.

The presence of water in the structure of the material influences its physical and electrical properties as well as its dimensional stability. Moisture absorbed by a resin before fabrication, unless removed by drying prior to processing, can cause serious degradation of properties. Visual flaws and severe hydrolysis during the molding process can occur.

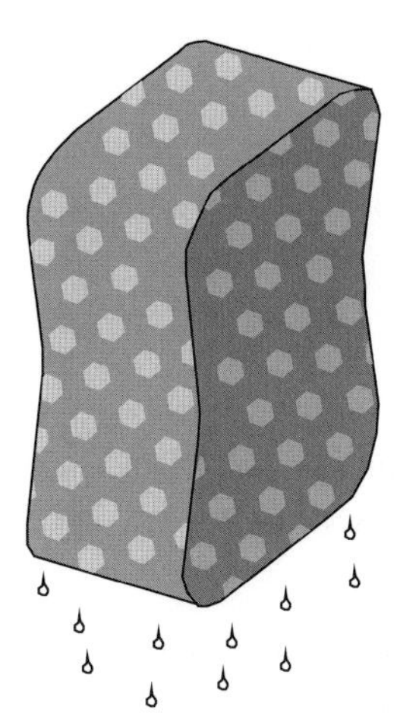

Figure 1.30
Water absorption

Dried-as-molded (DAM) finished parts (with no water content) will have higher stress levels (up to 40% higher than the 50% RH condition), better dimensional stability, and higher electrical insulation. The lower the water absorption rate, the better these properties are kept with time. On the other hand, if the part is expected to absorb a lot of mechanical energy, then a material with a higher water absorption rate will be chosen. This will take advantage of the material's enhanced elasticity at lower levels and dimensional stability.

1.7.11 Mold Shrinkage

Mold shrinkage is the amount of contraction from mold dimensions that a part exhibits after removal from the mold and cooling to room temperature (Fig. 1.31). Material shrinkage starts the very moment plastic is injected into the tool, so good engineering mold design will choose the best gate position, diameter, cycle time, and smoother flow paths to avoid delaminations and voids.

In order to improve structural and thermal performance as well as other mechanical and chemical properties of polymers, glass fibers and other types of fibers or fillers are used to reinforce the material. Glass fibers have a melt temperature exceeding

1,500–2,000°F. Therefore they will never melt or soften during normal or even extreme polymer processing conditions. The shape of glass fibers is cylindrical, having typical dimensions of 3 mm for its length and a few microns in diameter. The glass fiber in a molted resin can be visualized as a small canoe floating on a river.

Glass fibers align themselves with the flow direction of the resin, just as canoes align themselves with the flow of the river. The flow direction defines the anisotropic shrinkage property of glass-reinforced polymers.

The larger the thickness of the part is, the more likely it is that the glass fibers will have a random distribution in the thick area. Typically, for a wall thickness of 6 mm (0.25 in.), glass fibers aligned with the direction of flow are about 5% or less of the total. The random distribution of glass fibers in thick ribs creates unpredictable shrinkage (sometimes warpage as well), making it almost impossible to have predictable part dimensions.

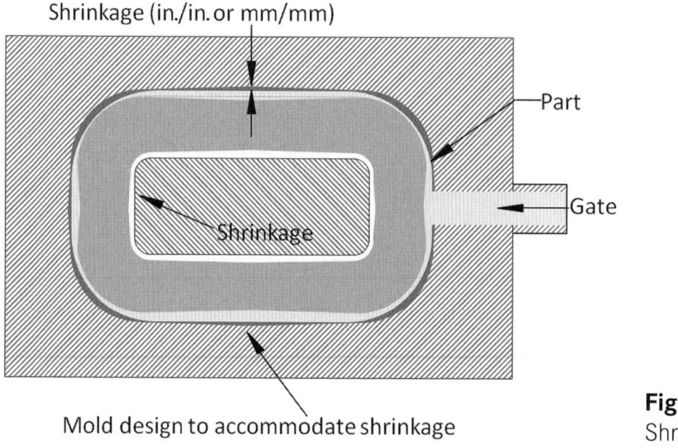

Figure 1.31
Shrinkage

Postmold shrinkage usually is completed the moment the finished part is taken out of the mold and left to cool down to room temperature. Factors such as complex shapes of the finished part and varying contraction rates relative to part dimensions can cause internal stress to build up during the cool-down process. This can cause warping to occur. An annealing process is often used after the molding is complete to relieve the stress and prevent warping.

One of the most important aspects of achieving dimensional predictability of plastic parts is the gate. The gate in injection molding represents an area through which the molten polymer reaches the tool cavity. The gating scheme has a dramatic impact on quality and dimensional stability and repeatability of an injection-molded part.

As a rule, gates should be placed in the thickest area of the part so as not to be visible, for aesthetic reasons, to the end user, due to a gate vestige remaining on the part after the runner system is removed.

1.8 Mechanical Properties

1.8.1 Normal Stress

In a standard tension test, a test bar specimen is subjected to an axial tension load. During this test, relationships between load, deflection, and stress are determined (Fig. 1.32).

Direct stress is the ratio of applied load to the initial cross-sectional area expressed in pounds per square inch (psi) or mega-Pascals (1 MPa = 1 N/mm²).

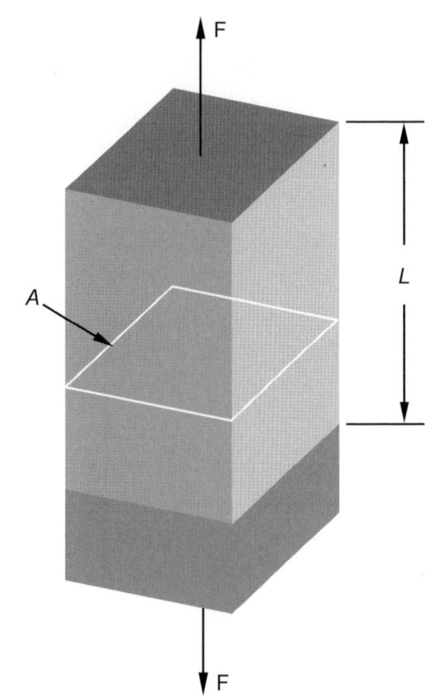

Figure 1.32
Normal stress

The test bar used in Fig. 1.32 was loaded in *tension*. If the load is reversed, the test bar will be in *compression*. Similar equations can be established.

1.8.2 Normal Strain

A test bar specimen subjected to an applied load changes its length. If the initial length of the test bar is L and it changes in length by the amount ΔL, the strain produced is defined as

$$\varepsilon = \frac{\Delta L}{L} \tag{1.3}$$

Strain is therefore a measure of the deformation of the material, a ratio that is dimensionless (Fig. 1.33).

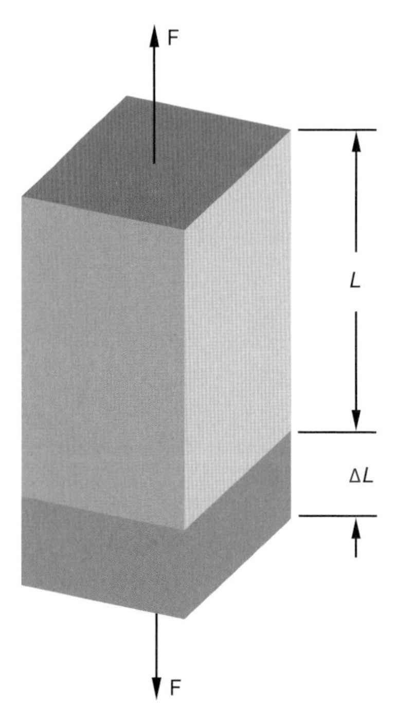

Figure 1.33
Strain

1.8.3 Stress-Strain Curve

We have seen that a material subjected to a load develops initial stress and changes its dimensions. The variation of these two parameters is proportional, and Hooke's Law (see Section 3.8) is the statement of that proportionality.

This *constant* value is measured in pounds per square inch or mega-Pascals and is known as *Young's modulus*. The plot of the stress-strain relation gives a curve f(σ, ε) we refer to as the *stress-strain curve* (Fig. 1.34).

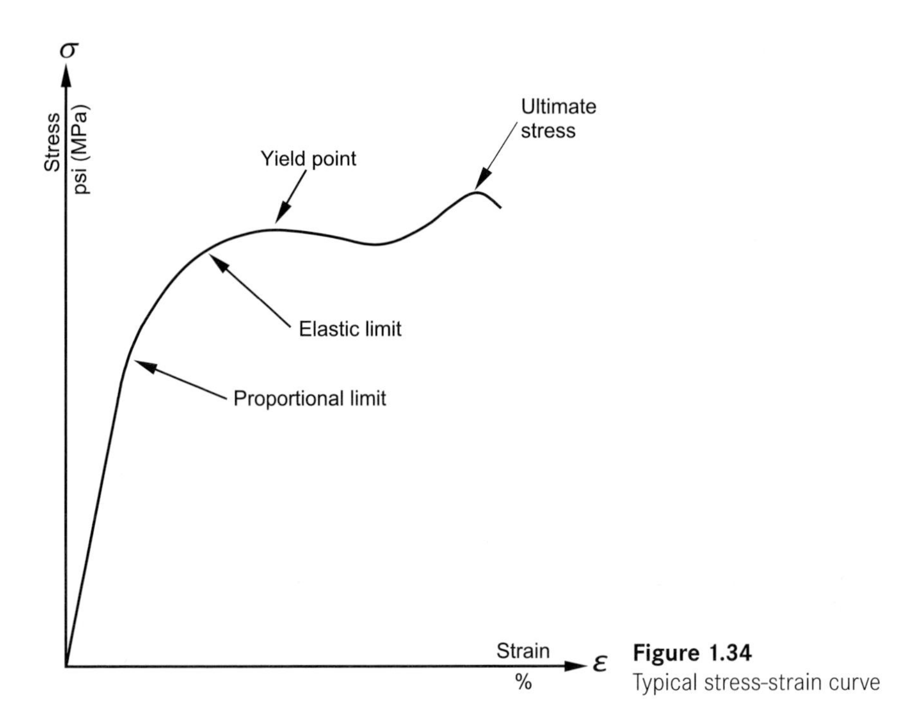

Figure 1.34
Typical stress-strain curve

Stress-strain curves are useful in giving a general picture of the strength and stiffness of the resin and allowing accurate comparisons for material selection. They also provide general information about ductility and toughness of the polymer. Compared to metal curves, plastic stress-strain curves show more pronounced visco-elastic properties (Fig. 1.35).

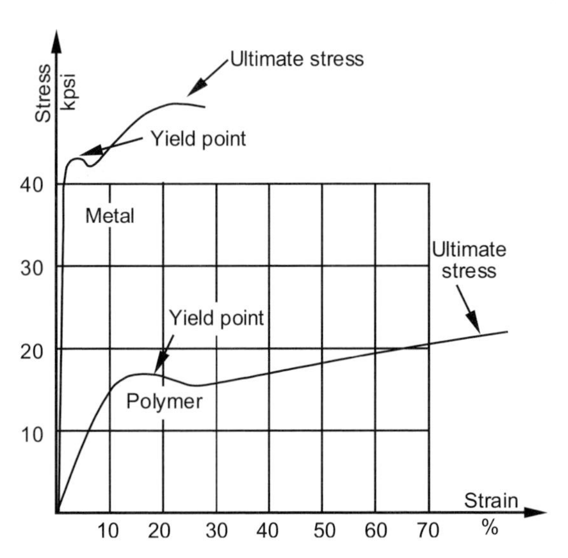

Figure 1.35
Stress-strain curve, metal vs. plastic

1.9 Creep

Plastic parts exhibit two important properties that can occur over long-term loading: creep and stress relaxation.

1.9.1 Introduction

When a constant load is applied to a plastic part, it induces an internal stress. Over time, the plastic will slowly deform to redistribute the internal energy within the part. A test that measures this change is performed by applying a constant stress over time. This time-related flow is called *creep.*

1.9.2 Creep Experiments

Figure 1.36 shows a creep experiment in which a specimen bar is held vertically from one end. The original length of the bar is L. When a weight is hooked at the free end of the bar, the load will immediately increase the length by an amount expressed as ΔL at time = 0.

If the weight is left on the part for some time—for example, 1 year or 5 years—the end of this time period is called time = end. During this time the specimen bar will elongate further. This further increase in length, which is brought about by the time factor rather than the weight, is called *creep.*

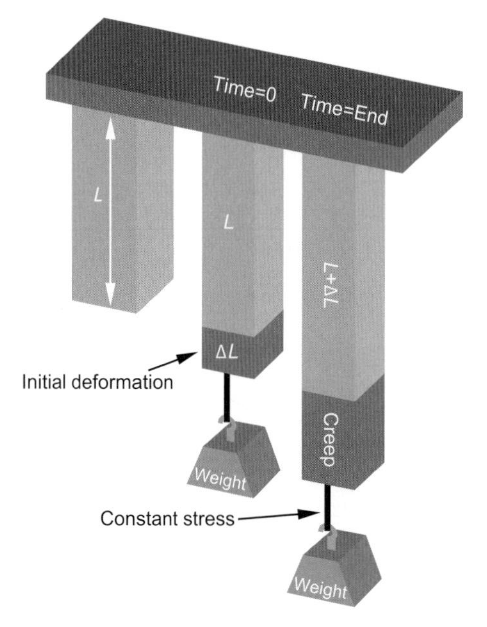

Figure 1.36
Creep experiment

1.9.3 Creep Curves

One of the best ways to demonstrate creep properties is through the use of isochronous stress-strain curves (Fig. 1.37). Most of these are generated from several test samples, each under a different degree of constant stress. After the appropriate load is applied, the elongation of each sample is measured at various time intervals. The data points for each time interval are connected to create isochronous stress-strain curves. Because material properties are also temperature dependent, the temperature must be kept constant throughout the experiment.

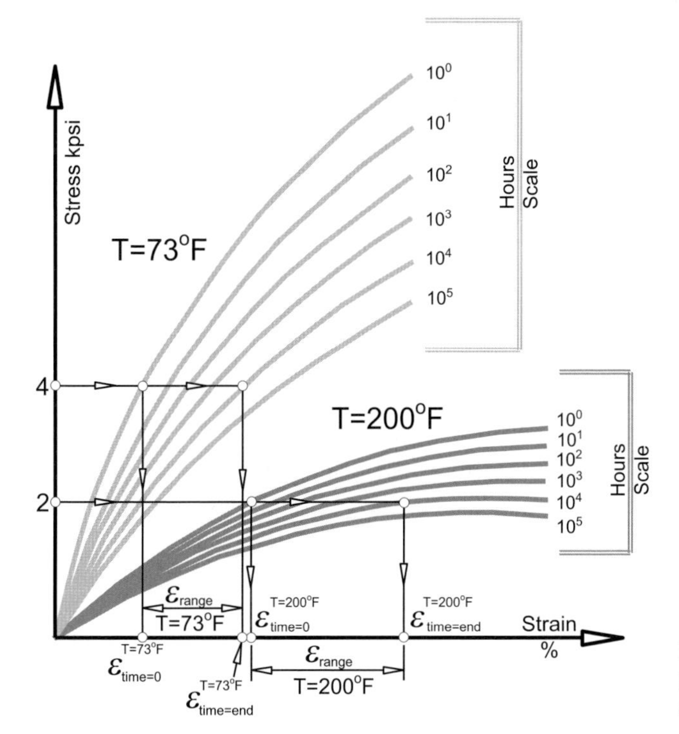

Figure 1.37 Isochronous stress-strain curves

Creep modulus is the modulus of a material at a given stress level and temperature over a specified period of time. Creep modulus is expressed as

$$E_C = \frac{\text{Stress}}{\text{Total Strain at time = end}} \tag{1.4}$$

Creep modulus is also called *apparent modulus* [173, 174]. These curves are usually derived from constant-stress isochronous stress-strain curves. The curves are plots of the creep (apparent) modulus of the resin as a function of time (Figs. 1.38 and 1.39).

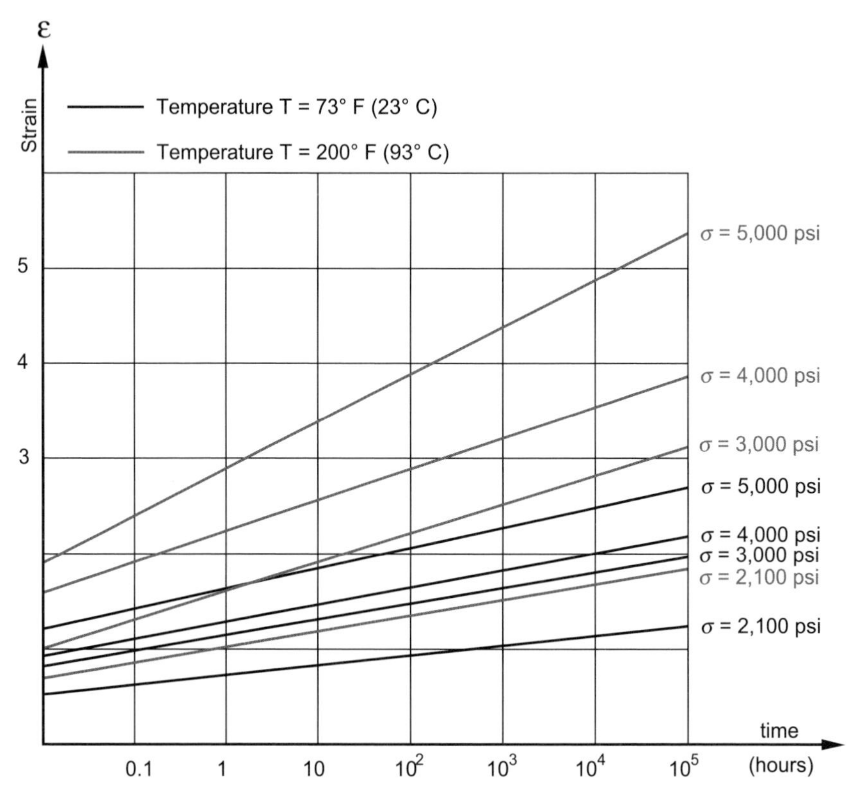

Figure 1.38 Constant stress, strain vs. logarithmic time

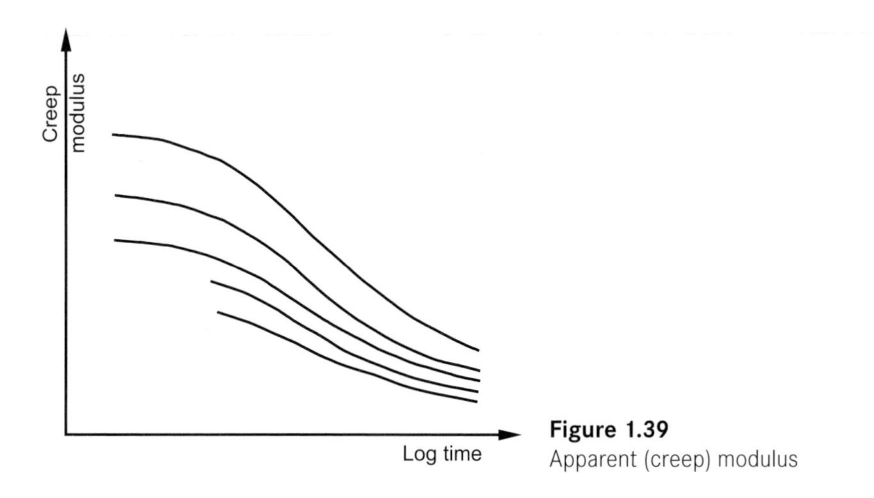

Figure 1.39
Apparent (creep) modulus

1.9.4 Stress-Relaxation

If the plastic part is subjected to a constant strain (or elongation) over time, the amount of stress necessary to maintain that constant elongation will decrease. This phenomenon is known as *stress-relaxation*.

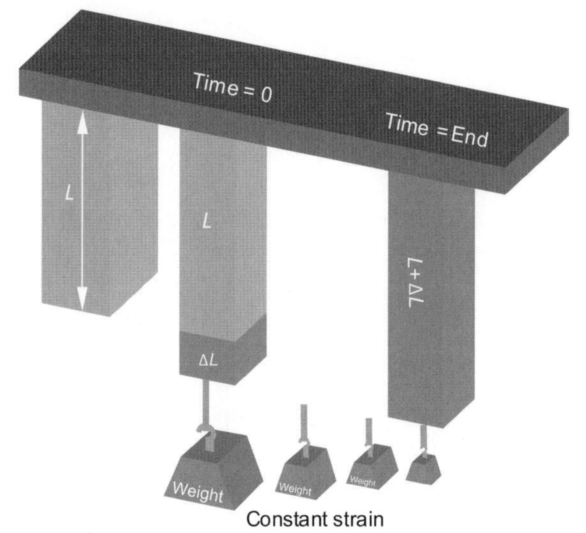

Figure 1.40
Stress-relaxation experiment

Figure 1.40 shows a stress-relaxation experiment. It is very similar to the creep experiment except that the weight of the load varies, decreasing over time as needed to maintain $L + \Delta L$ at a constant length. The variance of the weight in an idealized case should be continuous with instantaneous measurements. Because this is not always possible, creep curves can be used if stress-relaxation curves are not available. In most cases a margin of error of 5–10% is acceptable.

Stress-relaxation can be described as a gradual decrease in stress levels with time, under a constant deformation or strain.

1.10 Impact Properties

In the Izod impact test (Fig. 1.41), which is used mostly in North America, a specimen bar is clamped in a vertical position almost as a cantilever beam. The specimen is struck by a pendulum located a fixed distance from the bar. The pendulum weight is increased for subsequent strikes until the test bar breaks. This test can be done with a notched or unnotched specimen bar. It should be noted that the clamping load induces a variance to this test.

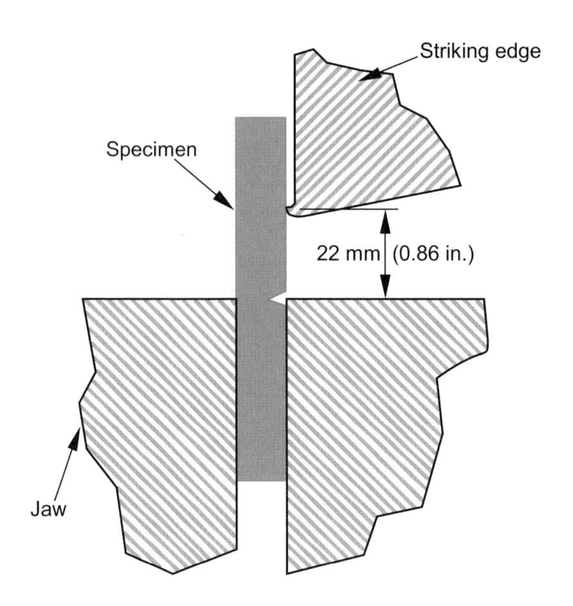

Figure 1.41
Izod impact test

The Charpy impact test (Fig. 1.42), which is used mainly in Europe, is similar to the Izod test, except that the bar is placed horizontally and is not clamped. This means there is no variance induced by clamp load.

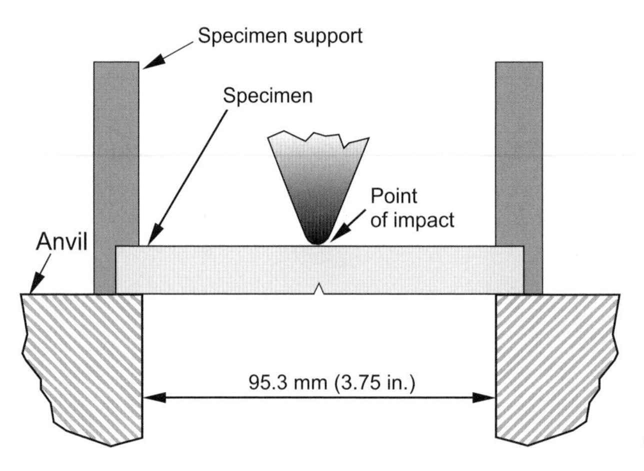

Figure 1.42
Charpy impact test

1.11 Thermal Properties

Polymers are sensitive to changes in temperature. The mechanical, electrical, and chemical properties of a polymer will all be influenced by temperature variations. The effects of changes in temperature on these properties are not always consistent; high temperature can affect the mechanical properties of some polymers a great deal, lowering stress levels and increasing strain levels. Higher temperatures will enhance the electrical properties of some plastics.

1.11.1 Melting Point

Thermoplastic polymers become more flexible with increases in temperature. Crystalline resins exhibit a well-defined melting point. The melting point property is important with respect to the molding process or assembly techniques.

Amorphous resins and liquid crystal polymers (LCP) do not have a sharp melting point. They exhibit this property over a larger range of temperatures.

1.11.2 Glass Transition Temperature

The *glass transition temperature* is the temperature at which polymers exhibit a significant change in properties. Usually, polymers have a more brittle response below T_g (glass transition temperature). Above it they behave with more elastic properties.

1.11.3 Heat Deflection Temperature

Heat deflection temperature or *HDT* is the temperature at which a test bar deflects 0.25 mm (0.01 in.) when subjected to a load of 0.45 MPa (66 psi) or 1.8 MPa (264 psi) in a recognized ISO test. Heat deflection temperature distinguishes between materials that lose their rigidity over a certain temperature range and the polymers able to sustain the above-mentioned loads at high temperatures. Heat deflection temperature essentially ranks polymers by their ability to sustain loads under heat for a short time.

1.11.4 Coefficient of Thermal Expansion

All polymers expand when heated and contract when cooled. Most resins show a change in dimensions for a certain temperature variation. The dimensional change exhibited by polymers is generally greater than that exhibited by metals for the same temperature variation.

Polymers expand and contract from 5 to 10 times more than metals. This repeated change in dimensions could cause internal stress to develop within the part. Stress concentrations can develop at the juncture of an assembly that comprises a polymer part and a metal part that have an area of contact. This can be overcome by using elastomeric gaskets at the interface between the two components. Also, using reinforcements, such as glass, mica, and carbon, in the base polymer can lower the polymer coefficient of linear thermal expansion and contraction. This will greatly reduce the difference between the thermal expansion of the plastic and the metal component.

Another way to accommodate the thermal differences of metal and plastic is to account for these differences in the part design itself if tolerance requirements allow.

The coefficient of linear thermal expansion or contraction (CLTE) is a material property. Table 1.6 shows the linear thermal expansion and contraction of several materials. This property is expressed as a ratio of change of linear dimension in a certain direction to the original dimension in the same direction for a given unit of temperature.

Table 1.6 Coefficient of Thermal Expansion for Common Metals and Polymers

Material	cm/cm/°C × 10^{-5}	in./in./°F × 10^{-5}
ABS	6.5–9.5	3.6–5.3
Acetal copolymer	6.1–8.5	3.3–4.7
Acetal copolymer 25% GR	2–4.4	1.1–2.4
Acetal homopolymer	10–11.3	5.5–6.2
Acetal homopolymer 20% GR	3.3–8.1	1.8–4.5
Aluminum	2.2	1.2
Epoxy	2–6	1.1–3.3
Brass	1.8	1.0
Bronze	1.8	1.0
Copper	0.9	1.6
Polyamide 6	8–8.3	4.4–4.6
Polyamide 6 GR	1.6–8	0.84–4.4
Polyamide 6/6	8	4.4
Polyamide 6/6 GR	1.5–5.4	0.8–3
Polyamide 6/12 GR	2.1–2.5	1.1–1.3
Polyamide 11	10	5.5

Material	cm/cm/°C × 10^{-5}	in./in./°F × 10^{-5}
Polyamide 12	–4	3.3–5.5
Polyamide imide	3	1.65
Polyamide imide GR	1.6	0.84
Polycarbonate	6.8	3.8
Polycarbonate GR	2.2	1.2
PBT	6–9.5	3.3–5.3
PBT GR	2.5	1.3
PET	6.5	3.6
PET GR	1.8–3	1–1.65
Low-density polyethylene (LDPE)	10–22	5.5–12.2
High-density polyethylene (HDPE)	5.9–11	3.2–6.1
Polyimide	4.5–5.6	2.5–3.1
Polyphenylene oxide	3.8–7	2.1–3.8
Polyphenylene sulfide	2.7–4.9	1.5–2.7
Polypropylene	8.1–10	–1
Polypropylene GR	2.1–6.2	1.1–3.4
Polystyrene	5–8.3	2.7–4.6
Polysulfone	5.6	3.1
Polyurethane thermoset	–10	5.5–11.1
Polyurethane thermoplastic	3.4	1.8
PVC rigid	5–10	2.7–5.5
PVC flexible	7–25	3.8–13.4
Steel	1.1	0.6
Zinc	3.1	1.7

1.11.5 Thermal Conductivity

Thermal conductivity represents the rate at which heat is transferred by conduction through a given unit area of a given material when the temperature difference or gradient is normal to the cross-sectional area. The quantity of heat that travels through a unit volume of a polymer in a given time, when the temperature gradient is one degree (1°), is referred to as a coefficient of thermal conductivity.

Plastics have a low thermal conductivity coefficient, compared with other materials, such as metals.

Recently, through new discoveries, polymers can be made thermally conductive (see Section 1.2.4.1).

1.11.6 Thermal Influence on Mechanical Properties

As we have seen so far, temperature, or more specifically a variance in temperature, affects many of a polymer's properties. From a mechanical design-engineering standpoint, we are most interested in the mechanical properties.

All polymers will exhibit similar changes on the stress-strain curve when subjected to the same test conditions.

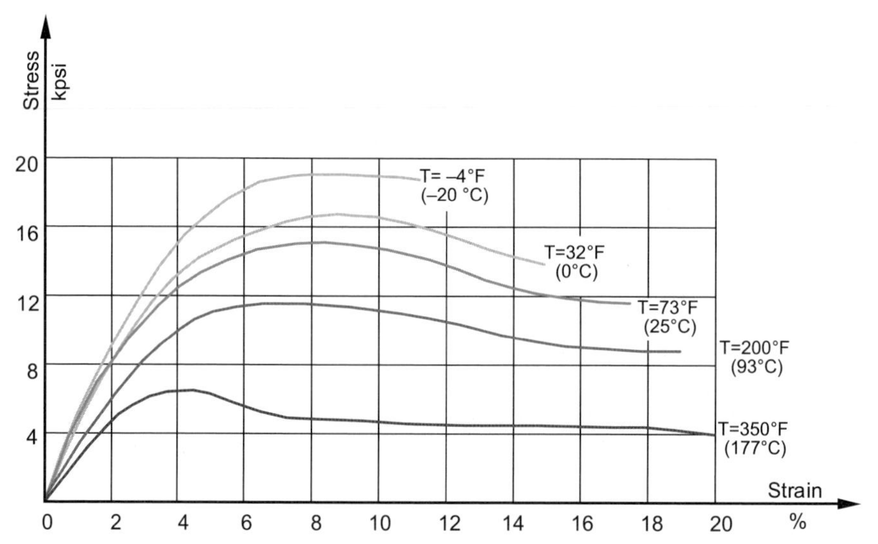

Figure 1.43 Temperature influence on the stress-strain curves

At -20°C (Fig. 1.43) the stress-strain curve of a polymer exhibits the highest strength values. The high strength occurs because the polymer contracts, packing the molecules closer together. The movement of the molecules is restricted by space, thereby lowering the strain values and making the polymer more brittle.

As temperature increases, the polymer expands, allowing more space for molecular movement, as shown in the higher temperature curves in Fig. 1.43. The material becomes more elastic and unable to sustain the same high stress value it possessed at -20°C.

1.11.7 Case History: Planetary Gear Life Durability

The area between the rib cage and pelvis of the human spine is called the "lumbar spine." In automotive seating, the portion of the seat back that contacts the lumbar spine area is known as "lumbar support" [178]. The purpose of a lumbar support system is to adjust the seat-back curvature to the most ergonomically correct position for the occupant's body. The planetary gear assembly we are discussing in this section is part of the lumbar support system of an automotive seat back, designed and manufactured by Leggett & Platt Company, Schukra Automotive Division.

In recent years this component of the seat back has seen a number of new developments, which allow the automobile driver or the passengers to adjust the lumbar portion of the seat up or down, according to the occupant's body height, and in or out, according to the occupant's spine curvature. In addition, a number of original automotive equipment manufacturers, especially for luxury cars, provide a variety of features such as seat-back heating and cooling, seat-back ventilation and, most recently, seat-back massage.

Figure 1.44 Actuator assembly of the lumbar support system having two push/pull cables and the electrical motor visible *(Courtesy of Leggett & Platt Company, Schukra Automotive Division)*

To adjust the lumbar support, the occupant can simply press a button to electronically activate the adjustable lumbar support system. The mechanical adjustment is achieved by transmitting the motion via an actuator. The actuator (Fig. 1.44) is composed of a two-stage gearbox. The first stage of the gearbox contains a worm gear and a worm wheel, while the second stage consists of a planetary gear train (Fig. 1.45).

The actuator transforms the rotational motion of the electrical motor into a push/pull linear motion by wrapping and unwrapping the metal cable, which activates the in/out motion of the lumbar support corresponding to the occupant's lumbar spine curvature.

Figure 1.45 Actuator assembly of the lumbar support system with the cover removed. The location of the two push/pull cable mountings is visible above. The 12-tooth sun gear is in the center. The three 14-tooth planet gears engage with the sun gear in the center as well as with the surrounding ring gear *(Courtesy of Leggett & Platt Company, Schukra Automotive Division)*

The project has passed through all developmental stages, reaching production. All necessary performance requirements were achieved. Only minor modifications were made to the eight-cavity injection-molding tool used to manufacture the planetary gears. Acetal homopolymer, a semicrystalline resin, was specified for planetary gears, the ring gear, and the wheel gear. Polyamide 6,6, also a semicrystalline resin, was specified for the worm gear, which engages the acetal homopolymer wheel gear when the actuator is in the "on" position.

As stated above, the actuator of the lumbar support system has a two-stage gearbox. The durability requirements call for the progressive loading of the actuator from 0 to 400 Newton. However, when the motor is in a stalled condition, the gearbox is exposed to the highest torque of 71.3 N · m. This torque is considered to be the worst-case loading of the gearbox. The first stage of the two-stage gearbox changes the axis of rotation of the motor by 90° using a worm and a worm gear (also called worm wheel) assembly. The worm is a single-start worm, and the worm gear has 88 teeth. The mesh efficiency of the two engaging gears is 34%. The second stage of the gearbox consists of one sun gear (center gear shown in Fig. 1.45), which is surrounded by three 14-tooth planet gears. The planet gears also engage the 42-tooth ring gear.

Individual gear calculations of the two-stage gearbox of the initial design are based on the Lewis parabola method, using the highest point of single contact (HPSC), with both gears having the minimum material condition. The following input assumptions were used:

- Number of gear teeth = 14
- Metric module = 1.00
- Pressure angle = 20°
- Tooth arc thickness = 1.83 mm
- Hob addendum = 1.33 mm
- Hob tip radius = 0.43 mm
- Roll angle for loading = 42.556°
- Tooth face width = 4.5 mm
- Gear input torque = 953.5 N·mm
- Root diameter = 12.052 mm
- Y =Lewis form factor = 0.37039
- Load tangent to base circle = 144.956 N
- Load normal to tooth center line = 119.872 N
- Parabola height = 1.72
- Parabola base = 1.995
- Bending moment arm = 1.422 N·mm
- Load radius at center of tooth = 7.954 N

The total tooth bending stress obtained was

$$\sigma_{\text{Bending}} = 71.92 \text{ MPa} \tag{1.5}$$

After about six months of production, during validation for a new automotive program, the planet gears started breaking at around 9,000 cycles instead of the regular specified requirement of almost 19,000 durability cycles without failure. Most failures were rupture of the tooth base. There were also some planet gears that were cracking through the rim.

The first analysis conducted for the failed parts was microstructural analysis (MSA). The equipment for MSA consists of a microscope with a magnification level of 100×, a polarized light source, a camera, a microtome cutter, and a glass support for

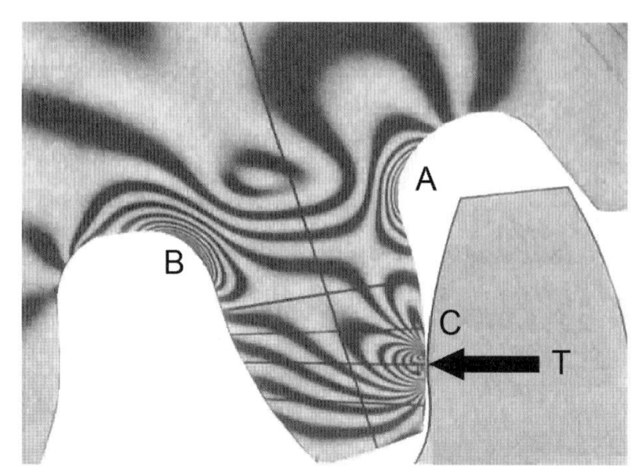

Figure 1.46 Photoelastic analysis of two gears in contact shows the two types of high stress in the teeth: A and B areas correspond respectively to tensile and compressive stresses due to the bending of the tooth. The compressive stress has a greater magnitude due to the radial component of the tooth force *T*. While at C location, contact stresses are present due to roll and slide of the two approximately cylindrical surfaces during the tooth contact *(Courtesy of ETS Inc.)*

microscopy. The analysis, also known as microtoming (see Fig. 1.46), is a technique that allows the analyst to use optical microscopy to investigate why glass-reinforced and unreinforced semicrystalline polymers fail. These polymers include the Delrin® 111P homopolymer acetal from DuPont, which had been used for the failed gears.

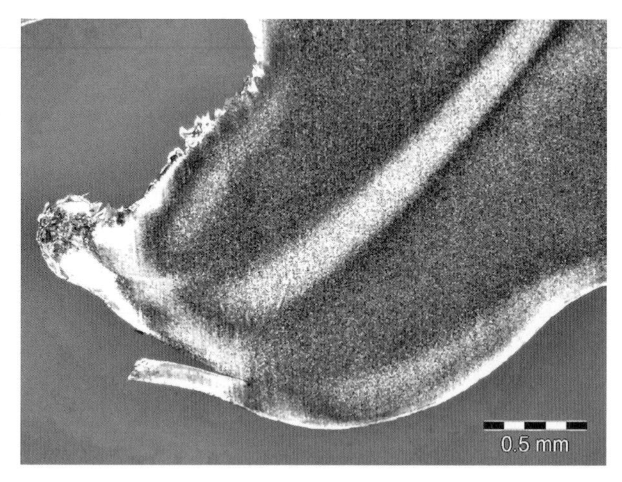

Figure 1.47 Microstructural analysis (MSA) of failed planet gear. Weld line (also known as knit line) is visible in the middle of the failed gear through bending tooth. Magnification factor is 70x

Microtoming provides a thin section for microscopic inspection within minutes. The cutting knife used to slice the specimen should be angled around 40 to 45° with respect to the part. Tungsten carbide tips provide the most precise cutting. The preferred thickness of the cut sample should vary between 0.01 and 0.025 mm (0.0004 and 0.001 in.). A drop of Canada balsam is spread on the microscope glass slide, which measures 3 in. by 1 in. Using tweezers, the thin slice obtained is then mounted onto the slide and flattened out because it has a tendency to curl. Once the specimen is flattened, the cover glass, with balsam on one of its surfaces, is placed on top, forming a sandwich assembly, with the specimen in the middle. The resulting assembly is heated to 50 to 60°C for a few minutes, and then it is allowed to cool down under a half-pound weight. Now it is ready to be analyzed.

MSA analysis shows that the resin has been well processed with homogenous melt throughout the part and correct melt temperature. The knit line is also visible (see Fig. 1.47), but not the cause of tooth failure. The microtome image shows high loads deforming the tooth, which is bent and then ruptured.

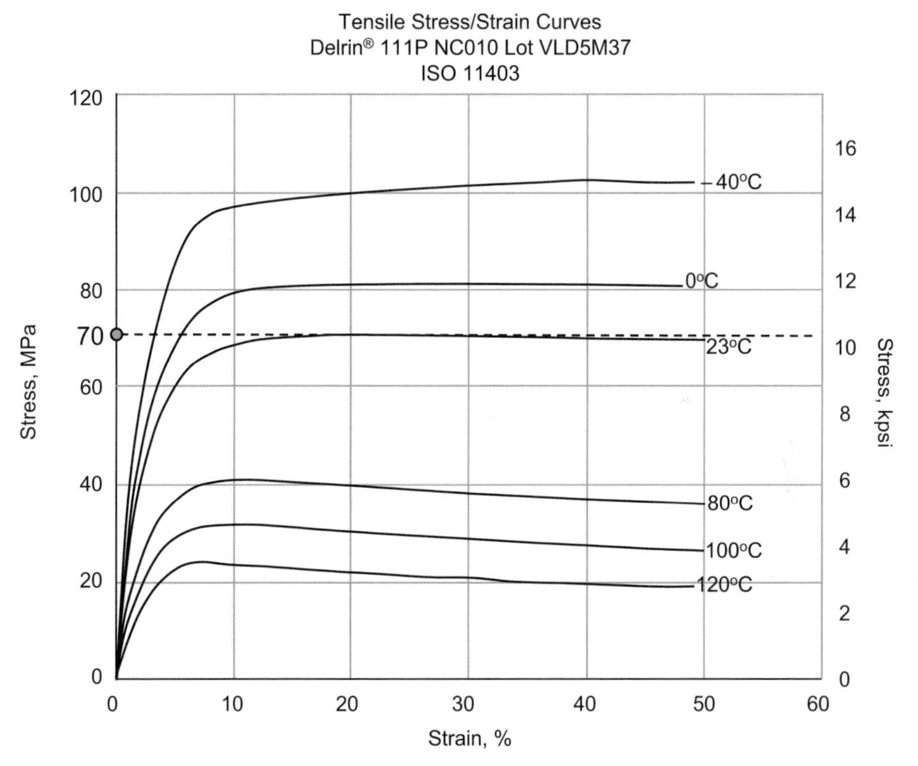

Figure 1.48 Stress/strain curves at six different temperatures (–40°C, 0°C, +23°C, +80°C, +100°C, +120°C) of Delrin® 111P, natural color *(Courtesy of ETS Inc.)*

Stress-strain curves for Delrin® 111P NC010 at various temperatures are shown in Fig. 1.46. It can be observed that when a tensile bar ASTM or ISO specimen is exposed to a stress of 71 MPa at room temperature (23°C or 73°F) the specimen will break. For example, if the temperature at which the test is conducted decreases to –40°C, the stress under which the Delrin® 111P will break increases to 105 MPa.

However, even if the temperature was kept precisely at room temperature, the friction between actuator components generated additional heat. Using an infrared laser gun probe, it was established that the actual temperature during actuator operation was 35°C, roughly 50% higher than room temperature.

The temperature of 23°C is borderline for the planet gears when exposed to tooth bending stresses of 71 MPa (see Eq. 1.5), representing the stalled condition of the electric motor. The actuator operating temperature is even higher (35°C). When the electric motor stalls at an operating temperature of 35°C, the mechanical strength of the acetal homopolymer drops below 70 MPa, to somewhere around 58 MPa. (For a detailed explanation of the graphic interpolation procedure used here, see Chapter 6.)

Although the failure mechanism of the planet gears is complex, 60 to 70% of the failure mechanism can be directly attributed to the fact that the planetary gears are exposed to higher stresses than they are capable of sustaining. The remaining 30 to 40% of the failure mechanism can be attributed to the ovality of the ring gear and to the fact that it is undersized. The combination of undersized ring gear diameter and its ovality (two opposite gates are used to mold the part) create an increased load on the planetary gear teeth, which fail in an unpredictable fashion.

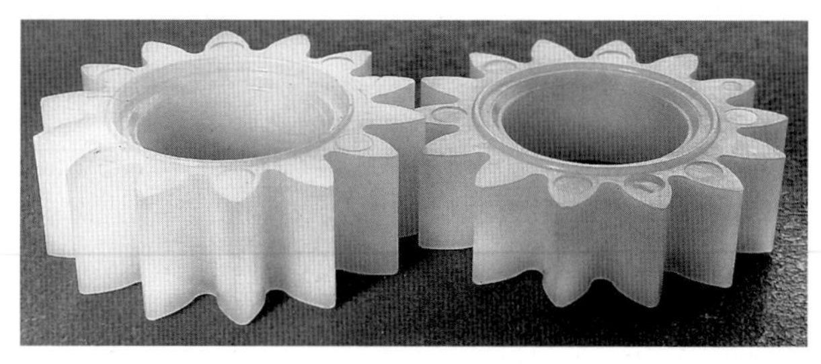

Figure 1.49 Planet gear: (right) old design with a tooth width of 4.4 mm and (left) new design with a tooth width of 5.4 mm *(Courtesy of Leggett & Platt Company, Schukra Automotive Division)*

The high tooth bending stresses have been successfully addressed when the planet gear tooth width increased from 4.4 mm to 5.4 mm (see Fig. 1.49). That is a 23% increase in the contact area between the engaged gears, corresponding to a similar drop in the tooth-bending stress.

Please note that when designing plastic parts made of polymers that have a significant variation of mechanical properties due to temperature gradient, such as acetals, the operating temperature of the component should always be checked [179].

2 Understanding Safety Factors

2.1 What Is a Safety Factor

The safety factor is a measurement of a product's ability to perform throughout its life expectancy. Ideally, after the life expectancy has been reached, the product should fail.

The safety factor of a bridge, for example, might be quantified with the number 10, while the safety factor of an aircraft could be 4. A bridge might be expected to perform over hundreds or even thousands of years, while an airplane's life expectancy is considerably less. It should be noted that an aircraft with a safety factor of 10 would never leave the ground. Also, the precision of the product must be taken into account. Aircraft parts are built to very precise tolerances, while a bridge's tolerances are comparatively high.

The safety factor is a coefficient the wise engineer has to take into consideration when designing a part. This coefficient gives assurance that a part will not fail under any operating conditions. In addition, it is proof that the material has been chosen correctly for those operating conditions. It also covers imperfections from processing of this material until it becomes a finished part. Safety factors can be divided into several categories:

- design factor
- material property safety factor
- processing safety factor
- operating condition safety factor

It is important to note that these categories are interrelated, and that continuing feedback to the design stage with regard to these safety factors is necessary.

2.2 Using the Safety Factors

2.2.1 Design Safety Factors

This is the most important category, largely because the designer requires input from all other categories. Material properties, processing, and operating conditions must all be considered by the designer.

At the design stage the engineer will start by choosing a material and designing the part to withstand external loading. If the finished product is expected to work under various adverse operating conditions, the engineer will perform stress-analysis calculus in order to ensure that the finished part will not fail throughout its life cycle. Depending upon the type of loading that will be applied to the finished part, design safety factors can be broken down into the following categories:

- design static safety factor
- design dynamic safety factor
- design time-related safety factor

Safety factors from other categories can be added at this stage.

NOTE: In the following design factor formulae, stress was chosen for convenience. Related strain or forces can also be used.

2.2.1.1 Design Static Safety Factor

When external loads are applied statically, a relatively simple calculation is performed and the safety factor is related to material allowable/permissible stress.

2.2.1.2 Design Dynamic Safety Factor

For cases where external loads are applied intermittently or in cycles, and the material is subjected to fatigue, safety factors will be higher than static ones and design stresses will be lower.

2.2.1.3 Design Time-Related Safety Factor

Creep and stress-relaxation are the most common time-related effects on thermoplastic polymers and are a major factor in determining the life expectancy of the product. In order to forecast the life expectancy, we can apply the theory that *initial design safety factor decreases with time.*

$$n = \frac{\sigma_{\text{Ultimate}\ \text{time=end}}}{\sigma_{\text{Allowable}\ \text{time=end}}} \tag{2.1}$$

In Fig. 2.1 the initial design (time = 0) safety factor n is greater than 1. Safety factor n reaches 1 when the product fails (time = end).

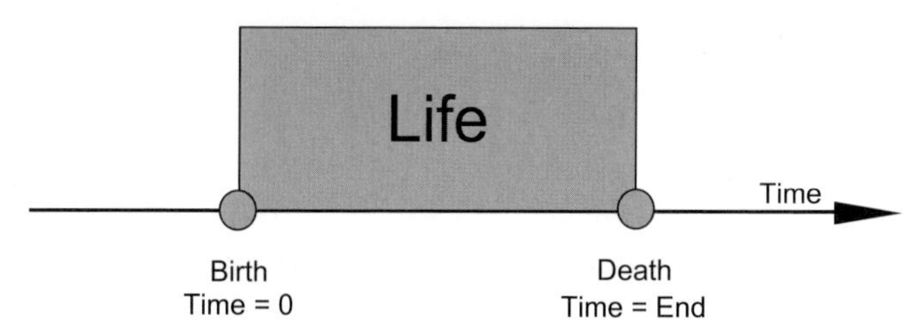

Figure 2.1 Creep

Product life expectancies vary between industries. Some products are expected to be completely rebuilt or replaced after 10 years; others may be replaced after less than one year. The safety factor is therefore determined by the norms (specifications) of the industry.

For a new product in a new industry with no established product life expectancies, testing the product and measuring the stress level for failure can determine a safety factor. These values, measured against original material properties, will provide a good starting point for establishing time-related safety factors.

2.2.2 Material Properties Safety Factor

If a safety factor can be estimated from experience, testing, or any other reliable means, then the maximum allowable stress is defined as:

$$\sigma_{\text{Allowable}} = \frac{\sigma_{\text{Yield}}}{n} \tag{2.2}$$

For snap-fitting and press-fitting calculations the material safety factor is based on yield stress (or strain) because it is more accurate for the elastic region.

$$n_{\text{SnapFit/PressFit}} = \frac{\sigma_{\text{Yield}}}{\sigma_{\text{Allowable}}} \tag{2.3}$$

For living-hinge calculations the material safety factor is based on ultimate stress.

$$n_{\text{Living Hinge}} = \frac{\sigma_{\text{Ultimate}}}{\sigma_{\text{Allowable}}} \tag{2.4}$$

This is more accurate for visco-elastic properties where large displacements and plastic deformation often occur.

The two safety factors mentioned take into account the following characteristics:

- imperfections of the material
- inclusions
- voids
- RH content

- reinforcement
- thermal treatment

The above factors are generally used as a sort of checklist to help in the selection of polymeric materials. In the predesign stages, it can reduce the virtually limitless array of possibilities to a more manageable–but not short–list of possibilities.

2.2.3 Processing Safety Factors

The processing safety factor covers imperfections caused during the injection-molding process. Some of the potential imperfections due to the process are

- knit (weld) lines
- cycle time
- voids
- stress concentrators

2.2.4 Operating Condition Safety Factor

Special operating conditions like extreme heat or cold or high relative humidity, ultraviolet exposure, saline water immersion, or a corrosive agent environment–in other words, weatherability conditions–can be covered by an *operating condition safety factor.*

3 Strength of Material for Plastics

3.1 Tensile Strength

Tensile strength is a material's ability to withstand an axial load.

In an ISO test of tensile strength, a specimen bar (Fig. 3.1) is placed in a tensile testing machine. Both ends of the specimen are clamped into the machine's jaws, which pull both ends of the bar. Stress is automatically plotted against the strain. The axial load is applied to the specimen when the machine pulls the ends of the specimen bar in opposite directions at a slow and constant rate of speed. Two different pull speeds are used: 0.2 in. per minute (5 mm/min) to approximate the material's behavior in a hand assembly operation; and 2.0 in. per minute (50 mm/min) to simulate semiautomatic or automatic assembly procedures. The bar is marked with gauge marks on either side of the midpoint of the narrow, middle portion of the bar.

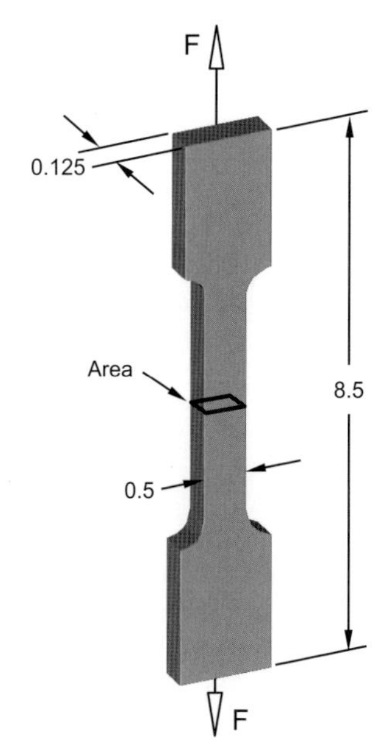

Figure 3.1
ISO test specimen bar

As the pulling progresses, the specimen bar elongates at a uniform rate that is proportionate to the rate at which the load or pulling force increases. The load divided by the cross-sectional area of the specimen within the gauge marks represents the *unit stress resistance* of the plastic material to the pulling or tensile force.

The stress (σ, sigma) is expressed in pounds per square inch (psi) or in mega-Pascal (MPa). One MPa equals 1 Newton per square millimeter (N/mm^2). To convert psi into MPa, multiply by 0.0069169. To convert MPa into psi, multiply by 144.573.

$$\sigma = \frac{F}{A} = \frac{\text{Tensile load}}{\text{Area}} \tag{3.1}$$

3.1.1 Proportional Limit

The proportional relationship of force to elongation, or of stress to strain, continues until the elongation no longer complies with the Hooke's law of proportionality. The greatest stress that a plastic material can sustain without any deviation from the law of proportionality is called *proportional stress limit* (Fig. 3.2).

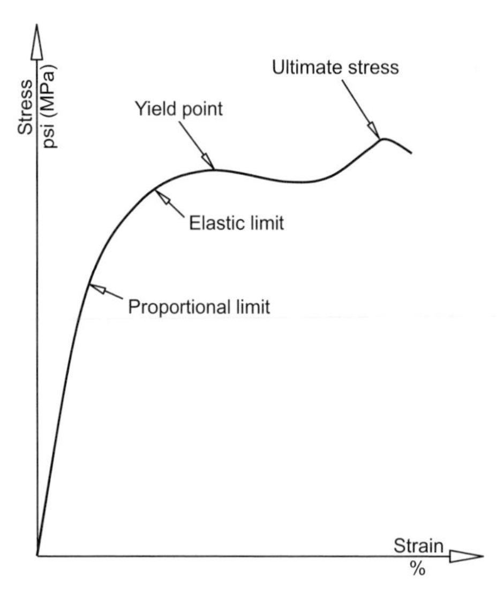

Figure 3.2
Typical stress/strain diagram for plastic materials

3.1.2 Elastic Stress Limit

Beyond the proportional stress limit, plastic material exhibits an increase in elongation at a faster rate. *Elastic stress limit* is the greatest stress a material can withstand without sustaining any permanent strain after the load is released (Fig. 3.2).

3.1.3 Yield Stress

Beyond the elastic stress limit, further movement of the test machine jaws in opposite directions causes a permanent elongation or deformation of the specimen. There is a point beyond which the plastic material stretches briefly without a noticeable increase in load. This point is known as the *yield point*. Most unreinforced materials have a distinct yield point. Reinforced plastic materials exhibit a *yield region*.

It is important to note that the results of this test will vary between individual specimens of the same material. If ten specimens made out of a reinforced plastic material were given this test, it is unlikely that two specimens would have the same yield point. This variance is induced by the bond between the reinforcement and the matrix material.

3.1.4 Ultimate Stress

Ultimate stress is the maximum stress a material takes before failure.

Beyond the plastic material's elastic limit, continued pulling causes the specimen to neck down across its width. This is accompanied by a further acceleration of the axial elongation (deformation), which is now largely confined within the short necked-down section.

The pulling force eventually reaches a maximum value and then falls rapidly, with little additional elongation of the specimen before failure occurs. In failing, the specimen test bar breaks in two within the necking-down portion. The maximum pulling load, expressed as stress in psi or in N/mm^2 of the original cross-sectional area, is the plastic material's ultimate tensile strength ($\sigma_{Ultimate}$).

The two halves of the specimen are then placed back together, and the distance between the two marks is measured. The increase in length gives the elongation, expressed as a percent. The cross section at the point of failure is measured to obtain the reduction in area, which is also expressed as a percent. Both the elongation percentage and the reduction in area percentage suggest the material ductility.

In structural plastic part design it is essential to ensure that the stresses that would result from loading will be within the elastic range. If the elastic limit is exceeded, permanent deformation takes place due to plastic flow or slippage along molecular slip planes. This will result in permanent plastic deformations.

3.2 Compressive Stress

Compressive stress is the compressive force divided by cross-sectional area, measured in psi or MPa.

It is general practice in plastic part design to assume that the compressive strength of a plastic material is equal to its tensile strength. This can also apply to some structural design calculations, where Young's modulus (modulus of elasticity) in tension is used, even though the loading is compressive.

The ultimate compressive strength of thermoplastic materials is often greater than the ultimate tensile strength. In other words, most plastics can withstand more compressive surface pressure than tensile load.

The compressive test is similar to that of tensile properties. A test specimen is compressed to rupture between two parallel platens. The test specimen has a cylindrical shape, measuring 1 in. (25.4 mm) in length and 0.5 in. (12.7 mm) in diameter. The load is applied to the specimen from two directions in axial opposition. The ultimate compressive strength is measured when the specimen fails by crushing.

A stress/strain diagram is developed during the test, and values are obtained for four distinct regions: the proportional region, the elastic region, the yield region, and the ultimate (or breakage) region.

The structural analysis of thermoplastic parts is more complex when the material is in compression. Failure develops under the influence of a bending moment that increases as the deflection increases. A plastic part's geometric shape is a significant factor in its capacity to withstand compressive loads.

$$\sigma = \frac{P}{A} = \frac{\text{Compressive force}}{\text{Area}} \tag{3.2}$$

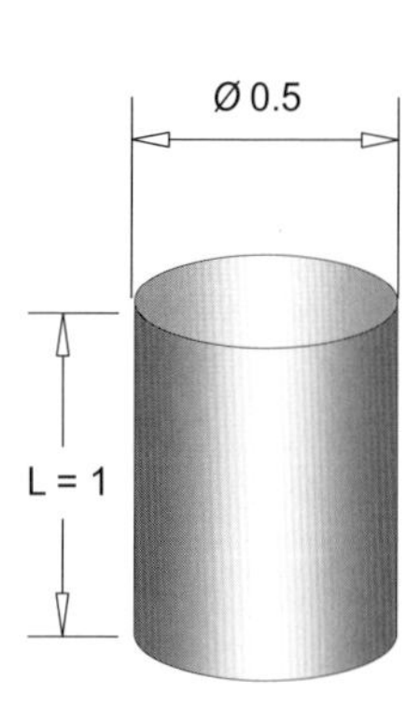

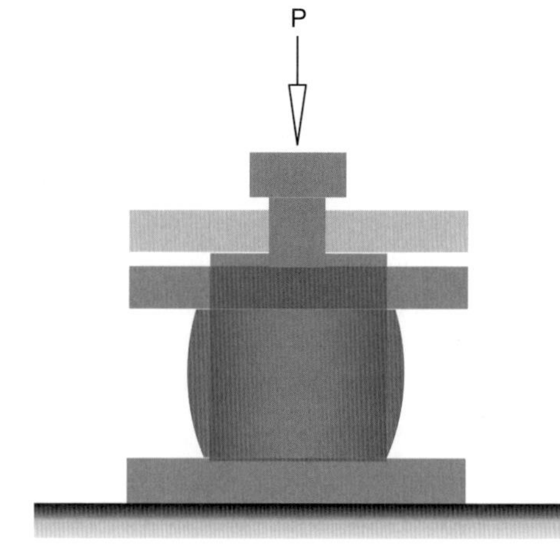

Figure 3.3 Compressive test specimen

The stress/strain curve in compression is similar to the tensile stress/strain diagram, except the values of stresses in the compression test are greater for the corresponding elongation levels. This is because it takes much more compressive stress than tensile stress to deform a plastic.

3.3 Shear Stress

Shear stress is the shear load divided by the area resisting shear. Tangential to the area, shear stress is measured in psi or MPa.

There is no recognized standard method of testing for shear strength (τ, tau) of a thermoplastic or thermoset material. Pure shear loads are seldom encountered in structural part design. Usually, shear stresses develop as a by-product of principal stresses or where transverse forces are present.

The ultimate shear strength is commonly observed by actually shearing a plastic plaque in a punch-and-die setup. A ram applies varying pressures to the specimen. The ram's speed is kept constant so only the pressures vary. The minimum axial load that produces a punch-through is recorded. This is used to calculate the ultimate shear stress.

Exact ultimate shear stress is difficult to assess, but it can be successfully approximated as 0.75 of the ultimate tensile stress of the material.

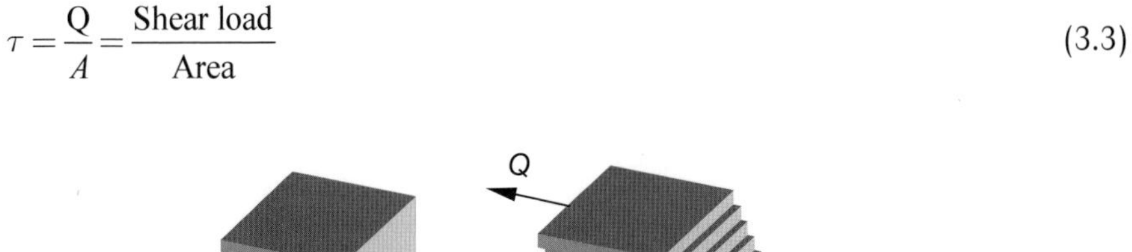

$$\tau = \frac{Q}{A} = \frac{\text{Shear load}}{\text{Area}} \tag{3.3}$$

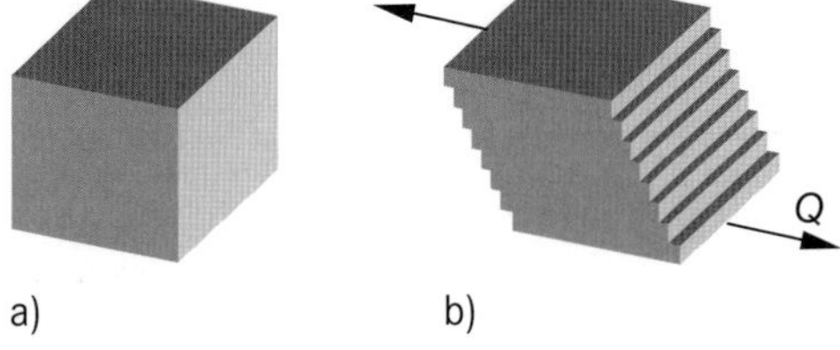

Figure 3.4 Shear stress sample specimen example: (a) before; (b) after

3.4 Torsion Stress

Torsional loading is the application of a force that tends to cause the member to twist about its axis (Fig. 3.5).

Torsion is referred to in terms of torsional moment or torque, which is the product of the externally applied load and the moment arm. The moment arm represents the distance from the centerline of rotation to the line of force and perpendicular to it.

The principal deflection caused by torsion is measured by the angle of twist or by the vertical movement of one side.

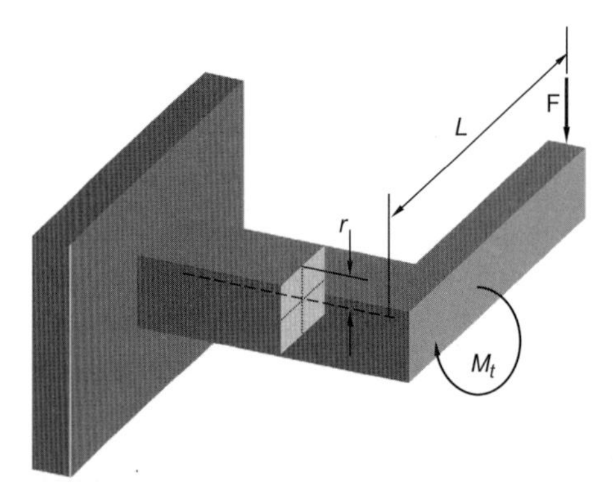

Figure 3.5
Torsion stress

When a shaft is subjected to a torsional moment or torque, the resulting shear stress is

$$\tau = \frac{M_t}{J} = \frac{M_t r}{I} \tag{3.4}$$

The torsional moment is M_t:

$$M_t = FR \tag{3.5}$$

The following notation has been used:

J = polar moment of inertia

I = moment of inertia

R = moment arm

r = radius of gyration (distance from the center of section to the outer fiber)

F = load applied

3.5 Elongations

Elongation (ε, epsilon) is the deformation of a thermoplastic or thermoset material when a load is applied at the ends of the specimen test bar in opposite axial directions.

The recorded deformation, depending upon the nature of the applied load (axial, shear, or torsional), can be measured in variation of length or in variation of angle.

Strain is a ratio of the increase in elongation by the initial dimension of a material. Again, strain is dimensionless.

Depending on the nature of the applied load, strains can be tensile, compressive, or shear.

$$\varepsilon = \frac{\Delta L}{L}\ (\%) \tag{3.6}$$

3.5.1 Tensile Strain

A test specimen bar similar to that described in Section 3.1 is used to determine the *tensile strain*. The ultimate tensile strain is determined when the test specimen, being pulled apart by its ends, elongates. Just before the specimen breaks, the ultimate tensile strain is recorded.

The elongation of the specimen represents the strain (ε, epsilon) induced in the material and is expressed in inches per inch of length (in./in.) or in millimeters per millimeter (mm/mm). This is a dimensionless measure. Percent notations such as $\varepsilon = 3\%$ can also be used.

Figure 3.2 shows stress and strain plotted in a simplified graph.

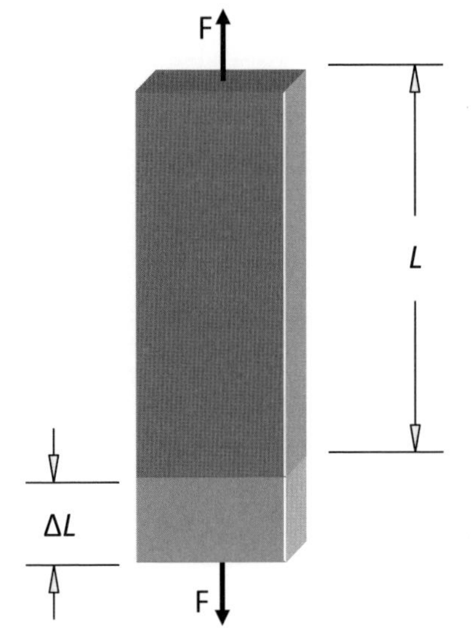

Figure 3.6
Tensile specimen loaded showing dimensional change in length. The difference between the original length (L) and the elongated length is ΔL

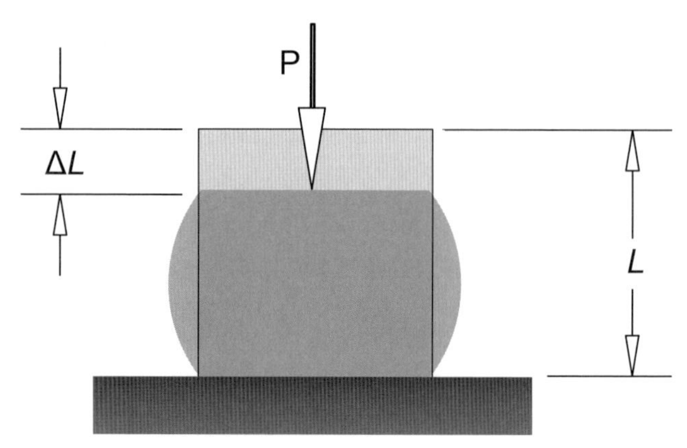

Figure 3.7 Compressive specimen showing dimensional change in length. L is the original length; P is the compressive force. ΔL is the dimensional change in length

3.5.2 Compressive Strain

The *compressive strain* test employs a setup similar to the one described in Section 3.2. The ultimate compressive strain is measured at the instant just before the test specimen fails by crushing.

3.5.3 Shear Strain

Shear strain is a measure of the angle of deformation γ (gamma). As is the case with shear stress, there is also no recognized standard test for shear strain.

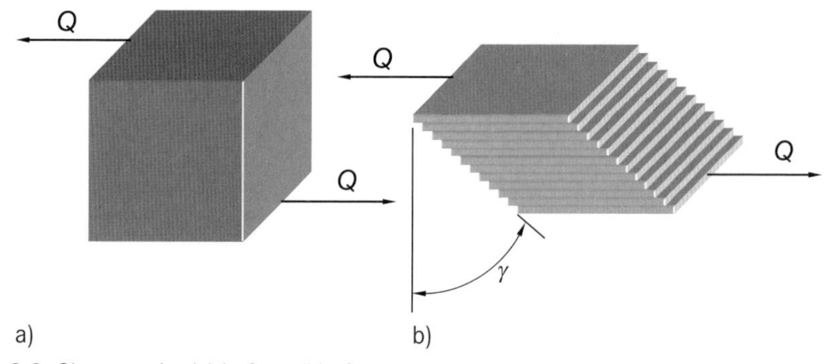

Figure 3.8 Shear strain: (a) before; (b) after

3.6 True Stress and Strain vs. Engineering Stress and Strain

Engineering strain is the ratio of the total deformation over initial length.

Engineering stress is the ratio of the force applied at the end of the test specimen by initial constant area.

True stress is the ratio of the instantaneous force over instantaneous area.

Equation 3.7 shows that the true stress is a function of engineering stress multiplied by a factor based on engineering strain.

$$\sigma_{\text{True}} = \sigma(1+\varepsilon) \tag{3.7}$$

True strain is the ratio of instantaneous deformation over instantaneous length.

Equation 3.8 shows that the true strain is a logarithmic function of engineering strain.

$$\varepsilon_{\text{True}} = \ln(1+\varepsilon) \tag{3.8}$$

By using the ultimate strain and stress values, we can easily determine the true ultimate stress as

$$\sigma_{\text{Ultimate True}} = \sigma_{\text{Ultimate}}(1+\varepsilon_{\text{Ultimate}}) \tag{3.9}$$

Similarly, by replacing engineering strain for a given point in 3.8 with engineering ultimate strain, we can easily find the value of the true ultimate strain as

$$\varepsilon_{\text{Ultimate True}} = \ln(1+\varepsilon_{\text{Ultimate}}) \tag{3.10}$$

Both true stress and true strain are required input as material data in a variety of finite element analyses where nonlinear material analysis is needed.

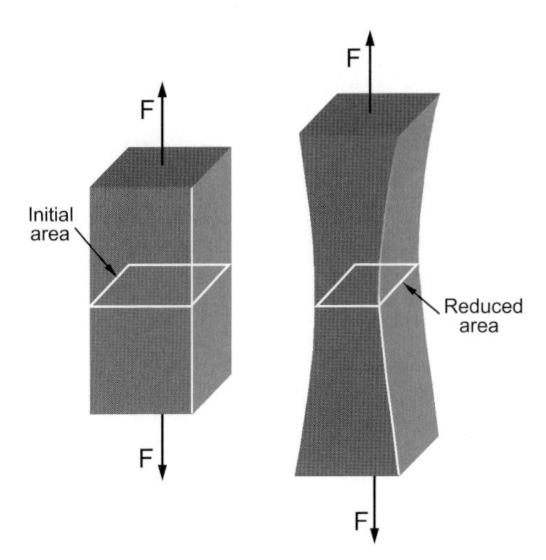

Figure 3.9
True stress necking-down effect

3.7 Poisson's Ratio

Provided the material deformation is within the elastic range, the ratio of lateral to longitudinal strains is constant, and the coefficient is called *Poisson's ratio* (ν, nu).

$$\nu = \frac{\text{Lateral strain}}{\text{Longitudinal strain}} \quad (3.11)$$

In other words, stretching produces an elastic contraction in the two lateral directions. If an elastic strain produces no change in volume, the two lateral strains will be equal to half the tensile strain times minus one (–1).

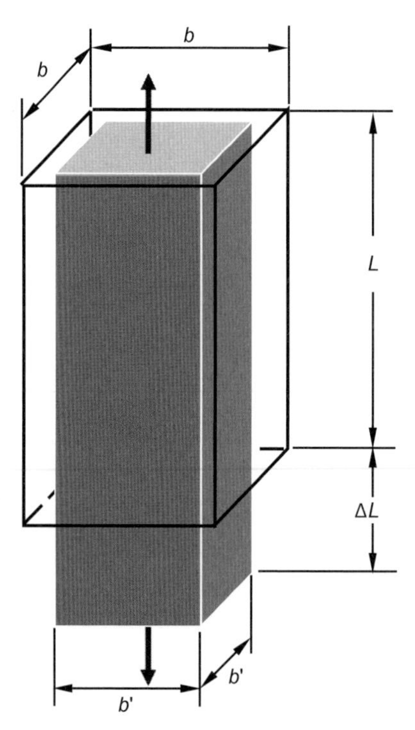

Figure 3.10
Dimensional change in only two of three directions

Under a tensile load, a test specimen increases (decreases for a compressive test) in length by the amount ΔL and decreases in width (increases for a compressive test) by the amount Δb. The related strains are:

$$\varepsilon_{\text{Longitudinal}} = \frac{\Delta L}{L}$$

$$\varepsilon_{\text{Lateral}} = \frac{\Delta b}{b} \quad (3.12)$$

Poisson's ratio varies between zero (0), where no lateral contraction is present, to half (0.5) for which the contraction in width equals the elongation. In practice there are no materials with Poisson's ratio zero or half.

Table 3.1 Typical Poisson's Ratio Values for Different Materials

Material Type	Poisson's Ratio at 0.2 in./min (5 mm/min) Strain Rate
ABS	0.4155
Aluminum	0.34
Brass	0.37
Cast iron	0.25
Copper	0.35
High-density polyethylene	0.35
Lead	0.45
Polyamide	0.38
Polycarbonate	0.38
13% glass-reinforced polyamide	0.347
Polypropylene	0.431
Polysulfone	0.37
Steel	0.29

The lateral variation in dimensions during the pull-down test is

$$\Delta b = b - b' \tag{3.13}$$

Therefore, the ratio of lateral dimensional change to the longitudinal dimensional change is

$$\nu = \frac{\frac{\Delta b}{b}}{\frac{\Delta L}{L}} \tag{3.14}$$

Or, by rewriting, the Poisson's ratio is

$$\nu = \frac{\varepsilon_{\text{Lateral}}}{\varepsilon_{\text{Longitudinal}}} \tag{3.15}$$

3.8 Modulus of Elasticity

3.8.1 Young's Modulus

The *Young's modulus* or *elastic modulus* is typically defined as the slope of the stress/strain curve at the origin.

The ratio between stress and strain is constant, obeying Hooke's Law, within the elasticity range of any material. This ratio is called Young's modulus and is measured in MPa or psi.

$$E = \frac{\sigma}{\varepsilon} = \frac{\text{Stress}}{\text{Strain}} = \text{Constant} \tag{3.16}$$

Hooke's Law is generally applicable for most metals, thermoplastics, and thermosets, within the limit of proportionality.

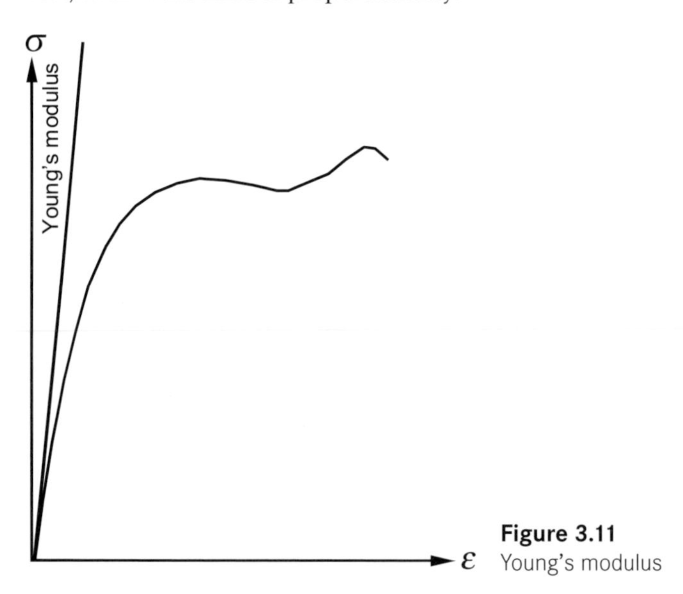

Figure 3.11 Young's modulus

3.8.2 Tangent Modulus

The instantaneous tangent over the elasticity range of a stress/strain curve for a thermoplastic or thermoset material gives a better approximation of the relation between stress and strain. The stress/strain curve of most plastic materials has a curved, elastic range (see Fig. 3.12). In these cases, the use of Young's modulus is difficult and less accurate.

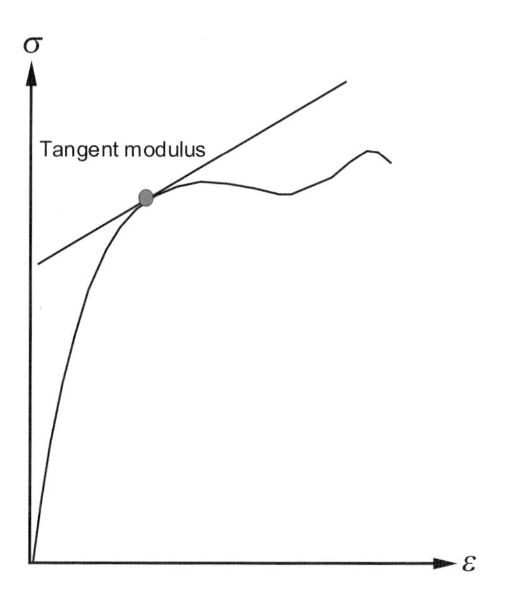

Figure 3.12
Tangent modulus

3.8.3 Secant Modulus

When the material stress/strain is nonlinear, as shown in Fig. 3.13, the use of Young's modulus describes the material to be less flexible than it actually is. This can result in the assignment of lower permissible or allowable deflections and higher loads for a specific stress level.

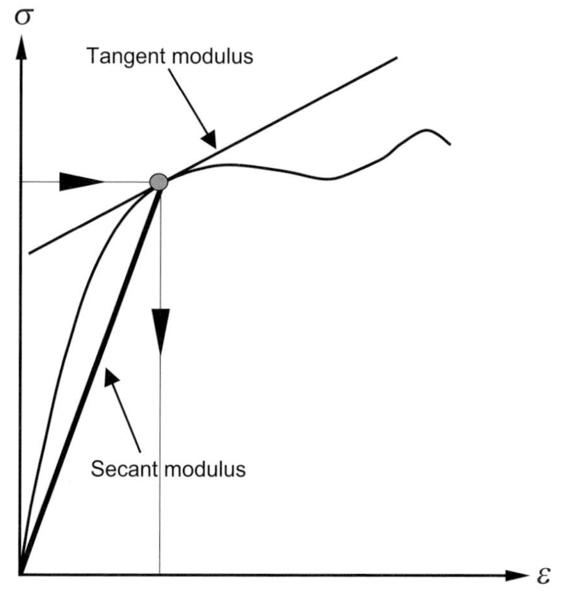

Figure 3.13
Secant modulus

The *secant modulus* represents the actual stiffness of a material at a specific strain or stress condition. It provides a more accurate load and deformation calculation.

The secant modulus is a good approximation of the stress/strain curve at exactly one point. If the calculation is conducted at another point on the curve, however, the degree of error can be excessive. This method is popular throughout the design community because it effectively copes with the visco-elastic properties of thermoplastic and thermoset materials.

Secant modulus appears to be a simple way to linearize a highly nonlinear stress/strain curve.

3.8.4 Creep (Apparent) Modulus

When a constant load is applied to a part, it induces an internal stress in that part. Over long periods of time, parts made of thermoplastic and thermoset materials will slowly deform to redistribute the internal stresses.

If stress condition and temperature are known and creep curves are also available at the respective temperature, then an apparent or creeping modulus can be calculated by employing the creep curves. The *creep modulus* formula is

$$E_c = \frac{\sigma}{\varepsilon_c} \tag{3.17}$$

The calculated stress condition (value) is σ, and ε_c is the value of the strain level corresponding to the given stress level from the creep curve at the expected time and temperature. *Apparent* or *creep modulus* is used to predict how a part will behave years in the future, bringing into account its projected stress level over its life expectancy.

3.8.5 Shear Modulus

Shear modulus, or *modulus of rigidity*, is directly comparable to the modulus of elasticity or Young's modulus.

The Hooke's equation for shear is

$$G = \frac{\tau}{\gamma} = \text{Constant} \tag{3.18}$$

The shear stress is τ (tau), and γ (gamma) is the angle of deformation. The ratio of the two is constant only within the elasticity region of a given material.

The equation that links Young's modulus to the shear modulus is

$$\frac{E}{G} = 2(1+\nu) \tag{3.19}$$

In 3.19, v represents the Poisson's ratio.

3.8.6 Flexural Modulus

The resistance to deformation in bending within the elastic range of a polymer is known as *stiffness in bending*. *Modulus of elasticity in bending*, also called *flexural modulus*, is a measure of this property. To illustrate this, a horizontal beam is supported at each end. A weight is placed at the center of the beam, exerting a vertical load. When the load causes the beam to bend, the beam exhibits two different stresses: compression stress above the neutral axis, and below it, tension stresses (Fig. 3.14).

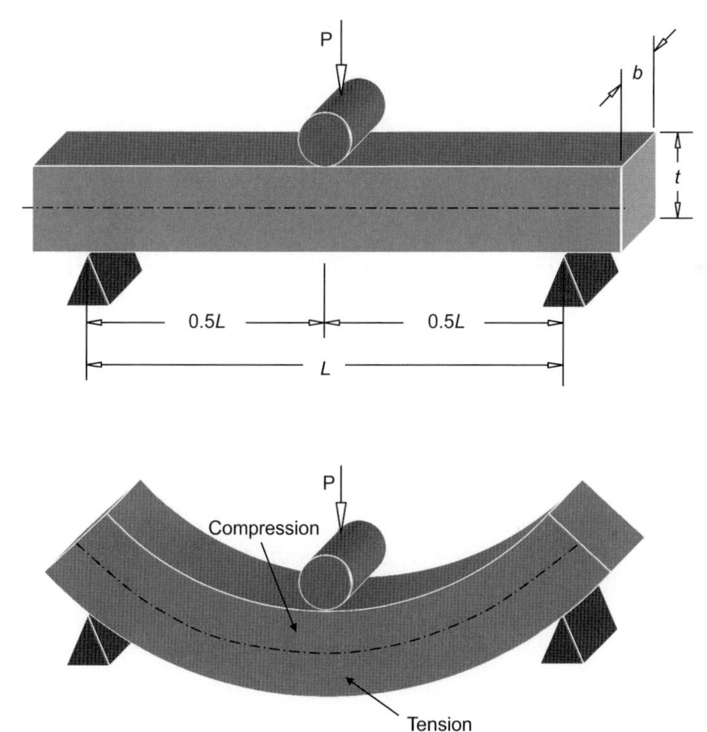

Figure 3.14 Flexural modulus

The relation between Young's modulus in tension, compression, and flexural modulus is

$$E_{\text{Flexural}} = \frac{4E_{\text{Tension}}E_{\text{Compression}}}{\left(\sqrt{E_{\text{Tension}}} + \sqrt{E_{\text{Compression}}}\right)^2} \quad (3.20)$$

3.8.7 The Use of Various Moduli

In the interest of accuracy and to minimize calculations and computing time, the examples presented in this book will use two moduli. Secant modulus will be used for press fitting, living hinges, and snap fitting. Creeping or apparent modulus will also be used for the press-fitting examples because stress relaxation curves for this specific material were not available. Also, creep curves were employed to approximate the material behavior over long periods of time.

3.9 Stress Relations

3.9.1 Introduction

In a three-dimensional (3-D) material element, six stresses can be defined. There will be three normal stresses in the *X*, *Y*, and *Z* directions (σ_X, σ_Y, σ_Z) and three shear stresses (τ_X, τ_Y, τ_Z). Corresponding to the six stresses there will be six strains (ε_X, ε_Y, ε_Z, γ_X, γ_Y, γ_Z).

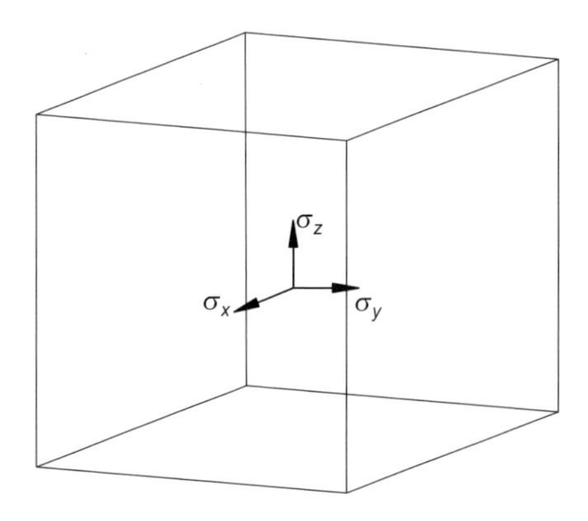

Figure 3.15
Three-dimensional material element

Complex loading is applied to the 3-D element. The vectorial summation of the three stresses will provide the *equivalent* level of stress for the element.

3.9.2 Experiment

In a simple tensile or compressive experiment a specimen test bar is loaded in tension or compression. One of the four parameters can be determined when the specimen is stretched or compressed to the limits of the polymer material:

Maximum normal stress = σ_{Ultimate}

Maximum normal strain = $\varepsilon_{\text{Ultimate}}$

Maximum shear stress = τ_{Ultimate}

Maximum deformation energy = W

3.9.3 Equivalent Stress

The variable $\sigma_{\text{Equivalent}}$ is defined as *equivalent stress* of three stress components σ_X, σ_Y, and σ_Z under complex loading, and its value is compared to one of the four (σ_{Ultimate}, $\varepsilon_{\text{Ultimate}}$, τ_{Ultimate}, and W) parameters of the tensile experiment.

Four classical analysis theories, also called theories of failure, are based on the comparison of equivalent stress with one of the following: ultimate stress, ultimate strain, ultimate shear stress and, finally, energy.

3.9.4 Maximum Normal Stress

The ultimate normal stress is reached in a 3-D complex loading when equivalent stress equals the normal stress in one of the three principal directions:

$$\sigma_{\text{Equivalent}} = \sigma_Z \tag{3.21}$$

By comparing the equivalent stress to the ultimate stress:

$$\sigma_{\text{Equivalent}} = \sigma_{\text{Maximum}} \tag{3.22}$$

Then:

$$\sigma_{\text{Equivalent}} = \sigma_Z = \sigma_{\text{Maximum}} \tag{3.23}$$

Failure will take place in a 3-D complex loading when $\sigma_{\text{Equivalent}}$ equals σ_{Ultimate} as in a simple tensile stress test. The results are not sufficiently accurate, but the calculations are rather simple.

Maximum stress (or Rankine) theory assumes that yielding of the element subject to combined stresses occurs when loading has reached a value such that one of the principal stresses becomes equal to the yield stress in simple tension or compression.

NOTE: For cases where the pressure is applied to an element from three directions, the element internal stress levels in the σ_X, σ_Y, and σ_Z directions go beyond the ultimate stress σ_{Ultimate} value as for a simple tensile stress test, and the theory doesn't apply.

3.9.5 Maximum Normal Strain

Conforming to Hooke's Law:

$$E_{\text{Equivalent}} = \frac{\sigma_{\text{Equivalent}}}{E} \tag{3.24}$$

The *maximum normal strain* theory (St. Venant's theory) assumes that failure by yielding in case of combined stresses occurs when the maximum value of the principal strain equals the value of the strain at yielding in a simple tension or compression.

The maximum normal strain is reached in a 3-D complex loading when the ultimate normal strain in a simple tensile test equals the equivalent strain in a complex 3-D loading test:

$$\varepsilon_{\text{Maximum}} = \frac{\sigma_{\text{Equivalent}}}{E} = \frac{\sigma_X - \nu(\sigma_Y + \sigma_Z)}{E} = \varepsilon_{\text{Equivalent}} \tag{3.25}$$

or, after simplifications:

$$\sigma_{\text{Equivalent}} = \sigma_X - \nu(\sigma_Y + \sigma_Z) \tag{3.26}$$

3.9.6 Maximum Shear Stress

The *maximum shear stress* theory is also known as Tresca or Coulomb theory. This theory assumes that yielding is reached when the maximum shear stress in the polymer becomes equal to the maximum shear stress at the yield point in a simple tension test.

Some materials (such as concrete) show 45°-oriented cracks when they are in normal tension or compression. In these cases, shear stress reaches the highest value ahead of any other stresses.

The shear stress is defined as being half the equivalent stress for a 3-D complex loading:

$$\tau_{\text{Equivalent}} = \frac{\sigma_{\text{Equivalent}}}{2} \tag{3.27}$$

Ultimate shear stress is reached in a 3-D complex loading case when the ultimate shear stress in a simple tensile or compressive test equals the equivalent shear stress level for a complex 3-D loading case test:

$$\tau_{\text{Maximum}} = \tau_Y = \frac{\sigma_X - \sigma_Z}{2} = \tau_{\text{Equivalent}} \tag{3.28}$$

or, expressing it as a function of normal and equivalent stresses:

$$\sigma_{\text{Equivalent}} = \sigma_X - \sigma_Z \tag{3.29}$$

NOTE: In a 2-D plane stress situation, the above equation is called *Mohr's circle method.*

3.9.7 Maximum Deformation Energy

The energy absorbed by a material element subjected to a 3-D complex situation loading is known as *maximum deformation energy*. The energy theory assumes that the energy absorbed in the material element is the same as the energy a similar material element will absorb in a simple tensile test up to the breaking point of the test specimen.

Therefore the equation for the maximum deformation energy is

$$W = \frac{\sigma_X^2 + \sigma_Y^2 + \sigma_Z^2}{2E} - \frac{\nu\left(\sigma_X\sigma_Y + \sigma_Y\sigma_Z + \sigma_Z\sigma_X\right)}{E} = \frac{\sigma_{\text{Equivalent}}^2}{2E} \tag{3.30}$$

and

$$W = W_{\text{Shape}} + W_{\text{Volume}} \tag{3.31}$$

The total maximum deformation energy is the sum of shape deformation energy, W_{Shape}, and the volume deformation energy, W_{Volume}:

$$W_{\text{Volume}} = \frac{1-2\nu}{6E}\left(\sigma_X + \sigma_Y + \sigma_Z\right)^2 \tag{3.32}$$

The shape deformation energy is

$$W_{\text{Shape}} = \frac{1-2\nu}{6E}\left[\left(\sigma_X - \sigma_Y\right)^2 + \left(\sigma_Y - \sigma_Z\right)^2 + \left(\sigma_Z - \sigma_X\right)^2\right] \tag{3.33}$$

The definition of the average or mean stress is

$$\sigma = \frac{\sigma_X + \sigma_Y + \sigma_Z}{3} \tag{3.34}$$

For positive values of the mean stress levels, materials deform both in shape and in volume. These thermoplastic and thermoset materials are referred to as tough materials. For tough materials, the equivalent stress, expressed as a function of principal stresses, is

$$\sigma_{\text{Equivalent}} = \sqrt{\sigma_X^2 + \sigma_Y^2 + \sigma_Z^2 - 2\nu\left(\sigma_X\sigma_Y + \sigma_Y\sigma_Z + \sigma_Z\sigma_X\right)} \quad (3.35)$$

When the levels of the mean stress are negative, the material's behavior is moderately tough. Thermoplastic and thermoset materials, for this particular case, deform in shape rather than in volume, which remains constant. In this case the equivalent stress will have the form

$$\sigma_{\text{Equivalent}} = \sqrt{\frac{1}{2}\left[\left(\sigma_X - \sigma_Y\right)^2 + \left(\sigma_Y - \sigma_Z\right)^2 + \left(\sigma_Z - \sigma_Y\right)^2\right]} \quad (3.36)$$

The above equation is known as the *von Mises'* equation.

3.10 ABCs of Plastic Part Design

3.10.1 Constant Wall

The ideal wall thickness for a plastic component is 3 mm or 0.125 in. This wall stock value is used by all thermoplastic and thermoset material suppliers worldwide. It is also used by the American Society for Testing and Materials (ASTM) and by the International Organization for Standardization (ISO).

If the designed part, assuming that the ideal wall thickness for the plastic component can't be maintained, has abrupt changes—thin to thick—there will be limitations in manufacturing the component and how well the part will correspond to its dimensional tolerance. From a manufacturing point of view, by using Gate 1 (as shown in Fig. 3.16(a)), filling the tool cavity with molten resin will be almost impossible because the polymer will most likely freeze in the second thin area. In order to fill such a component, the gate should be placed only in the thick area of the component, as the Gate 2 location indicates. Polymers flow very well from thick to thin. As far as part dimensions are concerned, the abrupt change from thin to thick induces warpage in the component, making it difficult to reach the required tolerance. It should be noted that the amount of warpage will be significantly higher in semicrystalline resin when compared to amorphous thermoplastics.

A better choice for thickness variation within the component would be to gradually taper the transition from thin to thick over three wall thicknesses, as shown in Fig. 3.16(b). However, the best choice would be to core out the heavy wall stock and maintain the same wall thickness consistently throughout the part whenever possible.

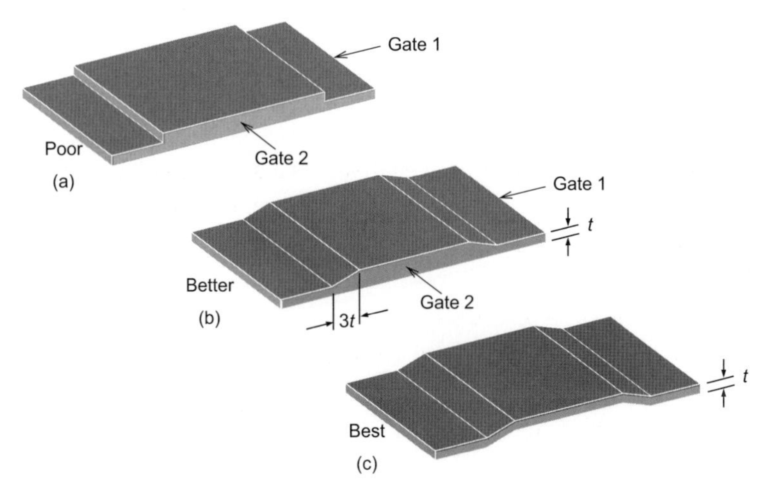

Figure 3.16 Wall thickness for plastic part design: (a) poor, (b) better, and (c) best

As stated above, the best wall thickness for a plastic part is 3 mm (0.125 in.). The component thickness can deviate from the ideal. The wall can vary between 0.5 mm (0.020 in.) and a maximum of 6.5 mm (0.25 in.). If the upper limit is reached the parts are inconsistent; they warp and create internal voids and air pockets, which will fail the component in an unpredictable and nonrepeatable way. Processing property data for polymers that are molded with very thin wall thickness, below 0.5 mm, is not available. The molder in this case has to develop the data pertinent to molding such thin parts.

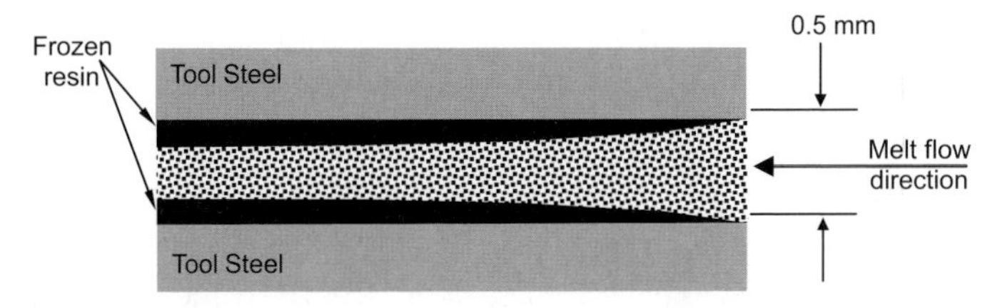

Figure 3.17 Component with wall thickness below 0.5 mm (0.02 in.)

Only high and very high melt flow resins are, in general, used in components having wall thicknesses below 0.5 mm. Figure 3.17 shows how the frozen layer is formed in the vicinity of both the tool cavity and the tool core. The larger the part and the longer the distance the melt has to travel to completely fill the tool, the more likely the molten polymer is to freeze without filling the mold. If the operator increases the injection pressure to prevent freezing, the tool could exceed the 2 to 5 ton per square inch (275 to 690 bar) clamping load and split the tool open, with pronounced flash along the parting line.

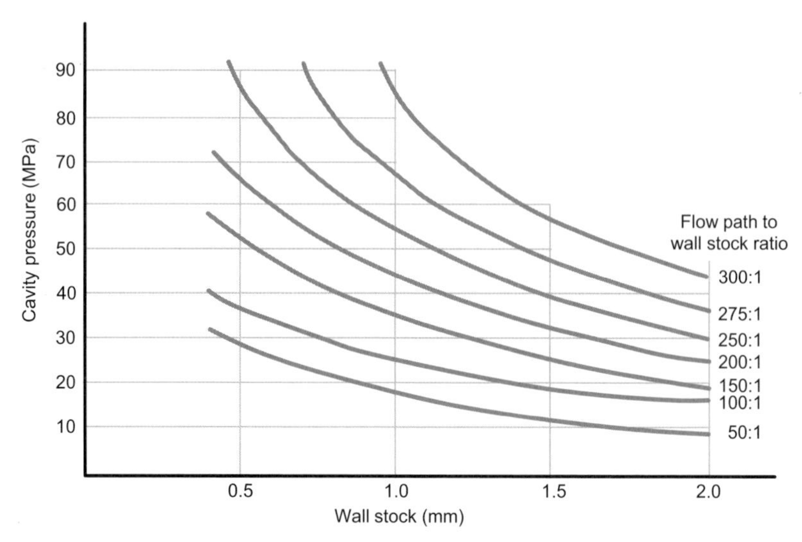

Figure 3.18 Mold cavity pressure versus clamping load as function of the component wall stock

When a manufacturer is molding thin components, the polymer behaves quite differently, as shown in Fig. 3.18. Below 0.5 mm (0.02 in.) the cavity pressure grows asymptotic, making it very difficult to predict whether the mold will split or not, thus creating large amounts of flash at the parting lines.

3.10.2 Fillets

Components made from thermoplastic and thermoset polymers should be designed with no sharp corners. Use fillet radii everywhere around the plastic part where sharp corners are present. Sharp corners induce stress concentrations, which are prone to air entrapments, air voids, and sink marks, all weakening the structural integrity of the part.

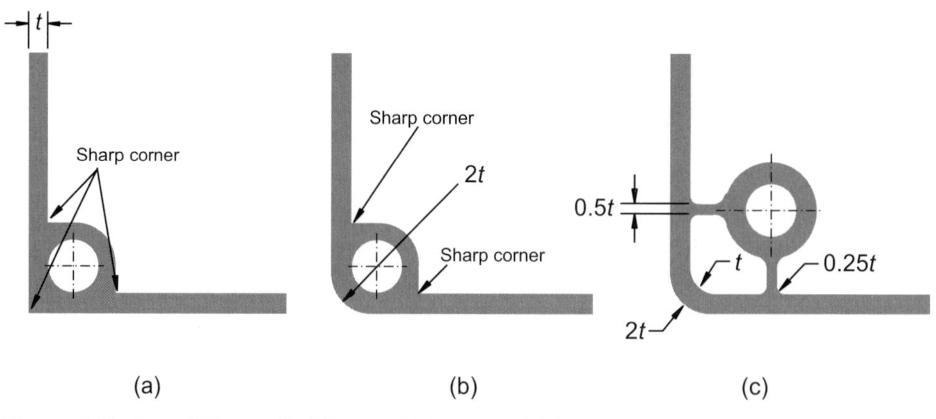

Figure 3.19 Use of fillets radii: (a) poor, (b) better, and (c) best

Designing a corner junction of a 3 mm (0.125 in.) wall stock, as shown in Fig. 3.19(a), is a poor choice. First of all, it has three sharp corners, and each of them creates stress concentrations, air voids, and sink marks. Second, the sharp corners are in an area larger than the typical wall thickness *t*, thus remaining hot even after the part is ejected from the tool. It will continue to cool down to ambient, thus potentially rupturing the still-molten polymer well inside the wall and sinking the surrounding walls. The component will not be acceptable from an aesthetic point of view and also may fail structurally in an unpredictable manner.

The design choice depicted in Fig. 3.19(b) is better. The corner has been blended with a 2*t* radius, thus eliminating the stress concentration induced by the sharp corner. But it still has two additional sharp corners, which, again, having more molten polymer in that specific thick area, will cool later, thus increasing the likelihood that air voids and sink marks will occur.

The option shown in Fig. 3.19(c) is the best choice. It eliminates all of the sharp corners and heavy wall stock, making it ideal for an injection-molding process.

Sharp corners are at the center of a tragic story. Back in the summer of 1952 the British aircraft manufacturer De Havilland of Hatfield, Hertfordshire, UK, introduced Comet 1, the world's first commercial jet airplane. The Comet featured an aerodynamically clean design with four De Havilland Ghost turbojet engines buried in the wings, a pressurized fuselage, and large square windows. The airplane offered a relatively quiet, comfortable passenger cabin and showed signs of being a commercial success at its debut.

Figure 3.20 Comet 1 passenger jet airplane manufactured by De Havilland of Hatfield, Hertfordshire, UK, having square windows

Less than one year after they were introduced, Comets started to fall out of the sky like apples from the tree. Extensive investigation revealed a devastating error. The design flaw was identified to be the dangerous stresses at the corners of the square windows. The constant stress of pressurization and depressurization would weaken an area of the fuselage near the Comet's square-shaped windows. The Comet's thin-skin exterior would become so stressed that high-pressure cabin air would burst through the slightest fracture, ripping a large slice in the aircraft's wall. All Comets were grounded until the line of jets could be redesigned. De Havilland never recovered; they lost the race to become a major player in designing and manufacturing commercial jet aircraft.

Figure 3.21 Wreckage from De Havilland Comet 1 aircraft showing failure at the corners of the square windows due to repeat cabin pressurization and depressurization while in flight or on the ground

The story of the Comet is a tragic lesson that should remind us to incorporate fillet radii in components when designing plastic parts.

3.10.3 Boss Design

Boss is a design feature that allows the use of mechanical fasteners to assemble components. A poor boss design is depicted in Fig. 3.22(a). The metal pin forming the inside of the boss is not deep enough. This creates a heavy wall stock at the bottom of the boss, which has hot polymer that can't cool off during the injection-molding cycle. It will continue to cool after it is ejected from the mold, thus creating air voids and sink marks.

A good design choice is shown in Fig. 3.22(b). The pin forming the inside of the boss penetrates deeper, thus preventing the formation of air voids and sink marks on the opposite side. However, if the opposite side is a class A (very smooth) surface and the polymer used to make the component is a highly shrinkable resin, such as polypropylene or polyethylene, the sink mark can't be avoided. The only option remaining is depicted in Fig. 3.22(c), in which a through-hole was used to remove the sink mark.

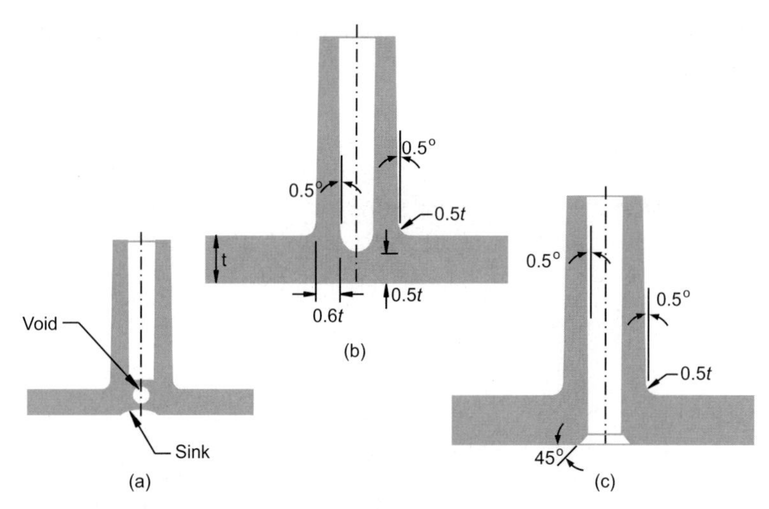

Figure 3.22 Boss design: (a) poor, (b) good, and (c) good

3.10.4 Rib Design

Ribs are added to part design to improve the overall strength characteristics of the component. Rib height is determined by the designer's skills. In general, ribs having heights below 1.5*t*, where *t* represents the wall thickness of the part, add very little as far as component strength is concerned. But incorporating ribs in the mold can increase tooling costs significantly (see Fig. 3.23). Also, a rib of a height exceeding five times the wall thickness (5*t*) could create difficulties during the injection-molding manufacturing process. The deep ribs could create vacuum in the tool, making it very cumbersome to execute part ejection from the mold. This problem will be more pronounced if there is any grain or surface roughness on the rib itself. Vents should be added inside the tool, at the tip of the ribs, to prevent vacuum formation.

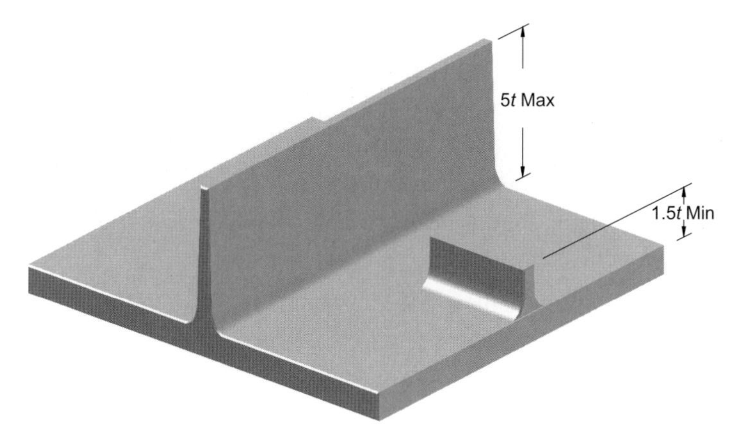

Figure 3.23 Rib height recommendation

The rib width depends mostly on the type of thermoplastic polymer used to manufacture the component. As discussed earlier, there are two major thermoplastic resin groups: amorphous and crystalline. Amorphous polymers shrink much less than crystalline; therefore the rib thickness for these types of resin can be greater (see Fig. 3.24).

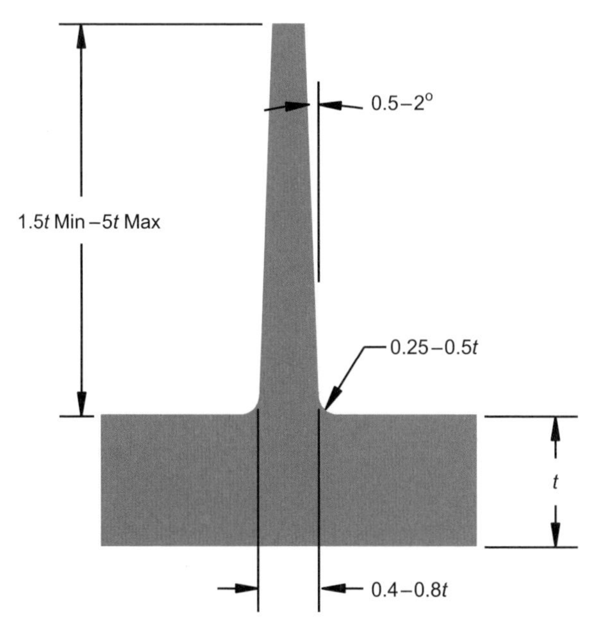

Figure 3.24
Rib width or thickness

Components made of crystalline polymers, which shrink more than amorphous resins, should have the rib width or thickness very small. Polypropylene (PP), for example, which shrinks 0.030 mm/mm or in./in. or more, should have a very thin rib–about 40% of the component wall stock. On the other hand, if the part is made of polycarbonate, which shrinks less than PP, it should have a rib thickness of about 60% of the part wall thickness. Both filled and reinforced amorphous and crystalline polymers shrink very little when compared to their unfilled or unreinforced counterparts. For these types of filled polymers the rib thickness could be as much as 80% of the wall stock.

Draft angles of at least half a degree per side are required in order to eject the part from the mold, assuming that the tool cavity is properly polished. For complex part geometry, the draft angle could reach two degrees per side. If the part and ribs are textured or grained, the draft angle should be increased by two degrees for each 0.025 mm (0.001 in.) of grained depth.

The radius connecting the rib base to the part should be small: between $0.25t$ and as much as $0.5t$. Junctions having sharp corners should not be used.

3.10.5 Case History: Ribs

Assume that a structural calculation was completed and requires the use of a thermoplastic component having the shape of a plate: 250 mm (10 in.) long by 9 mm (0.34 in.) thick (Fig. 3.25).

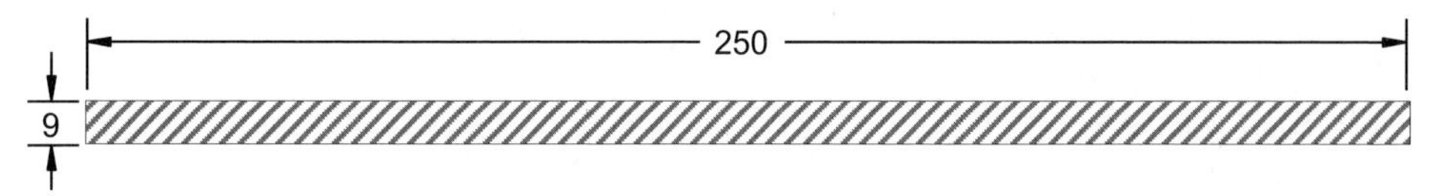

Figure 3.25 Impractical design difficult to mold

A designer familiar with the limits of the injection-molding process will immediately notice that molding such a plate in large volumes is not possible. Parts will warp and shrink differently, no two parts having the same tolerance.

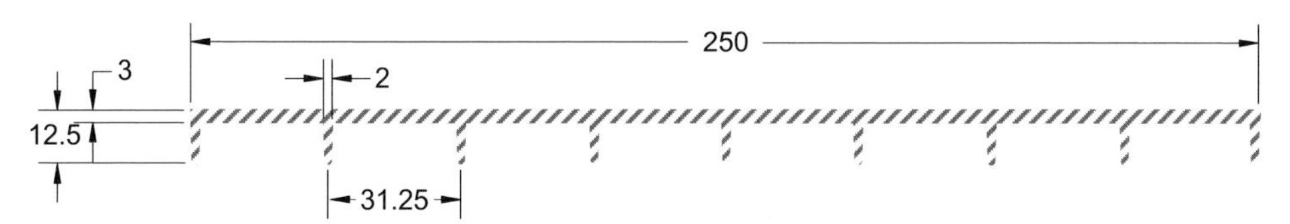

Figure 3.26 Redesigned component much easier to mold

To eliminate such issues, the component was redesigned, making it structurally identical to the previous one (see Fig. 3.26). The wall thickness of the redesigned component is 3 mm (0.125 in.), which is ideal for the injection-molding process. It also employed nine ribs each 2 mm (0.08 in.) thick, where they attach to the part, which represents 65% of t (t being the part wall stock of 3 mm). Ribs are 9.5 mm (0.375 in.) deep, which falls within the recommended range (see Fig. 3.23). The rib 2° draft angle per side represents an amount of 0.332 mm, as per Eq. 3.37:

$$x = 9.5 \tan 2^{\circ} = 0.332\,\text{mm} \tag{3.37}$$

Subtracting $2x$ from the rib width at base, the value of b, the small base of the trapezoid (see Fig. 3.27), is calculated:

$$b = 2 - 2 \cdot 0.332 = 1.336\,\text{mm} \tag{3.38}$$

thus obtaining the rib width at the tip as 1.336 mm.

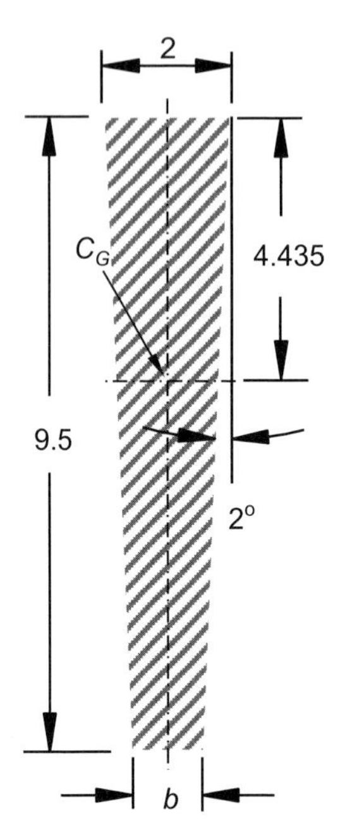

Figure 3.27
Rib detail

Next, the trapezoid centroid, or its center of gravity, is calculated. On the horizontal axis, the centroid is located in the middle of the trapezoid. However, on the vertical axis, because of the trapezoidal shape, the location of the center of gravity shifts toward the large base (Eq. 3.39):

$$y_{1,\ldots,9} = \frac{9.5(2+2\cdot 1.336)}{3(2+1.336)} = 4.435\,\text{mm} \tag{3.39}$$

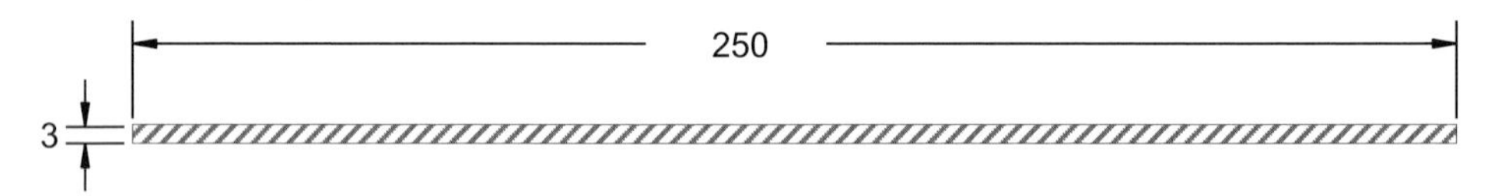

Figure 3.28 Plate detail

The center of gravity for the 3-mm-thick plate is determined in a similar manner. Because the plate is rectangular in shape, the centroid is in the center of the two axes of symmetry (Fig. 3.28). On the vertical axis, the center of gravity for the plate is 1.5 mm from each side.

The moment of inertia of each rib is

$$I_{1,\ldots,9} = \frac{9.5^3}{36}\frac{\left(2^2+4\cdot 2\cdot 1.336+1.336^2\right)}{2+1.336} = 117.73\,\text{mm}^4 \tag{3.40}$$

and the moment of inertia for the 3-mm-thick plate is

$$I_{10} = \frac{250 \cdot 3^3}{12} = 562.5\,\text{mm}^4 \tag{3.41}$$

Finally, the total moment of inertia can be calculated:

$$I_{\text{Total}} = 9I_1 + I_{10} = 680.23\,\text{mm}^4 \tag{3.42}$$

The area of the initial design was 2,250 mm^2 (3.39 in.^{2}). Through redesign, a better component was achieved, one that can be molded consistently in a repeatable fashion. The area of the redesigned part is just 922 mm^2 (1.43 in.^{2}). Overall, a component using 59% less polymer was achieved while at the same time having a shorter injection-molding cycle time. Significant improvements were reached by redesigning the part: reducing the amount of resin used, lowering cycle time, increasing productivity, and, probably most important, the manufacture of a consistent product.

Ribs are, in general, added to improve part performance and reduce component weight while at the same time allowing for manufacturability. Some examples are shown below. The parking brake handle (Fig. 3.29) uses a thermoplastic polyamide 6,6, highly reinforced with glass fiber to 50% levels. Its redesign relies on properly located ribs, which are loaded in compression, thus preventing creep from taking place prematurely in the thermoplastic polymer.

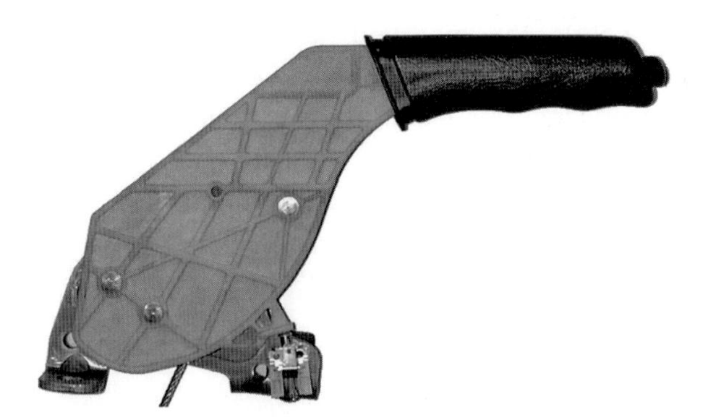

Figure 3.29 Parking brake made of 50% glass-reinforced polyamide. The rib layout is such that when the brake is engaged, the ribs are loaded in compression

Figure 3.30 shows how an aluminum bracket, required to mount the vehicle transmission to the chassis, was redesigned from polyamide 6,6, reinforced with 50% glass fibers. Many ribs ensure the proper stiffness of the plastic component while at the same time allowing for rather short injection-molding cycle times in a repeatable manufacturing setting.

Figure 3.30 The aluminum bracket above, used for the BMW 550i and 750i, was redesigned using a thermoplastic polyamide 6,6, 50% glass fiber reinforced, supplied by BASF, which has a large number of ribs to improve stiffness, reduce weight, and enable consistent injection molding in large volumes

3.11 Conclusions

To simplify things a little, the important theories to remember for the preferred procedure of quick calculations are *ultimate normal stress* and *ultimate shear stress.* The most accurate methods for our day-to-day needs are *maximum deformation energy* and especially the *von Mises' stress theory.*

4 Nonlinear Considerations

4.1 Material Considerations

4.1.1 Linear Material

Linear materials are those that fully recover to original dimensions once a load is removed. Materials displaying this type of behavior include most metals and certain stiff polymers.

To describe a linear material, one should employ one of the following moduli:

- Young's modulus
- secant modulus
- tangent modulus

It is important to consider linear material properties at the early stages of any design. This will help greatly in approximating the overall geometry needed to handle the required loads, pressures, or deflections. Most calculations at this stage are simple enough to be done by hand or with a pocket calculator.

4.1.2 Nonlinear Materials

Nonlinear materials exhibit visco-elastic properties. They have a dual behavior, showing both elastic and plastic deformations when subjected to an external load.

A material has nonlinear properties when the stress is no longer a linear function of the strain. The use of *linear* material approximations in the analysis of a nonlinear material model induces error. Depending on the point on the stress/strain curve at which the analysis is conducted, the magnitude of error can be quite large (see Fig. 4.1).

The basic assumption one can make about a nonlinear polymer is that it does not recover fully to its initial dimensions once a load is removed (it does not obey Hooke's law). Most rigid polymers, certain metals, all thermoplastic elastomers, and all thermoset elastomers are classified as nonlinear materials.

It is extremely cumbersome to conduct analysis using hand calculations for nonlinear material models. Instead the analysis can be performed using a computer with the appropriate software. Computer programs that can handle nonlinear material models as well as large deformation analyses are available, and engineers with proper training in this type of software can perform calculations quite easily, using finite element analysis or the boundary analysis method.

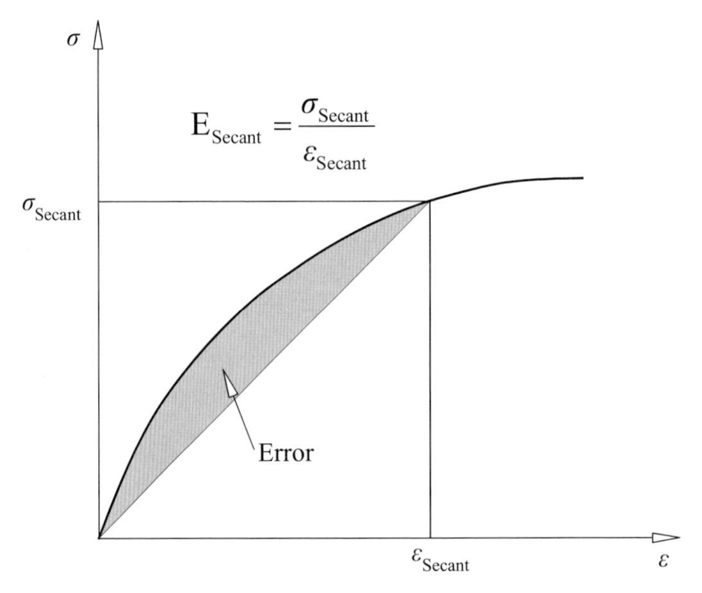

Figure 4.1 Error due to material nonlinearity

4.2 Geometry

The term *geometry* refers to all dimensions of the part that will make the final design. Based on the type of loading the part will be subjected to, two types of geometry can be used for analysis: *linear* geometry and *nonlinear* geometry.

4.2.1 Linear Geometry

Linear geometry includes most finite element types present in all finite element analysis (FEA) codes now commercially available. Modeling with linear geometry elements implies that rotations should not exceed 10°, and the ratio between the plate thickness and any of the model dimensions does not exceed an amount specified by the code.

Typical linear elements include bar elements, quad (plate) elements, and solid (brick) elements.

4.2.2 Nonlinear Geometry

Large deflections are referred to as those that exceed a certain structural dimension, such as the typical plate thickness. Large rotations are those that exceed 10°, when the sine function of the angle differs substantially from the angle itself.

Contact problems present another type of geometric nonlinearity.

A good example of contact nonlinearity is the relationship between a ball and a rigid surface. Other types of contact problems are threaded connections and the line of contact between the teeth of two engaged gears. When friction has to be taken into account, the analysis becomes difficult. It becomes necessary to solve a series of load increments that follow the actual load history.

Typical finite elements used for modeling nonlinear geometry include gap elements, rigid surface elements, spring elements, friction elements, contact elements, as well as others.

4.3 Finite Element Analysis (FEA)

4.3.1 FEA Method Application

Modeling techniques for simple static stress and deflection analyses are different from those of vibration and nonlinear analyses. The construction of a model is generally dependent on the expected response, whether it is structural yielding, creep, temperature profile throughout the part, or other responses.

When analyzing a symmetrical structural part, considerable computer time can be saved by modeling only half of the part, when the other half would be geometrically identical.

A typical ratio between the length and thickness for elements used should be less than 10. The modeling material used should imitate the polymer, which in the vast majority of cases is anisotropic. In general, polymeric material responds to applied loads with large displacements or large strains. The complete stress/strain curve for a given material is required for proper modeling (see Chapter 1, Fig. 1.34).

4.3.2 Using FEA Method

Three distinct steps are required to perform a complete finite element analysis.

The first stage in the FEA process is *preprocessing*. A 2-D or 3-D geometry needs to be created either in a computer-aided design (CAD) system or with the help of an FEA modeler. If the geometry for the part to be analyzed was created with a CAD package, the 2D or 3D description of the geometry must be translated or imported into the FEA modeler if it is from a different software company.

Next, the model creation begins. Elements are generated with a given density over the entire part surfaces. Then the nodes belonging to the elements are optimized in a way that places no two nodes at the same location. Depending on how the elements (for plate elements only) were generated, the normal of each element should point towards the outside or inside only. Normals to the elements should not be mixed and should all be pointing in the same direction.

Next, it is important to verify that the model is continuous, to ensure that there are no unaccounted-for gaps in the model. After one is certain that the mesh takes advantage of the part symmetry and there are no element discontinuities, boundary conditions can be applied. *Boundary conditions* represent any movement restrictions or enforced displacements that the model should obey.

Finally, the loads are applied. The loads represent static or dynamic forces, moments, or pressures that the part is subjected to. Loads can also mean temperature gradients, time-dependent loading, stress relaxation, and other factors.

The second step in applying an FEA method is *solving*, or "number crunching." This can be done using the internal code of the software or an external code.

The last step of the FEA method consists of *results interpretation* and *evaluation*. This is the most specialized and challenging part of the process, requiring highly experienced and qualified engineers. Results are reviewed in the form of displacements in any direction, or any principal directions X, Y, and Z, or as total displacements. The results can be evaluated according to strains, stresses, energies, thermal expansion or contractions, and other forces.

4.3.3 Most Common FEA Codes

There are three categories of general finite element analysis codes: they are for personal computers, for workstations, and for mainframes.

For personal computers (PC), some of the most common FEA programs are Algor from Algor Inc. of Pittsburgh, PA; Cosmos/M from Dassault Systèmes S.A. of Vélizy-Villacoublay, France; ANSYS from Ansys Inc. of Cecil Township, PA; and NISA from Cranes Software Inc. of Troy, MI.

The FEA software for workstation and mainframe environments are Abaqus from Dassault Systèmes S.A. of Vélizy-Villacoublay, France; ANSYS; MARC, MSC/NASTRAN and PATRAN from MSC Software Corporation of Newport Beach, CA; NISA; Cosmos/M; and Algor.

4.4 Conclusions

A finite element analysis or FEA is recommended for press-fitting assemblies involving polymers when the total deflection (strain) required to engage the two parts exceeds 10%. In the case of living hinges that are molded in, FEAs are usually required if the hinge length is 2 mm (0.080 in.) or less and the hinge bending angle is greater than 60°.

There are two types of cantilever beam snap fits: one-way snap fits, which require special tools for opening; and two-way snap fits, which can be opened and closed many times by hand. The beam theory does not apply in all of these cases. Where snap fits are used to assemble two or more parts and the angle of deflection of the cantilever beam is greater than 8°, finite element analysis with large displacement capabilities should be applied.

5 Welding Techniques for Plastics

There are many different methods for welding two parts together. All variables, such as materials, design, and conditions under which the finished product will be used, including cost of the process, must be considered when deciding which welding technique should be employed.

Polymers can be melted, and therefore welded, using relatively little energy. Heat, friction–even ultrasonic vibrations and radio frequencies–can be used to create the melting necessary for a polymer weld. Welding methods include ultrasonic welding, ultrasonic heat staking, hot plate welding, spin welding, vibration welding, and laser welding. Welding requires no additional materials with one exception: electromagnetic welding, which requires bonding agent consumables.

5.1 Ultrasonic Welding

The principle behind ultrasonic welding technology is based on vibration. One of the parts being assembled is vibrated against the other, stationary one. Heat generated through vibration melts the materials at the joint interface to accomplish the weld.

Thermoplastics are the only polymers suited for this process. Thermoset materials do not melt when reheated because of their intermolecular cross-links.

5.1.1 Ultrasonic Equipment

The type of equipment required for an ultrasonic welding process depends upon the size of the manufacturing operation. The ultrasonic welding equipment requirements of a large-volume production environment will be different from those of a small prototype operation. They will, however, be very similar in principle.

A typical ultrasonic welding system consists of a power supply, also referred to as an *ultrasonic generator*; a *converter*, also known as a *transducer*; a *booster*; and a *horn* (see Fig. 5.1). The horn is a metal bar designed to resonate at a certain frequency, delivering the actual energy to the parts to be welded. The converter, booster, and horn are mounted inside a frame, which can slide along the stand, allowing them to travel vertically under the power of a pneumatic cylinder. The pressure applied by the air cylinder can be preset for manual systems or fully controlled by a computer for automatic systems. The pressure, trigger pressure, stroke speed, and stroke travel are all adjustable through the control panel or by the computer. The two palm buttons are used by the operator to activate the machine.

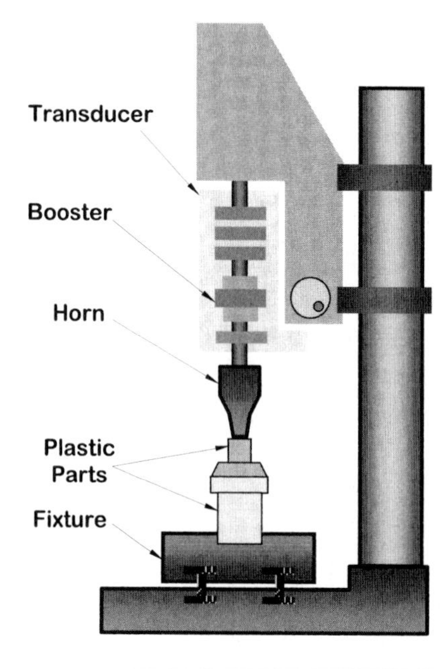

Figure 5.1
Ultrasonic welder

To generate the necessary amount of vibration required for a particular assembly, an electrical current is passed through a stack of crystalline ceramic material that possesses *piezoelectric* properties, which allow the material to change its size. The electric power supply has a frequency of 50 to 60 Hz. Once an electric current is applied, the material expands and contracts at a very high frequency, converting the electrical energy into mechanical energy or vibrations. These vibrations occur with frequencies ranging from 15 to 70 kHz. The most common output in ultrasonic welding systems provides frequencies of 20 to 40 kHz.

The distance the mechanical vibrations travel back and forth is called the *amplitude.* A typical converter of 20 kHz could have amplitudes of 0.013 to 0.02 mm (0.0006 to 0.0008 in.) between its maximum expansions and contractions.

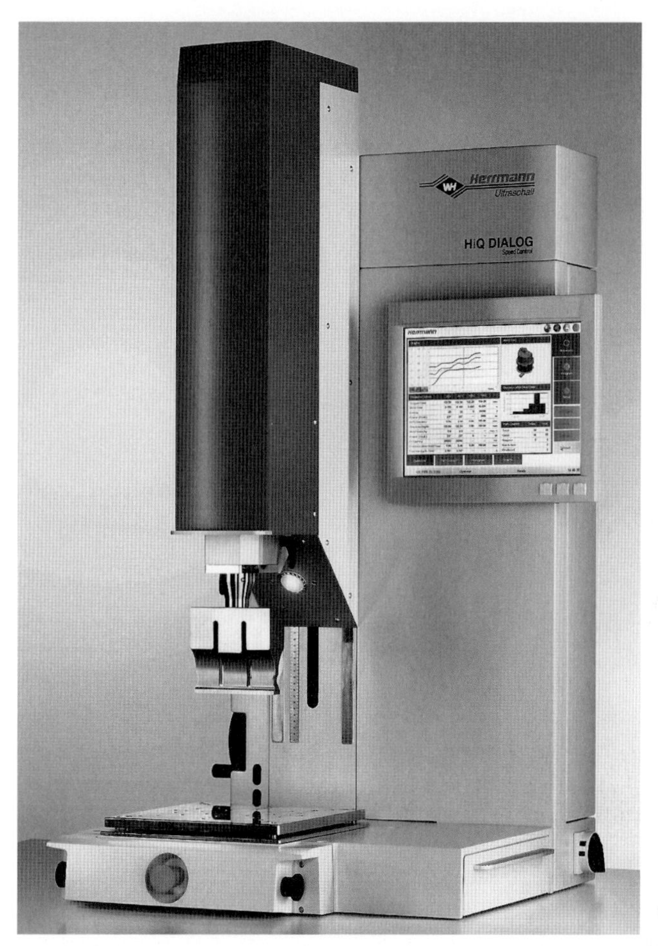

Figure 5.2 Ultrasonic welder HiQ Dialog also includes software to control and operate the welding process and machine functions, having the additional capability of welding visualization in two graphic modes: EasySelect and Expert mode *(Courtesy of Herrmann Ultrasonics)*

There are different types of ultrasonic welding systems for different applications. An integrated welder (see Fig. 5.1) is a self-contained unit, which has a power supply, actuator, and the acoustic components packaged as a stand-alone system. Advantages of this type of system include low investment cost and ease of service.

Modular systems include, in addition to the welder, a rotary indexing table and an in-line conveyor. These systems are ideal for assembling large numbers of parts. Also, their components are interchangeable and easy to upgrade.

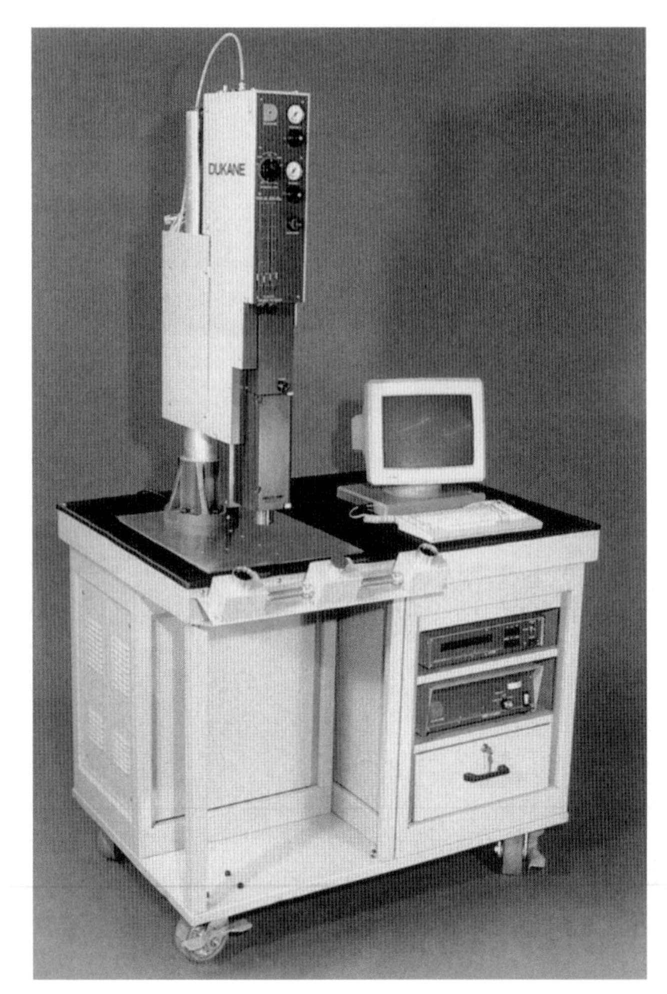

Figure 5.3 Mobile ultrasonic workstation. An aluminum top plate acts as the base plate for the press table. The generator and controller are on the recessed shelves *(Courtesy of Dukane Corporation)*

Modular systems are available in semiautomatic and automatic models. Automatic systems include a pick-and-place robot arm.

Power level is frequently determined by the cycle time or the material used in a given application. Power supplies are available from 150 to 3,200 W for the 20 kHz machines, and from 150 to 700 W for 40 kHz systems.

Controls are integrated into the power supply and may be analog or digital. Digital controls are computer controlled.

The frame or box contains the converter. The vibration produced by the converter must be amplified further in order to produce meaningful results when it reaches the horn face.

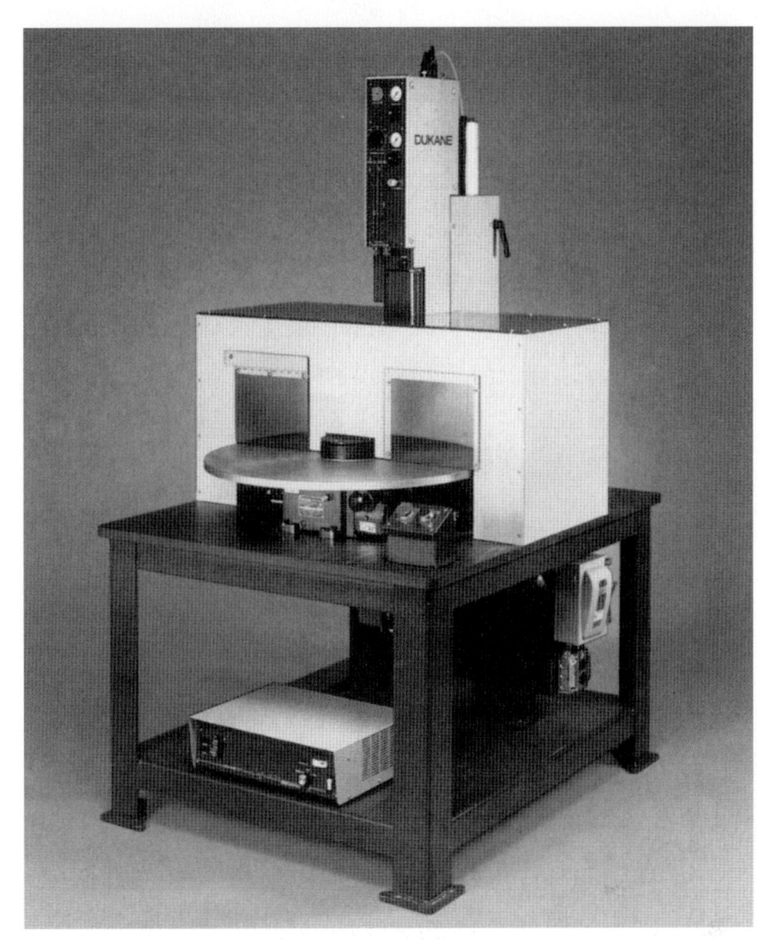

Figure 5.4 Semiautomatic ultrasonic press with rotary index table *(Courtesy of Dukane Corporation)*

5.1.2 Horn Design

When the horn receives high energy in the form of vibrations from the booster, it reaches its resonant frequency. At that time, the ends will expand and contract longitudinally about its center (also called the *nodal point* of the horn), alternately lengthening and shortening the horn's dimensions. The movement from the longest length value to the shortest length value at the horn face (the portion of the horn in contact with the part) is referred to as the *horn amplitude*. The face of the horn is usually machined to conform to the plastic part with which it comes in contact.

The horns are designed as resonant elements with a *half wavelength*. The materials for horns must have low acoustical impedance (low losses at ultrasonic frequencies) and high fatigue strength.

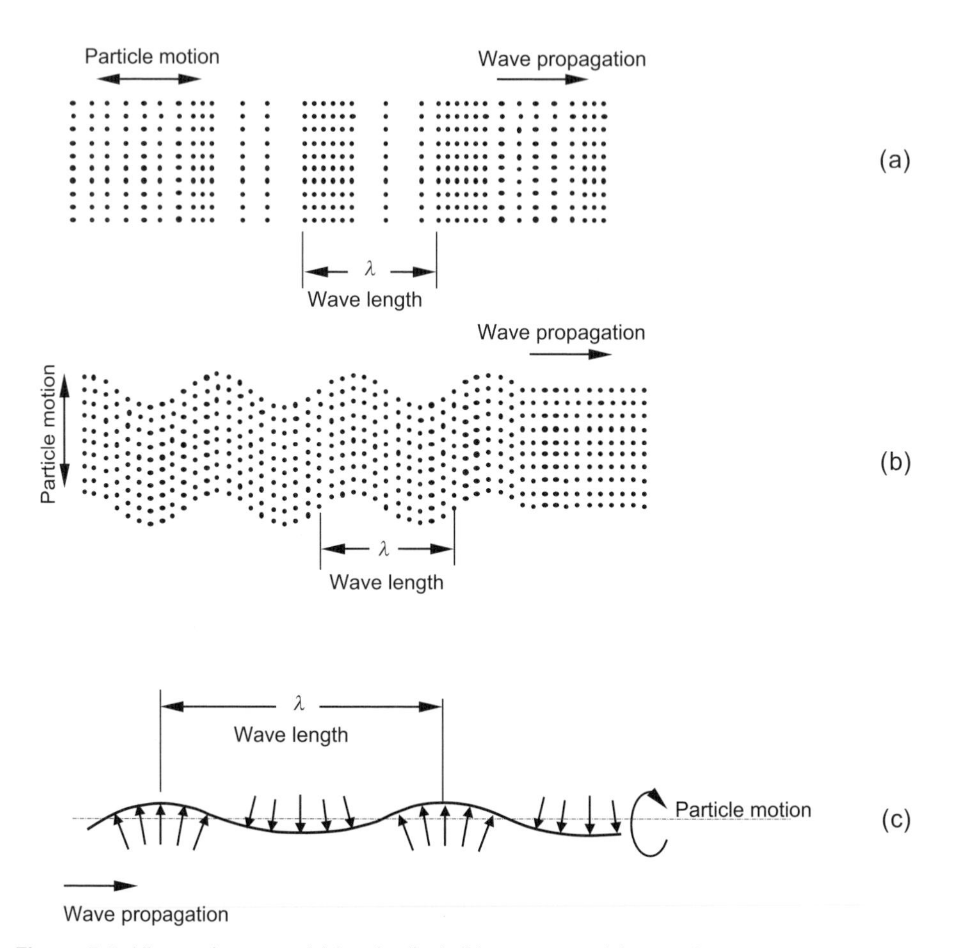

Figure 5.5 Ultrasonic waves: (a) longitudinal; (b) transverse; (c) curved

There are three types of vibrations produced in the ultrasonic welding process. The first type is the *longitudinal wave*. These waves are transmitted in a direction parallel to the horn axis, which is vertical to the stand. The oscillations are a function of the wavelength λ *(lambda)*, both in amplitude and direction. Longitudinal waves (see Fig. 5.5(a)) act as energy carriers to allow a proper weld, and they are the only desirable type of vibrations in an ultrasonic welding process.

A second type of vibration is the *transverse wave*. Typically, transverse waves are electromagnetic waves of very high frequency and can be generated only in the presence of shear stresses. Transverse waves move in a direction normal to the horn vertical axis (see Fig. 5.5(b)). Transverse waves should be avoided because they create vibrations only at the horn surface and not in the entire horn body. As a result, almost no energy is transmitted to the parts to be welded.

The third type of vibration consists of *curved waves*. These waves are detrimental to the ultrasonic welding process and will occur when the system components are out of balance due to misalignment, for example. This results in uneven pressure reaching the part, creating a nonuniform weld. During the transmission of ultrasonic waves from the transducer to the horn, curved waves returning from the horn to the piezoelectric material could crack the ceramic material. Curved waves generate high compression and tensile loads in the parts being welded (see Fig. 5.5(c)). In order to correct the system imbalance, asymmetrical masses can be placed on the misaligned components to bring the system back into balance. Horns should be designed to completely avoid transverse and curved waves.

Horns are commonly made from aluminum, titanium, Monel metal, stainless steel, and steel alloys. These materials have different properties, which are beneficial for different applications. An important consideration when selecting a horn material is that the material should not dissipate acoustical energy.

Titanium is one of the high-strength materials with the best acoustical properties, and it wears better than other horn materials.

Aluminum, although it does not wear as well as titanium, is the best of the low-strength materials. Aluminum has low amplitude and is appropriate for assembling large parts.

Steel materials are best for low-frequency losses or *premiation*. Steel has high wear but loses a great deal of its own frequency. It is good for low amplitudes and high wear such as ultrasonic metal inserting.

Carbide-faced titanium is recommended for high-amplitude horns and high-wear applications.

5.1.3 Ultrasonic Welding Techniques

In order to achieve a proper ultrasonic weld, the horn must be applied as close to the joint as possible. To help ensure an accurate weld, a *nest* or supporting fixture is required to hold the parts together. Fixtures have two purposes: to provide alignment between the parts and the horn, and to provide support to the weld area. The nest is made of chrome-aluminum or epoxy and steel.

Figure 5.6 shows a fixture that provides nesting for the parts as well as accurate location and securing of the part. The fixture holds the part in place by applying a vacuum for the duration of the weld cycle. Once the cycle is completed, the vacuum is reversed to create an air pressure, which ejects the final assembly from the nest.

The majority of fixtures are machined or cast. These manufacturing processes create fixtures that engage the lower part and hold it securely in a given position. Variations in thickness and flatness of the parts close to the joint area can adversely affect the welding process. To accommodate such variations, fixtures may be lined with rubber or rubbery material, such as silicone. Rubber or silicone strips allow the part to align in the fixture under nominal static loads and act as rigid constraints during the high-frequency vibration phase of the process. They also may help absorb random vibrations, which can often create cracking or melting in regions away from the joint area.

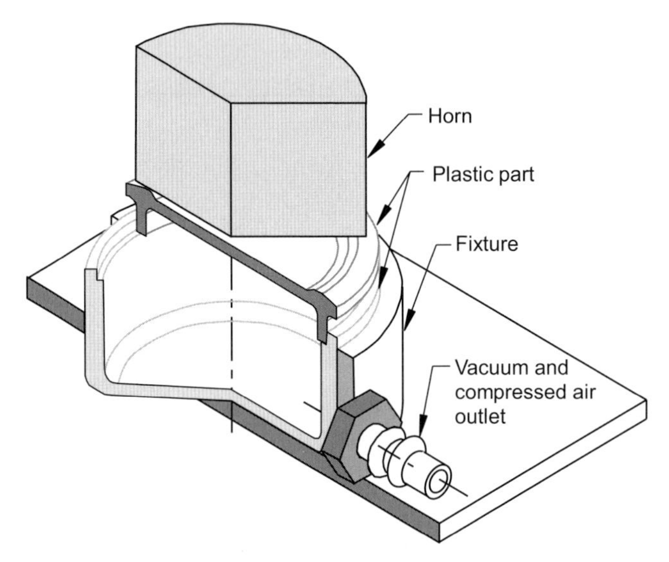

Figure 5.6
Air-assisted fixture design

There are various factors that influence the ultrasonic welding process. Polymer (material to be welded), part geometry, and wall stock (thickness) all affect the transmission of the mechanical energy to the joint interface. These factors also influence the design of the fixture.

The booster or amplifier regulates the vibration, keeping it at the appropriate level to melt the correct amount of resin in the weld area for the most efficient weld possible. Boosters are made of titanium or aluminum and are color-coded to identify the amount of amplitude they can generate.

The overall ultrasonic weld cycle (see Fig. 5.7) can vary from a fraction of a second up to a few seconds, depending on the part size and joint area. The hold time could be anywhere from 0.25 second to approximately 1 second, again depending on the size and shape of parts to be assembled.

The horn transfers the vibrational energy it receives from the booster to the parts to be assembled. The amount of amplitude the horn receives from the booster depends on the horn design. Different horn designs deliver different amplitudes.

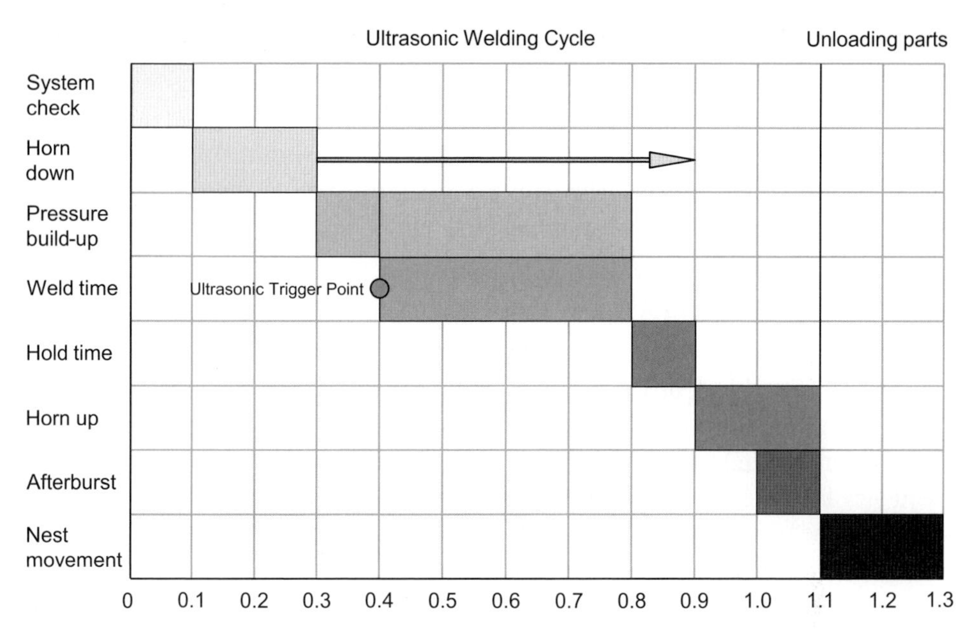

Figure 5.7 The welding cycle

Figure 5.8 shows a stepped horn arrangement, which is a convenient way of modifying the amplitude. By interconnecting horns, one can increase or decrease the amount of amplitude to which the last horn in a series can vibrate. The horn in the middle of the arrangement in Fig. 5.8 is also called the booster horn.

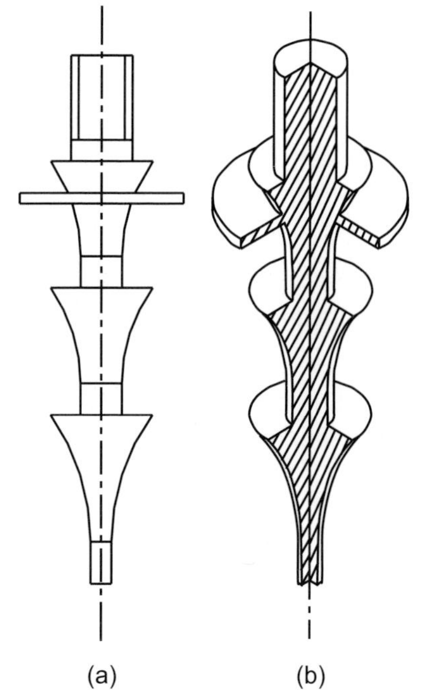

Figure 5.8
Stepped horn arrangement

It is important to avoid overstressing horns when interconnecting them. This could lead to failure of the system through fatigue.

The ratio between the amplitude generated by the converter (also called input amplitude) and the amplitude at the end of the horn in contact with the part welded (output amplitude) is called *gain*. The gain itself is a function of the transversal area between the converter (input section) and the horn face (output section).

If the cross-sectional area of the output end is less than that of the input end, the gain will be greater than 1 and the corresponding amplitude will increase.

There are different horn shapes for various welding applications. Stepped, conical, exponential, catenoidal, or Fourier horns can be connected at the stress antinodes, the point between two adjacent peaks in the wave pattern. Larger horns (greater than 75 mm or 3 in.) can be constructed with slots cut out to change the resonating frequency by more than a quarter of a wavelength. Each is designed to change the gain to a specific value.

5.1.4 Control Methods

Amplitude is the most important variable in determining the power output for the part to be welded. It is also very important in the horn design. The horn, as mentioned earlier, is a metal bar a half-wavelength long, dimensioned to resonate at a certain applied frequency.

Constant energy is the total amount of ultrasonic energy required by the mating parts and actually delivered to them (Fig. 5.9). There is a time window within the welding process when the total energy is applied to the joint area. The energy is delivered independent of any external influences such as voltage fluctuations. All other parameters, such as time or amount of travel—the downward vertical distance the horn moves during the process—are varied in order to determine their optimum values for a given joint design.

The relatively new computer-controlled ultrasonic systems greatly enhance the process by allowing direct control of the energy transmitted to the parts rather than controlling only the time.

Another technique of control is based on travel. Controlling travel is one way of controlling the weld quality. There are two ways of applying the control method: through partial travel and total travel.

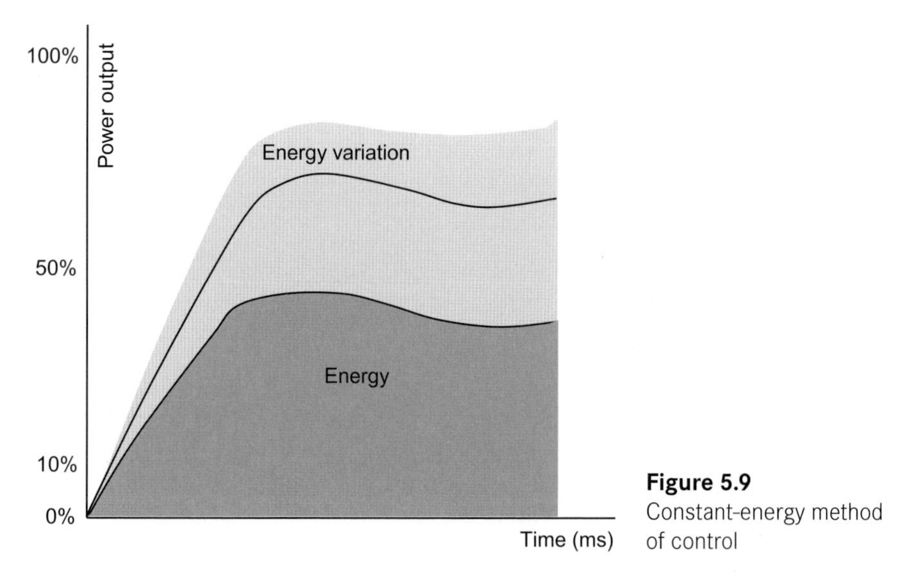

Figure 5.9 Constant-energy method of control

Partial travel implies that the horn is moving downward until it makes contact with the part to be welded. Once the horn makes contact, the circuit is closed by the digital readout sensor, and the ultrasonic is activated (see Fig. 5.10(a), ultrasonic activation point (UAP)). The pneumatic cylinder applies pressure until the top part reaches the ultrasonic deactivation point (UDP). The weld is completed.

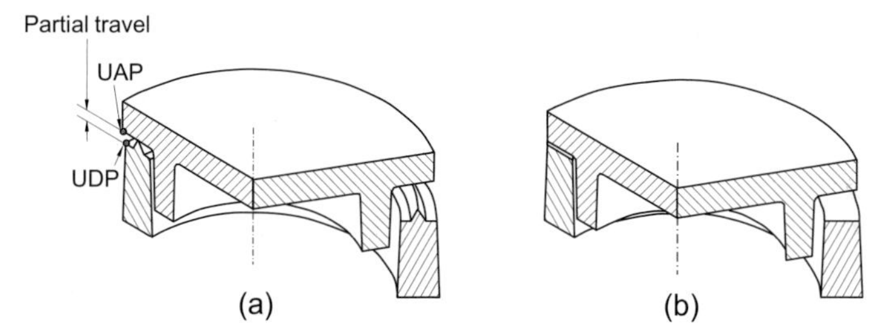

Figure 5.10 Partial-travel method of control (UAP represents the ultrasonic activation point and UDP represents the ultrasonic deactivation point): (a) before, and (b) after assembly

Sometimes the partial-travel method is not feasible, for example, when the parts are unstable. In these cases *total travel* or *absolute travel* is used. This is the best method when dimensional accuracy is the most important feature of the assembly. When total travel is used, the horn is deactivated only when a preset amount of travel is achieved.

The total-travel method (Fig. 5.11) uses a digital readout sensor mounted on the press to activate the horn before it makes contact with the part. The location of the sensor can be taken from a fixed reference point or by measuring the collapse of the plastic in the joint during the welding process (see Fig. 5.11(a), UAP point). The horn will stay triggered until it reaches the UDP position. Holding time starts once the preset travel is reached. A signal from the encoder is sent to the computer, which stops the flow of vibration energy to the horn.

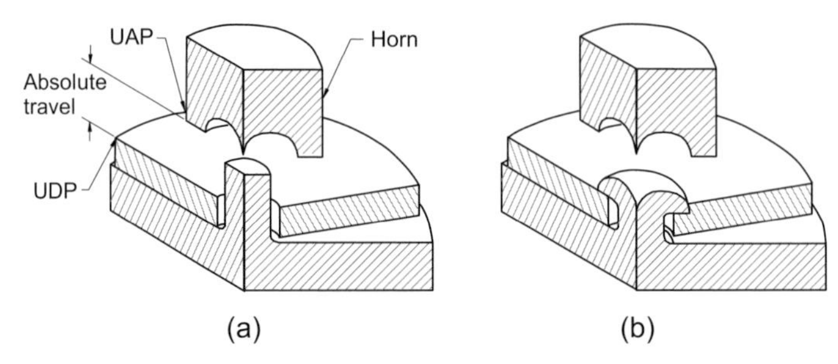

Figure 5.11 Absolute- or total-travel method of control (UAP represents ultrasonic activation point and UDP represents the ultrasonic deactivation point): (a) before, and (b) after assembly

While total travel is a preset distance, partial travel is usually limited to the height of the energy director.

Another method of controlling the weld is based on time. The *constant-time method of control* implies that the ultrasonic will be on for a predetermined time (usually 0.2 to 0.3 seconds) while the other parameter will be varied to determine optimum values.

For this method, time can be determined by an experienced operator through trial and error. Part size, materials, and other variables will influence the time selected.

The ultrasonic welding method produces joints with strength up to 90 or 95% of the virgin polymer. In some instances, hermetically sealed joints can be produced ultrasonically.

Another important factor in the welding process is the use of *far-field* and *near-field*.

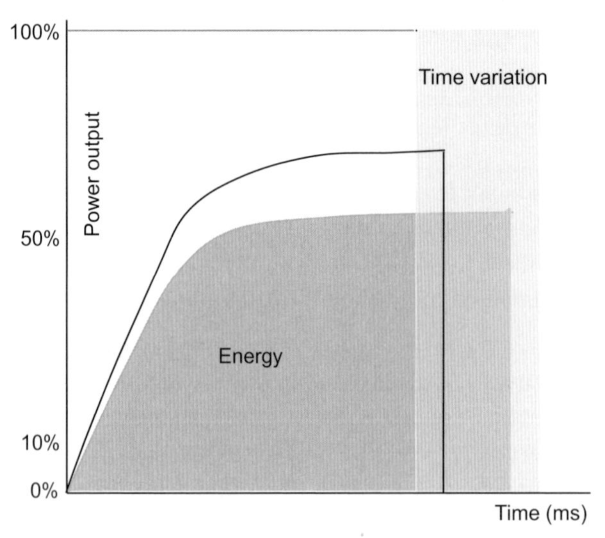

Figure 5.12 Constant-time method of control

The field refers to the distance between the joint weld area and the point at which the horn comes in contact with the part. When the distance is more than 7 to 8 mm (0.25 to 0.375 in.) it is referred to as a far-field weld. When the distance is less than 7 or 8 mm, it is a near-field weld. Special horns can be designed for assemblies requiring near- and far-field welds, but it is advisable to avoid combining near- and far-field welds in one assembly.

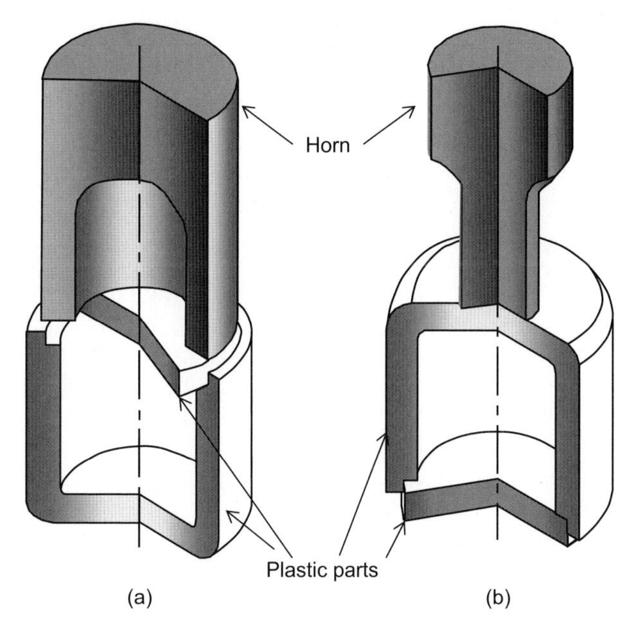

Figure 5.13
Horn position: (a) near-field, and (b) far-field

The frequency to be used in the weld depends on several factors, including the size of the part and the rigidity of the plastic. A general rule is that larger parts and softer plastics require lower frequencies. Sometimes, frequencies as low as 15 kHz may be used for very large parts in excess of 150 mm (over 6 in.). In these cases the horns are also quite large. Conversely, harder plastics and smaller parts require higher frequencies. Far and near fields also affect the choice of frequencies. Near-fields need higher frequencies; frequency decreases as the distance between the horn contact and the joint increases.

Also, the number and accuracy of controls determines the quality of the ultrasonic system. The timer controls the weld and hold time.

The melt temperature, Young's modulus, and overall structure usually determine the amount of vibration energy required for a specific weld. Rigid plastics exhibit the best weldability properties because they are good transmitters of vibration energy. Soft polymers, on the other hand, have a low value for Young's modulus or secant modulus. They dissipate the vibration energy, making the part difficult to weld. Softer polymers are, however, well suited for ultrasonic staking, forming, or spot welding.

Amorphous materials tend to soften gradually before melting and flow easily without solidifying prematurely.

Crystalline polymers do not readily transmit ultrasonic energy and therefore need higher energy levels than amorphous resins. Due to their sharp melting point they lend themselves more easily to controls, giving assemblies made from these materials a narrow margin of variation from part to part.

Mold-release agents such as zinc stearate, aluminum stearate, fluorocarbons, and silicones are not compatible with this process. If molding agents must be used in the molding process, a paintable grade should be selected. Incompatible agents can be removed with a Freon TF solution for crystalline polymers or a 50/50 solution of water and detergent.

5.1.4.1 Common Issues with Welding

As with any manufacturing process, problems may occur. The following is a list of common problems associated with ultrasonic welding, their probable causes, and possible solutions.

Overweld is usually caused by too much energy reaching the part. It can be corrected by reducing the pressure exerted by the pneumatic cylinder or by reducing the overall weld time. Other possible solutions include slowing the air cylinder motion and the use of a power control.

Nonuniform welds around the joint could have many possible causes. Warped parts could be one of them. The part dimensions, tolerances, and general processing conditions should be reviewed. A higher trigger pressure could be one of the solutions.

If the nonuniformity is created by the energy director's variance in height, the energy director should be redesigned. The problem could also be caused by a lack of parallelism between the horn, the nest, and the parts.

Sometimes flexure of the walls is the cause of nonuniformity. Ribs can be added to the part, or the fixture can be modified to prevent outboard flexure.

A knockout pin location in the joint area can result in an uneven weld. The knockout pin should be moved away from the joint area. Also, the knockout pin marks should be flush with the surface.

Insufficient support in the fixture can lead to nonuniformity. Improving support in critical areas, redesigning the nest, or switching from a flexible fixture to a rigid nest design may solve the problem. Sometimes the pressure from the pneumatic cylinder will cause larger sections of some parts to bend. This can be corrected by adding a rigid backup.

Tighter part tolerances or molding parameters are needed when the part tolerance is not within the part requirements.

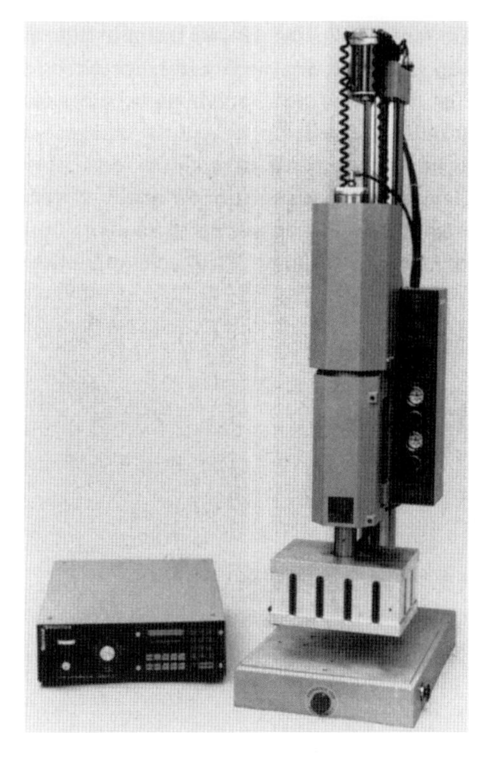

Figure 5.14
Ultrasonic welder has a power output of 4,000 watts and a frequency of 15 kHz. The horn dimensions are 300 mm by 350 mm (12 in. × 14 in.) *(Courtesy of Sonics and Materials, Inc.)*

Improper alignment is another possible cause of a nonuniform weld appearance. This can result from the part shifting during the weld cycle. Provisions for alignment in the mating parts need to be reviewed, and parallelism between the horn, part, and fixture should be rechecked.

A lack of intimate contact between the horn and the part can also cause a nonuniform weld. One has to make certain that there are no sink marks, raised symbols, or other inconsistencies to impede contact.

The presence of a mold-release agent on the part surface could also cause uneven welding. As mentioned earlier, parts should be cleaned prior to welding. If possible, a paintable mold-release grade should be used.

Fillers can also affect weld uniformity. If that is the case, processing conditions should again be reviewed, and the amount of filler should be reduced if possible. Also, the filler type—short fiber vs. long fiber—should be verified. It is also advisable to check for uniform filler distribution.

If the nonuniformity is a result of cavity-to-cavity variations (a cavity is an empty volume in a closed tool that becomes filled with polymer during the molding process), there will be a need to conduct a statistical study to determine if a pattern develops with certain cavity combinations. Both the cavity and the gate (the space provided in the tool for the molten polymer to reach the cavity) should be checked for excessive wear. This is particularly important for fiber-reinforced polymers, where wear is a major issue.

The percentage of regrind or degraded plastic in the material could be a problem. If so, the molding parameters should be verified and the percentage of regrind should be reduced. If the regrind is absolutely necessary, its quality should be consistent.

Drops in the pneumatic cylinder pressure should be combated by increasing the output pressure for the compressor. A surge tank with a safety valve may be added.

Changes in line voltage contribute to uniformity problems. This can be solved with a voltage regulator.

Marking is a welding deficiency that can have many different causes, a common one being an overheated horn. When this occurs, check for loose studs and a loose tip. Other possible solutions are simply to cool the horn, check that the coupling of the horn and the booster is correct, and ensure that no cracks are present in the horn. If the horn is made of titanium, the problem could be solved by switching to an aluminum horn. If the horn is made of steel, the amplitude should be reduced.

If marking is caused by localized high spots in the part, such as lettering or symbols, the horn will need to be redesigned in order to properly fit the part. Another solution may be to recess the lettering or symbols.

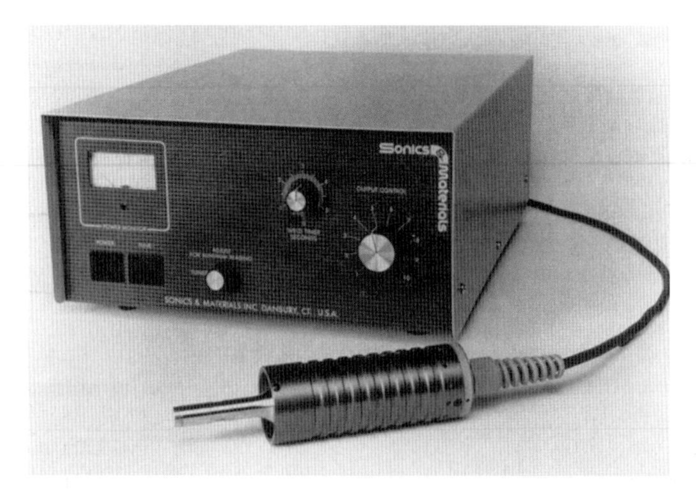

Figure 5.15 Ultrasonic handheld welder with replaceable horn (tip) *(Courtesy of Sonics and Materials, Inc.)*

Marks are often caused by the presence of aluminum oxide at the interface of the horn and the part. Aluminum oxide can be eliminated by using a chrome-plated horn and/or fixture or by applying a polyethylene film between the horn and the part.

Marking can also be created when a long weld cycle is used. Marking can be eliminated by reducing the overall weld cycle. This can be accomplished by lowering the amplitude or the pressure and adjusting the dynamic trigger pressure.

Flash in the weld can be the result of an energy director that is too large. Reducing the energy director size, reducing the weld time, and reducing pressure are all possible solutions. If the flash is caused by shear interference that is too great, the problem may be overcome by simply reducing the amount of interference. Flash can also be caused by poor part tolerances (too tight) and by nonuniform joint dimensions.

Misalignment of the welding assembly, which might suggest a poor initial design, could indicate the need for an alignment feature to be added to the parts. If improper support in the fixture is causing the misalignment, redesigning the fixture is recommended in order to provide proper support. Another option may be to shim the fixture. When misalignment is caused by wall flexure, with large sections deflecting, the addition of a rigid backup is suggested. Or the source of the misalignment problem may lie in improperly dimensioned joint design, in which case the parts must be redimensioned. Part tolerances and poor molding could also be the cause. Part tolerances should be tightened and molding conditions checked.

Internal components damaged during welding. This could be caused by excessive amplitude, which can be reduced by switching to a lower frequency. If long weld time is the cause, reduce weld time by adjusting the amplitude and/or pressure as well as the dynamic triggering pressure.

Internal damage can also be caused by too much energy entering the part. This can be corrected by reducing the pressure, weld time, or amplitude, or by using a power control. Also, proper mounting of the internal components should be verified. Sometimes a simple solution will be to isolate them from the housing or move them away from the area of high-energy concentration.

Melting or fracture of part sections outside the joint area. This problem is usually caused by sharp internal corners. In this case, a fillet should be used to radius or "round out" both the internal and the external corners. The correct internal radius should be equal to half the wall thickness. An external radius must be 1.5 times the wall stock.

If excessive amplitude is the cause, it can be reduced by changing to a lower booster. Fractures and melting can also come as a result of long weld time, in which case increased amplitude, increased pressure, or adjustments to the dynamic triggering pressure could correct the problem.

5.1.4.2 Joint Design

Joint design is a crucial component in successful ultrasonic welding. The design of the part and the materials used are important considerations in determining which joint design will be utilized.

5.1.4.3 Butt Joint Design

The butt joint design, also known as an energy director joint design or tongue-and-groove joint design, is appropriate for welding parts made mostly from amorphous resins, which lend themselves well to ultrasonic welding.

The joint should contain an energy director, which is a triangular protrusion or peak at or near the center of one of the faces. The peak provides line contact between the two surfaces to be welded. The volume (or area, as the calculations can be conducted in a 2-D plane if the part is symmetrical) of the triangular peak should equal the volume of the free space between the faces to be welded. This could be easily approximated with the areas of the two regions in a 2-D drawing. The melted material contained by the energy director or peak area should have an equal volume of space available to it in order to obtain a proper weld.

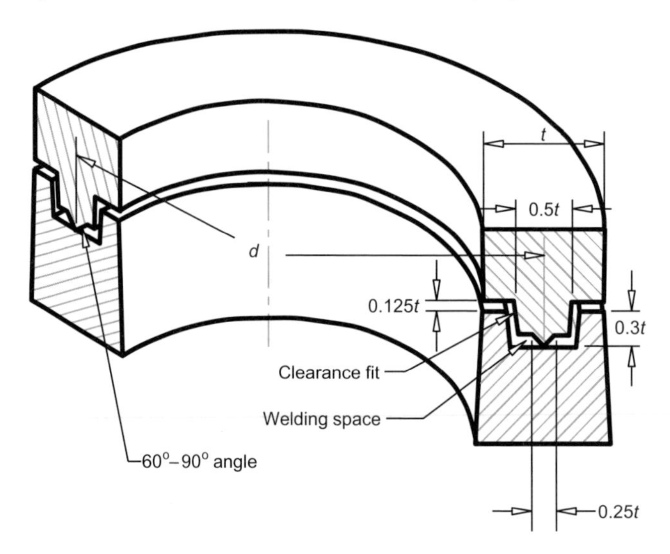

Figure 5.16
Butt joint, energy director, or tongue-and-groove joint design

The welding space or volume (the area in a 2-D cross-section drawing) should be at least three times the volume (or area) of an energy director for a butt joint design. This ratio increases as the angle decreases below 90°. Angles below 60° should not be considered.

The design shown in Fig. 5.16 provides a strong butt joint. This joint is difficult to mold, however, because of the clearance on both sides of the groove. The base of the energy director should be 0.25 times the wall stock. The angle should not exceed 90°. Usual values vary between 60° and 90°. The tongue width should be approximately half the wall thickness.

Equation 5.1 provides a formula to calculate the volume of polymer contained by the energy director for symmetrical welds (see Fig. 5.16).

$$V_{\text{Energy Director}} = 0.125t\left[\frac{\pi d^2}{4} - \frac{(d-0.25t)^2}{4}\right] \tag{5.1}$$

The following notation is used:

t = wall thickness (stock)

d = part diameter at the tip of the energy director

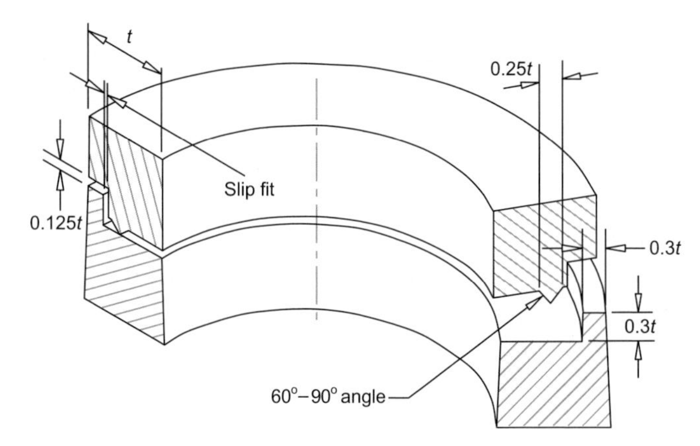

Figure 5.17
Butt joint: step design

To calculate the groove space needed:

$$V_{\text{Joint}} = 0.125t\left[\frac{(d+0.5t)^2}{4} - \frac{(d-0.5t)^2}{4}\right] \tag{5.2}$$

The butt joint step design is stronger than a pure tongue-and-groove design. The material flows into the slip fit clearance (Fig. 5.17), creating a seal that has good shear strength as well as tension strength. This design is based on an isosceles triangle. The height of the triangle should be a minimum of 0.5 mm (0.02 in.), and the base should be no less than 1 mm (0.04 in.).

Figure 5.18 shows other possibilities available in designing energy director joints or tongue-and-groove joints.

5.1.4.4 Shear Joint Design

Crystalline polymers require a joint design that provides a shearing action as the welding occurs. Figure 5.19 shows a typical shear joint. It should be noted that interference varies based on part dimensions. For small components with any dimensions in the *X*, *Y*, or *Z* direction less than 20 mm (0.75 in.), the interference should vary between 0.2 and 0.3 mm (0.008 to 0.012 in.). For medium-sized components with any dimensions between 20 and 40 mm (0.75 to 1.5 in.), the interference should increase to 0.3 up to 0.4 mm (0.012 up to 0.016 in.). Finally, for large part dimensions, which exceed 40 mm (1.5 in.), the interference should be between 0.4 and 0.5 mm (0.016 and 0.02 in.).

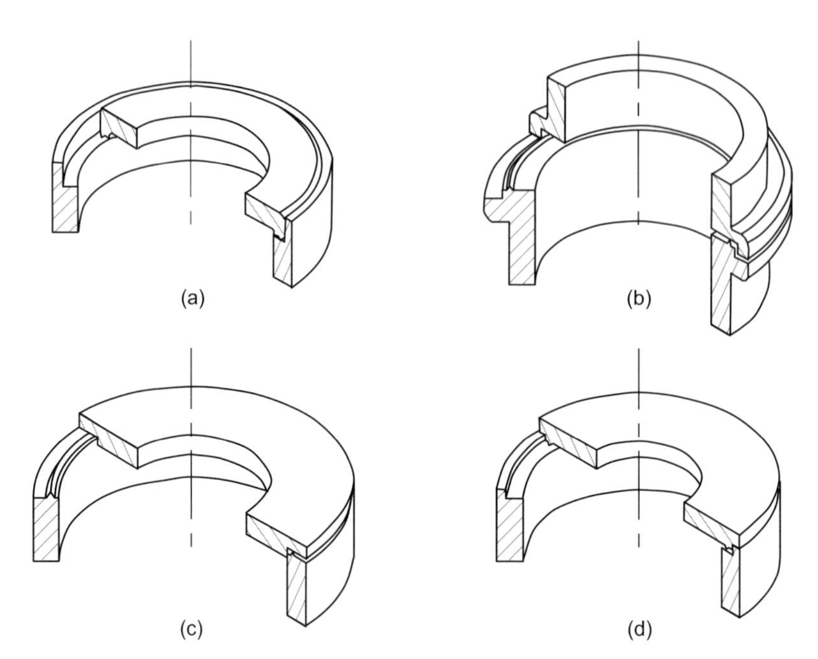

Figure 5.18 Variations of the butt joint design: (a) flat step; (b) double step; (c) flush step; (d) double flush step

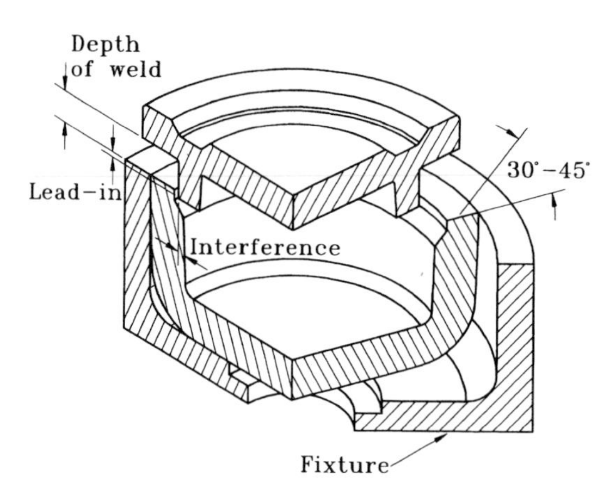

Figure 5.19
Shear joint design

The minimum lead-in recommended is between 0.5 and 0.6 mm (0.02 and 0.024 in.). The depth of the weld is related to the wall thickness and should be 1.25 to 1.5 times the wall stock.

The initial contact for this type of joint is limited to a small recess area in either of the parts. The recess helps the alignment of the parts during the welding process, which starts by melting the surfaces immediately on contact. Once the initial melting takes place, the parts continue to melt along the vertical walls, sliding together in a shearing process. The shearing action of the two melt surfaces eliminates possible leaks, resulting in good, leak-free seals.

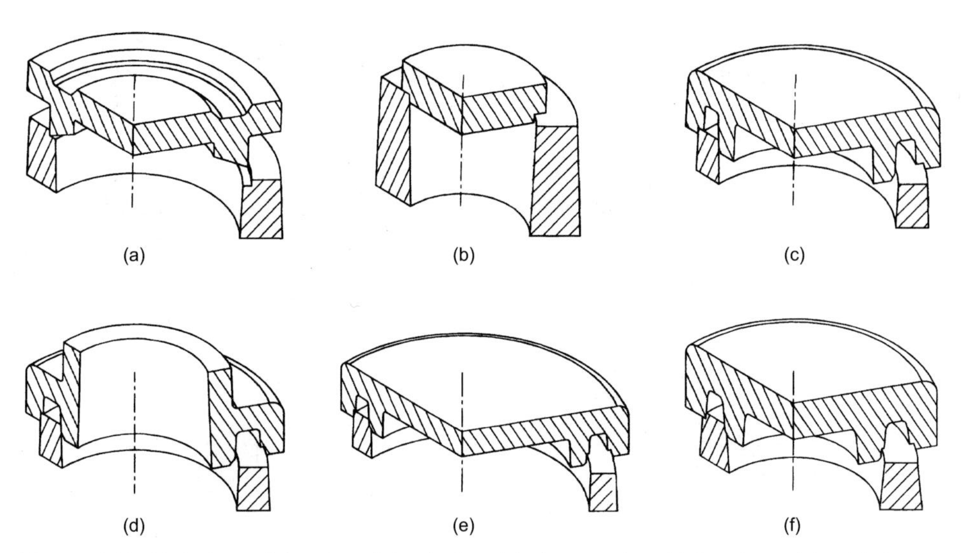

Figure 5.20 Variations of the shear joint design: (a) shear wedge; (b) flat shear; (c) guided shear; (d) control shear; (e) double shear; (f) double split shear

Figure 5.20 shows a few variations of the basic shear joint design. The joint designs shown in (d), (e), and (f) are mostly used for large parts (in excess of 80 mm). They help support the wall deflection that takes place during welding.

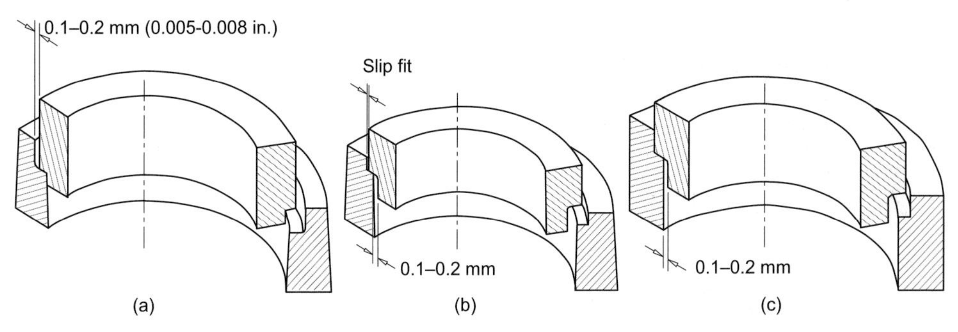

Figure 5.21 Shear joint designs with flash traps: (a) outside trap; (b) double trap; (c) inside trap

Figure 5.22 Various assembled parts using ultrasonic welding methods *(Courtesy of Dukane Corporation)*

5.1.4.5 Torsional Ultrasonic Welding

Available in the marketplace for a few years, the torsional ultrasonic welding, developed in Switzerland by Telsonic AG, brings new possibilities to the classic axial ultrasonic welding, detailed above. The new welding system, known by its trade name, Soniqtwist®, combines regular longitudinal or axial ultrasonic welding with a back-and-forth motion of less than 60 microns (µm) or circular motion (known as "twisting" motion) of just a few degrees in the same welding plane with identical frequency.

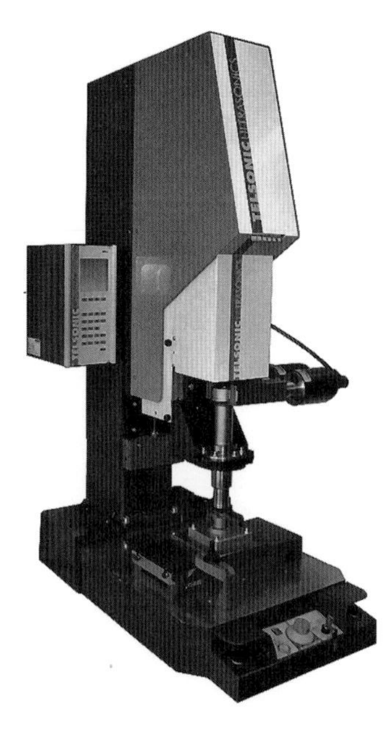

Figure 5.23
Torsional ultrasonic welder, Soniqtwist®

For the torsional ultrasonic welding process, the cycle times are comparable to regular axial welding. The system is used for sensitive components such as electronic assemblies, in which integrated circuits are enclosed polymer components assembled with a very thin film or membrane, and automotive crankshaft pulse sensors.

Components designed for this new process can use thinner wall thicknesses because no marks will permeate to the show surface. Typically, the design should incorporate an energy-director joint design even for crystalline polymers. For automotive components such as bumpers, fascias, and even cowl vents, the wall thickness of the actual part can be reduced by as much as 20% when compared to classic axial ultrasonic welding, without marking the automotive painted class-A surfaces, or even molded-in color class-A surfaces.

Figure 5.24 Detail of torsional ultrasonic sonotrode

The reduced wall stock results in significant reductions in weight—which nowadays is a major driver in the automotive marketplace due to electrification efforts and the overall injection-molding cycle time necessary to manufacture the part.

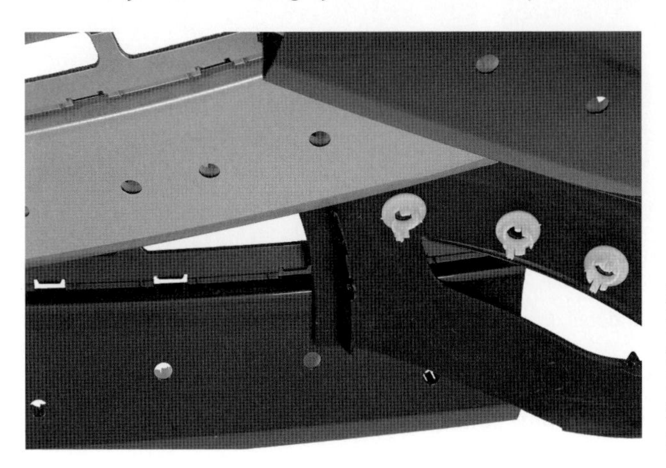

Figure 5.25 Painted automotive rocker panels having assembly locators mounted using the Soniqtwist® torsional ultrasonic machine

5.1.4.6 Case History: Welding Dissimilar Polymers

Going to the dentist is an unappealing task for most people, especially those prone to cavities. Cavities mean getting fillings, which traditionally entail a large injection in the gums and then a scary and noisy drill filing away at the molars. German dental equipment manufacturer DMG has developed the dental applicator (Fig. 5.26) called Icon® that should calm many people's fear of going to the dentist.

Figure 5.26
Dental applicator Icon® made by DMG of Hamburg, Germany

DMG's Icon product applies hydrochloric acid directly onto the weak area of the tooth and eats away at the enamel until it reaches the cavity. The therapy uses a light-cured resin that fills the enamel cavities and then is activated by blue light and seals the tooth surface. The technique works very well for early-stage cavities and makes the trip to the dentist a little more appealing.

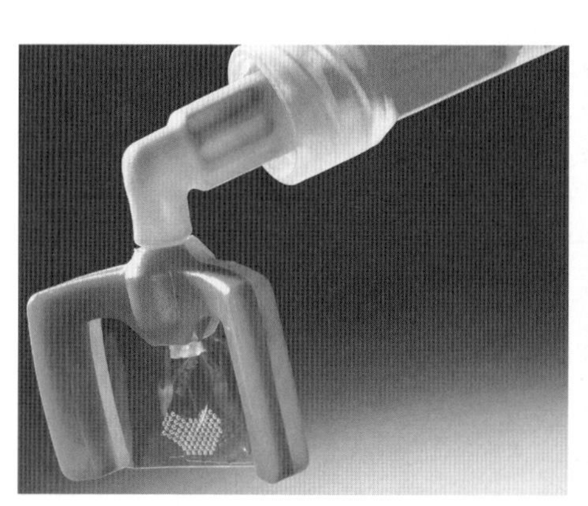

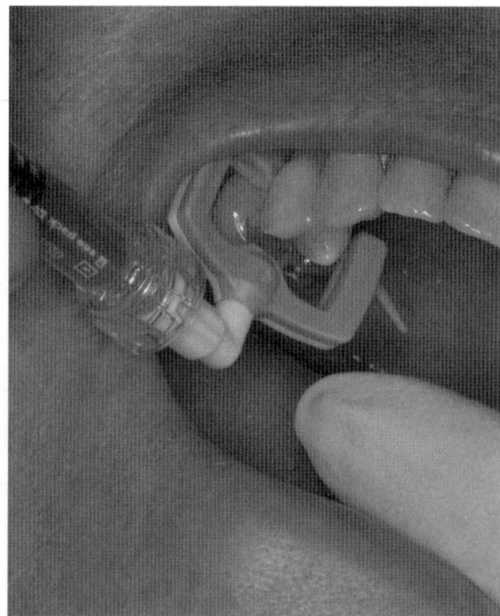

Figure 5.27 Dental applicator (a) detail, (b) detail in use *(Courtesy of Herrmann Ultrasonics)*

The dental applicator uses three polymers assembled with an axial ultrasonic welder. A semicrystalline polyethylene terephthalate (PET) double-layered PET film (called

Hostaphan® and manufactured by Mitsubishi Polyester Film), partially perforated and with a wall stock of about 0.05 mm, is clamped between two halves of a U-shaped frame made of amorphous polystyrene (PS), called Polystyrol®, from BASF (Fig. 5.27(a)).

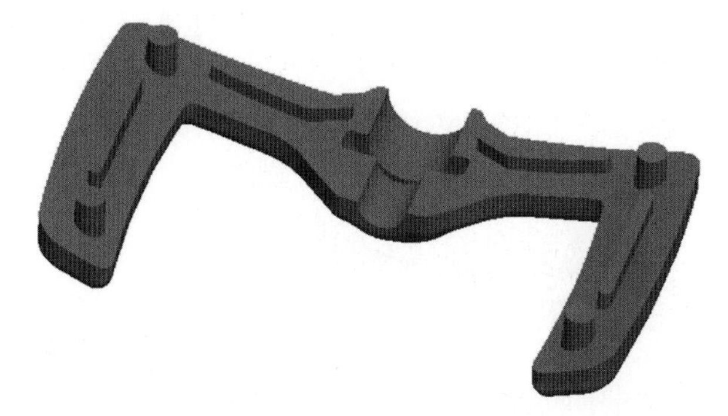

Figure 5.28 U-shaped frame having four studs and four energy directors *(Courtesy of Herrmann Ultrasonics)*

Ultrasonic welding produces a controlled melt built up to ensure a tight bond with minimal thermal load. The key to properly welding dissimilar crystalline-amorphous polymers (PET/PS) into a homogenous assembly is to program the weld force required to uphold the desired joining velocity. Amorphous resins, like polystyrene in general, are hard and rigid. They require low specific heat and small welding amplitudes, between 10 and 25 microns at a 35,000 cycles per second (Hz) frequency.

Figure 5.29 Frame assembly *(Courtesy of Herrmann Ultrasonics)*

On the other hand, crystalline polymers–like PET film–are softer, tougher, and can generally withstand higher temperatures, thus requiring higher specific heat to disrupt the resin structure. This is accomplished by employing larger amplitudes in the welding cycle.

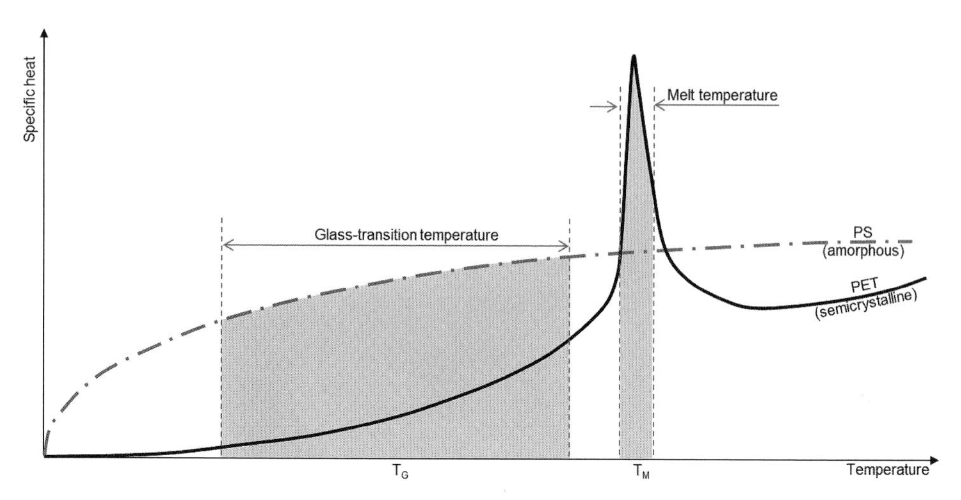

Figure 5.30 Specific heat versus temperature graph of the two polymers, polystyrene and polyester *(Courtesy of Herrmann Ultrasonics)*

Due to the complex shape of the U-shaped frame (convex/concave) as well as its small dimensions and the thin PET film, the dental applicator is a rather complex assembly. Warpage of the U-shaped components within the injection-molding process, as well as any postmold shrinkage, could pose serious welding problems due to tolerance stack-ups. To design a suitable weld joint that doesn't allow molten resin to escape laterally when melting the frame components while simultaneously clamping the film, was a challenge. The correct positioning of the very thin PET film during the production process was critical (Fig. 5.31).

The existing perforations in the PET film represent disruptions within the film surface and create areas known as "notch-effect points," where the mechanical vibrations imposed by the horn lead to stress peaks, which could lead to undesirable plasticizing of the polymer.

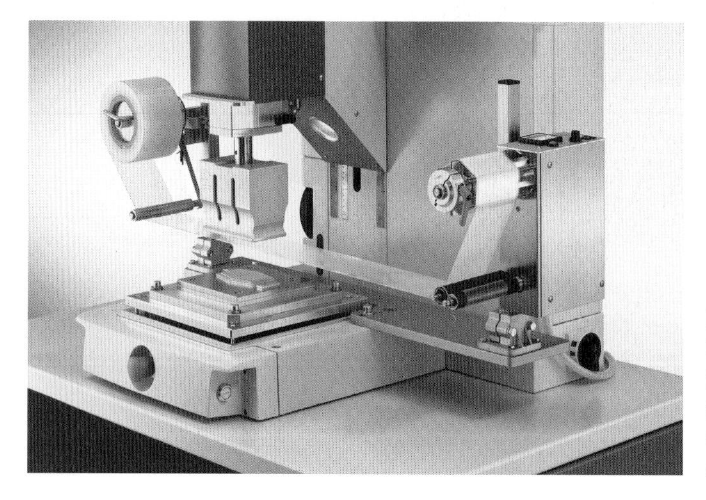

Figure 5.31 Ultrasonic welder with film indexer used to assemble the dental applicator *(Courtesy of Herrmann Ultrasonics)*

Herrmann Ultrasonics designed the U-shaped frame components in such a way that four locating pins and four small, thin-walled joints ensured the best weld joint strength. The energy director design is especially well suited for small parts made of amorphous polymers with thin walls.

Amorphous and semicrystalline resins have different melt temperatures and are difficult to join (Fig. 5.32). While the amorphous frame heats up quickly due to the mechanical vibrations provided by the horn, the PET film has a delayed response that protects the polyester from degrading thermally.

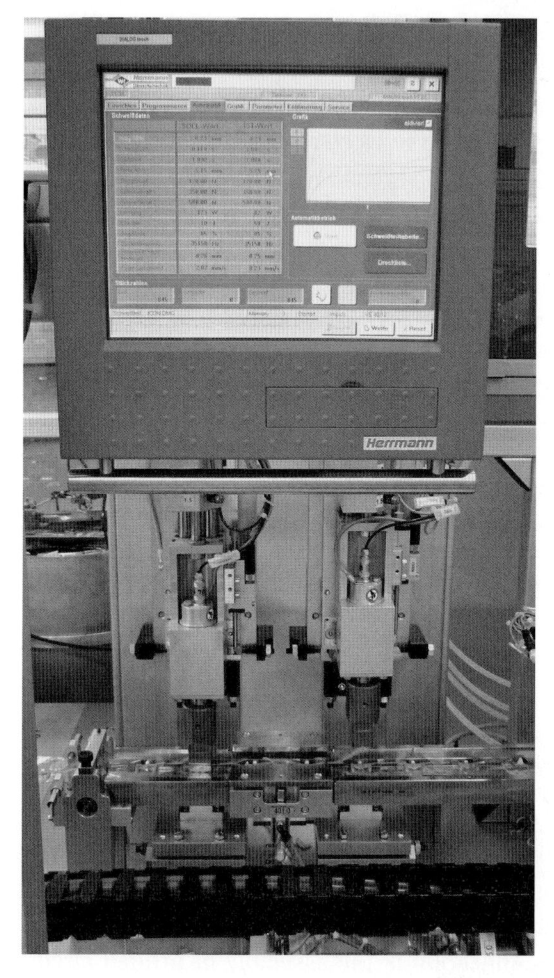

Figure 5.32
Ultrasonic welder using tow horns and digital display for welding parameters *(Courtesy of Herrmann Ultrasonics)*

Quality welds can be displayed visually. Graphical representation of the welding parameters such as power, joining velocity, and weld force allows high-quality ultrasonic welding to take place. A rapid and linear slope of the joining velocity curve is desired for complete melt buildup, together with repeatable results (see blue curve in Fig. 5.33). This guarantees exact repeatability of the welding process.

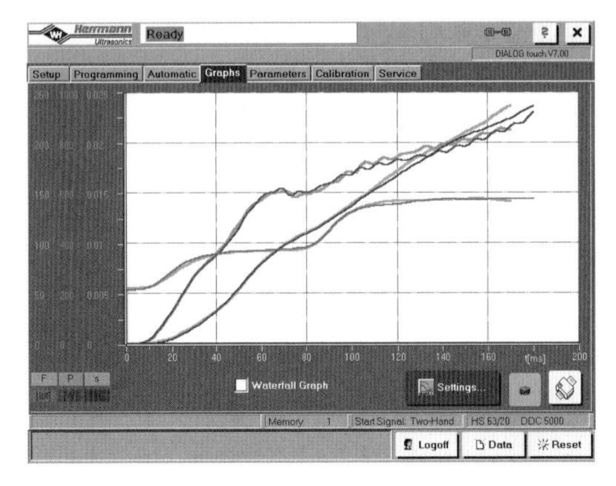

Figure 5.33
Ideal joining velocity linear slope is shown in blue *(Courtesy of Herrmann Ultrasonics)*

When welding the DMG Icon applicator parts, the welder (Fig. 5.32) switches from the first force exercised by the first horn to a second weld force in the last third portion of the process applied by the second horn. The melt generated with the weld force created by the first horn is compressed when switching to a second, higher weld power generated by the second horn. As a consequence, the velocity of the welding operation is retained until the end. Employing two horns in tandem allows the overall welding time to be shortened. Also, the load exercised on the polyester film by the mechanical vibrations of the horn is reduced, which is quite important in preventing any film damage. Furthermore, the melt, while cooling, is pressed more strongly against the polyester film during the hold time period, thus increasing the strength of the weld.

From an ultrasonic equipment standpoint, the prerequisites for weld accuracy are precise determination of the ultrasonic activation (also called "trigger point"), controlled melt buildup, and quick termination.

5.2 Ultrasonic (Heat) Staking

Ultrasonic or *heat staking* is an assembly method that uses controlled melting and forming of a boss or stud to capture or lock another component in place. The boss is made of plastic material, and the parts can be made of the same or different materials. An example of a heat staking application is the assembly of printed circuit boards.

The plastic stud or boss protrudes through a hole in the dissimilar material. The high frequency of the horn is transmitted to the head of the stud. The stud begins to melt and fills the volume of the horn cavity, which produces a head, locking the other component in place. The head is produced by the progressive melting of the post or boss under continuous and light pressure provided by the pneumatic cylinder.

Unlike ultrasonic welding, ultrasonic staking requires that *out-of-phase* vibrations be generated between the horn and the plastic surface of the stud or boss. Initial contact pressure of the horn must be light. This pressure, in combination with the out-of-phase vibration of the horn, provides an effective ultrasonic stake within the limited contact area.

5.2.1 Standard Stake Design

The diameter of the head produced by *standard stake design* is double the initial stud diameter. The recommended minimum stud diameter is 1.5 mm (0.0625 in.). The stud height, after the stud passes through the hole in the other component, is 1.5*d* to 1.7*d*, where *d* is the diameter of the stud.

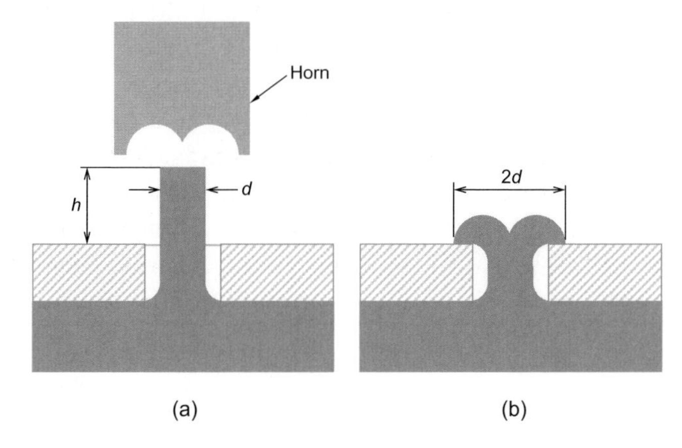

Figure 5.34
Standard (flare) stake design: (a) before, and (b) after assembly

The stud protruded through the dissimilar material has a volume V_{Cylinder}. This volume must equal the volume of the horn cavity. In this case, the volume of the horn cavity will be V_{Horn}. Therefore, by using the condition that the two volumes should be equal, one can determine the diameter and height of the plastic stud or boss.

$$V_{\text{Stud}} = V_{\text{Cylinder}} = \frac{\pi d^2 h}{4} \tag{5.3}$$

$$V_{\text{Horn}} = \frac{V_{\text{Torus}}}{2} = \frac{\pi^2 d^3}{8} \tag{5.4}$$

If the two volumes are equal, the relation between *d* and *h* can be determined:

$$V_{\text{Horn}} = V_{\text{Stud}} \tag{5.5}$$

$$h = 1.57d \tag{5.6}$$

Equation (5.6) shows the existing relation between the stud height and the stud diameter.

The flare stake design is recommended for nonabrasive thermoplastics.

5.2.2 Flush Stake Design

The *flush stake design* is recommended for assemblies in which a flush surface is required and where there is sufficient thickness in the held-down piece.

The volume of the recess is calculated by adding the three volumes that constitute it:

$$V_{\text{Recess}} = V_1 + V_2 + V_3 \tag{5.7}$$

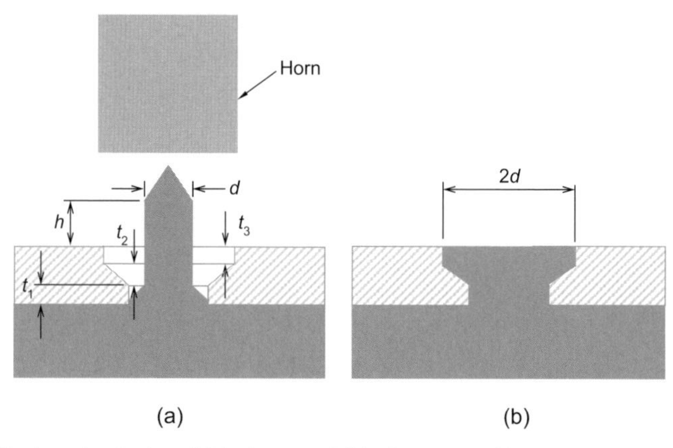

Figure 5.35 Flush stake design: (a) before, and (b) after assembly

If the relationship between the diameters of the three volumes is

$$V_{\text{Recess}} = \frac{3\pi d^2 t_1 + 7\pi d^2 t_2 + 12\pi d^2 t_3}{12} \tag{5.8}$$

and the height of each zone is

$$t = t_1 + t_2 + t_3$$

$$t_1 = 0.25 \text{ mm} = 0.01 \text{ in.}$$

$$t_2 = t_3 \tag{5.9}$$

Then the relationship between the various dimensions can be determined:

$$V_{\text{Stud}} = \frac{0.87\pi d^3 + \pi d^2 h}{4} \tag{5.10}$$

$$V_{\text{Stud}} = V_{\text{Recess}} \tag{5.11}$$

$$h = 2t - 0.87d - 0.58 \tag{5.12}$$

The recommended design for this type of weld is the tapered stud design commonly used for dome staking.

5.2.3 Spherical Stake Design

The spherical stake design is usually recommended for materials such as crystalline polymers, which exhibit a well-determined melting point. This design is also well suited for glass-reinforced materials. To calculate the protruded height of the stud, it is first necessary to write the equation for the total stud volume. This is based upon a cylinder volume to which the conical volume of the tip is added:

$$V_{\text{Stud}} = \frac{0.87\pi d^3 + \pi d^2 h}{4} \tag{5.13}$$

$$V_{\text{Horn}} = V_{\text{Semisphere}} = \frac{2\pi d^3}{3} \tag{5.14}$$

Because the stud volume must equal the horn cavity volume, the condition is

$$V_{\text{Horn}} = V_{\text{Stud}} \tag{5.15}$$

The height of the stud protruding above the flat surface is

$$h = 1.81d \tag{5.16}$$

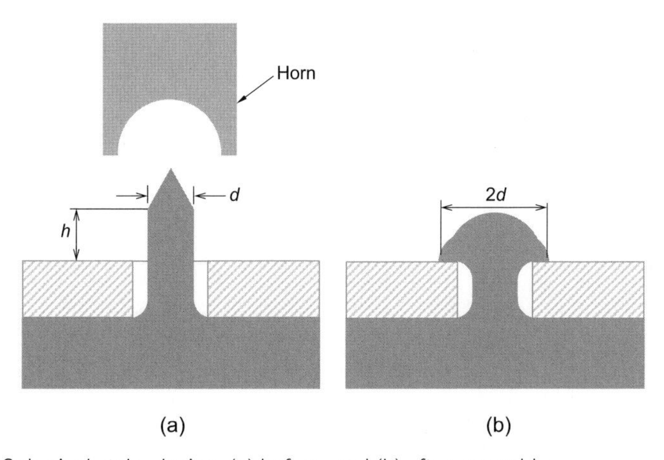

Figure 5.36 Spherical stake design: (a) before, and (b) after assembly

5.2.4 Hollow (Boss) Stake Design

The *hollow* or *boss stake design* is recommended for assemblies where no sink marks should be visible on the opposite side or when the opposite side is a show surface. If disassembly is required, this design allows the parts to be reassembled by using self-tapping screws.

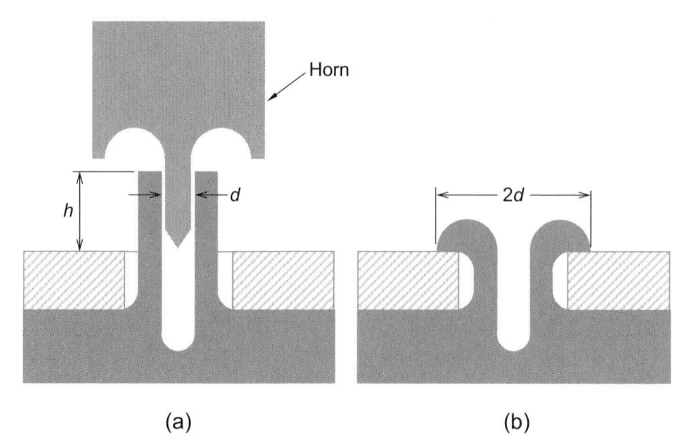

Figure 5.37 Hollow (boss) stake design: (a) before, and (b) after assembly

Similar calculations can be conducted in order to determine the optimum height and diameter relationship required for a proper design:

$$V_{\text{Horn}} = \frac{V_{\text{Torus}}}{2} = \frac{\pi^2 d^3}{4} \tag{5.17}$$

$$V_{\text{Stud}} = \frac{3\pi h d^2}{4} \tag{5.18}$$

$$V_{\text{Horn}} = V_{\text{Stud}} \tag{5.19}$$

And finally, the relationship between the diameter of the boss and its height is

$$h = 1.05d \tag{5.20}$$

5.2.5 Knurled Stake Design

Knurled stake design is used mostly where appearance is not an issue. This design is not recommended where material strength plays an important role in the assembly.

The knurled stake method is used in high-volume production. The integrity of ultrasonic stake assemblies depends upon the geometric relationship between the horn cavity and the stud, which must be volumetrically compatible. The process requires accurate control over the three principal process variables: amplitude of vibration, pressure, and weld time. If any variable is altered, the other two must be precisely adjusted or the process will fail. It should also be noted that ultrasonic staking requires overall pressures substantially lower than those for ultrasonic welding.

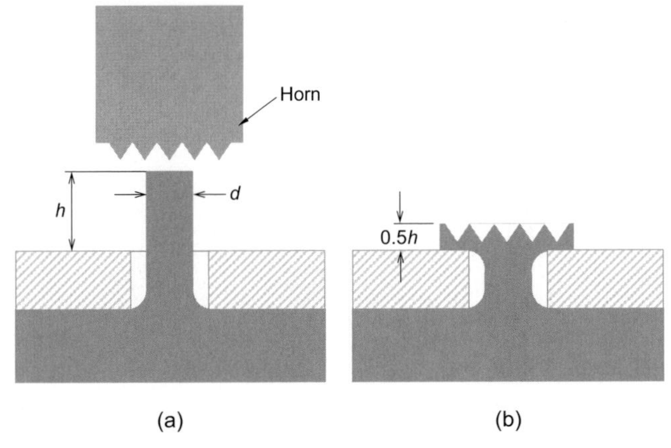

Figure 5.38 Knurled stake design: (a) before, and (b) after assembly

Proper stake design produces optimum head strength and appearance.

Several design configurations for stud or horn reviewed above are available to the designer. The design to be used depends upon the requirements of the application and the physical size of the stud or studs.

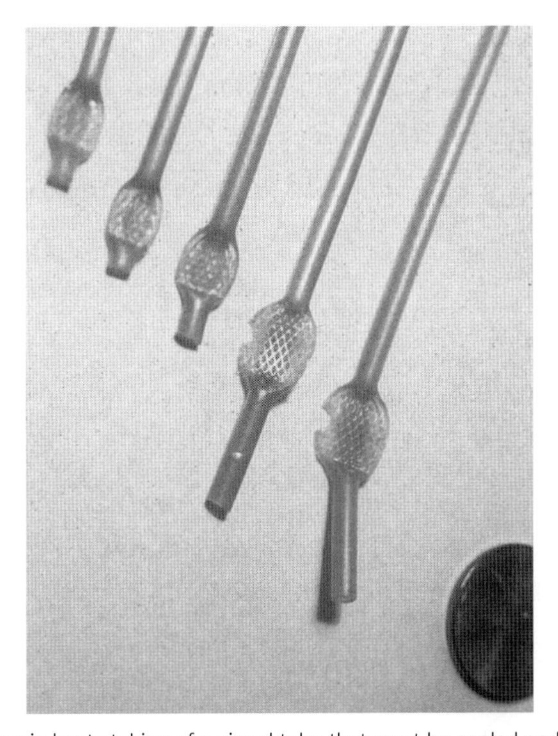

Figure 5.39 Ultrasonic heat staking of a signal tube that must be sealed and sustain a 0.2 MPa (30 psi) internal pressure. A 2,000 W system at 20 kHz frequency was employed with the following parameters: weld time 0.75 seconds, hold time 0.5 seconds, a 0.35 MPa (50 psi) horn pressure, and amplitude of 0.03 mm (0.0012 in.) *(Courtesy of Branson Ultrasonics, a business of Emerson Electric Company)*

For each of the designs presented, the same staking principle applies: the area of initial contact between the horn and the stud must be kept to a minimum, thus concentrating the output energy delivered by the horn to produce a rapid melt.

5.3 Ultrasonic Spot Welding

The *ultrasonic spot welding* technique is usually employed to join two layers of similar materials in a single location. This method does not require any type of energy directors or any preformed (pilot) holes.

The protruded area of the horn melts the first surface to be joined and then passes through it. The output energy delivered by the horn is released at the interface between the horn and the parts. This energy produces frictional heat that melts the material. As the horn travels to the lower surface, the displaced melted thermoplastic polymer flows between the two components, which are spot-welded together, forming a bond.

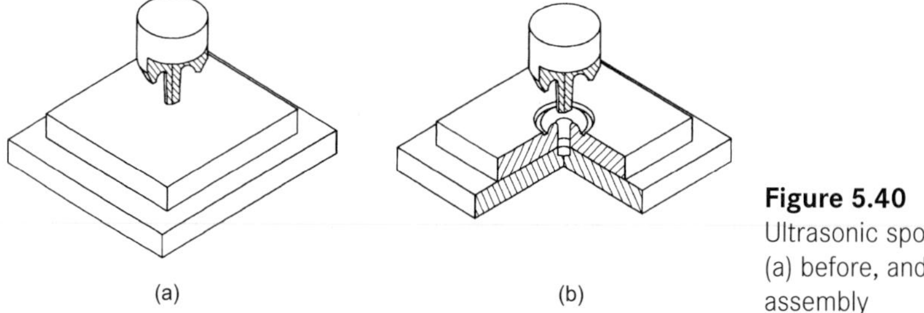

Figure 5.40
Ultrasonic spot weld design: (a) before, and (b) after assembly

5.4 Ultrasonic Swaging

Ultrasonic swaging is a process by which the plastic is melted and re-formed as a ridge to encapsulate or hold another component of the assembly. Calculations similar to those in the previous section can be carried out.

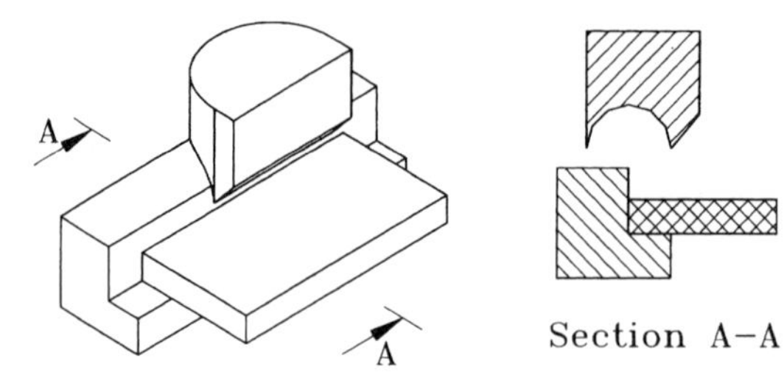

Figure 5.41
Ultrasonic swaging design: isometric view of the components and the horn, and section A–A through the axial centerline of the horn

5.5 Ultrasonic Stud Welding

Ultrasonic stud welding uses a shear joint design (Fig. 5.42) with typical dimensions of a = 0.4 mm (0.016 in.), b = 0.5d (half the stud diameter), c = 0.4 mm, and d = stud diameter. Each design allows for the escape of air during the welding process. Caution must be used when employing the stepped stud design, as air can become trapped, creating internal stresses in the joint.

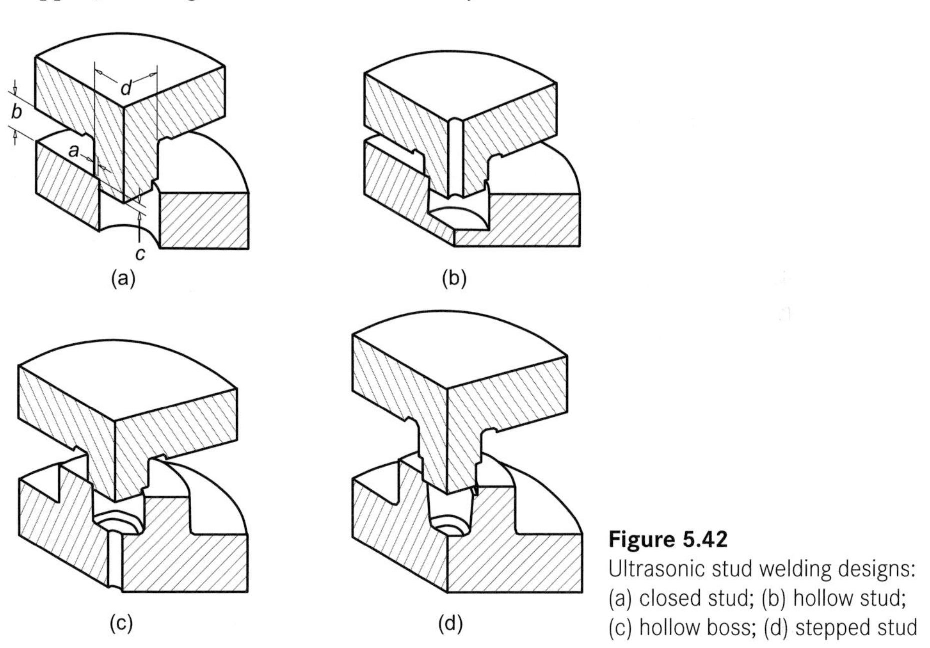

Figure 5.42 Ultrasonic stud welding designs: (a) closed stud; (b) hollow stud; (c) hollow boss; (d) stepped stud

5.6 Spin Welding

Spin welding is a friction-welding assembly method used for joining thermoplastic parts that exhibit a circular cross section. In the spin welding process, one part is rotated while the other is held stationary. Spin welding is accomplished by the frictional heat developed at the interface of two parts in contact. Once a melt film has been formed due to the heat generated by friction, the relative motion is stopped and the weld is allowed to solidify under pressure.

In addition to producing joints of exceptional quality and strength, this technique offers economic advantages because it is fast, simple, and inexpensive.

5.6.1 Process

Once the rotation has begun, the rotating part contacts the stationary part under a constant preset axial pressure, and heat is generated.

Depending on the type of machine used, the pressure needed to join the parts can be provided in two ways. The rotating part can be pressed down toward the stationary one, or the stationary part can be pushed upwards against the rotating part.

The heat is generated by the external friction between the parts and the internal friction created by the shearing of plastic layers in the parts. After melting takes place, the spinning stops and the parts are allowed to cool under a preset pressure. Sensors measure different process variables such as normal pressure, torque, rotations per minute, and axial tolerance between the centerline of the two parts. The variables that can be used to control the weld joint are time control, which refers to the duration of the friction, and energy control, or the amount of frictional energy that is applied.

The spin welding process can be divided into two methods: the *pivot method* and the *inertia method*.

Spin welding by the pivot method requires continuous rotation and exact timing to properly join the parts. The tool for the pivot method contains a spring-loaded pivot pin.

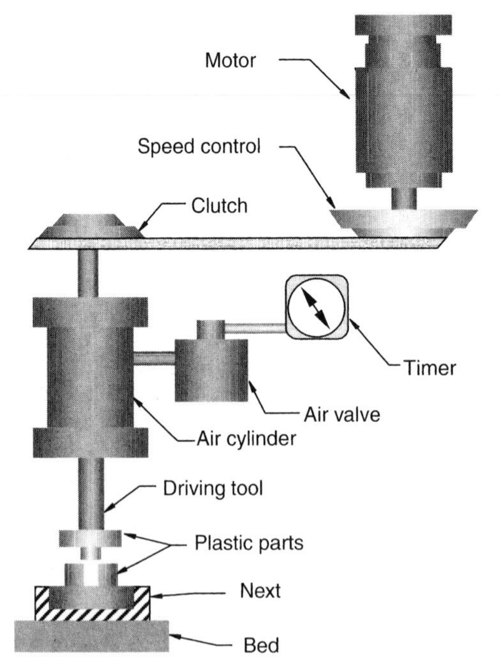

Figure 5.43
Spin welding equipment

As the tool advances, the pivot pin engages at the center of the button for alignment. The pivot pin is then compressed. The tool continues to advance until teeth grip the surface of the part. The part spins until the timer signals for the tool to retract. Both contact surfaces melt during the spinning process. As the tool retracts, the teeth disengage. Using a self-braking feature, the part stops spinning instantly due to joint friction. The joint remains pressurized until the pivot pin disengages. This provides sufficient time for the thin melt film in the joint to solidify.

The teeth of the pivot tool will leave indentations in the polymer surface of the finished part. Where this is not desired, a rubber-faced pivot tool may be used. Also, lugs may be molded onto the surface of the polymeric part to engage the mating lugs on the tool.

The inertia method also requires rotation and pressure. It depends upon the parts sealing themselves after initial rotation. An inertia tool utilizes the energy stored in a spinning flywheel to melt the contacting surfaces. This allows a constant amount of energy to be applied to each weld.

The spinning tool advances and makes contact with the mating part and the lugs engage. The flywheel or mass comes to a stop within 0.5 seconds, losing all its kinetic energy in friction. The generated friction causes melt to form in the joint. After the flywheel stops, the melted material solidifies within a fraction of a second. The tool is then retracted, completing the cycle.

When using the inertia technique in an automatic operation, it is necessary to provide a quick-acting clutch to bring the flywheel up to speed. The clutch can be placed above the pneumatic cylinder. This way, as the tool advances, the clutch disengages, allowing the driving head and shaft to spin freely. Retraction of the tool after welding will re-engage the clutch. The same technique can be used by placing the clutch above the chuck, which will accelerate the speed of the powered spindle.

The key points of this technique are that the rotating mass or flywheel should stop in less than one second, and a bead of melt should be formed around the weld. The size or weight of the flywheel can be approximated roughly by using 5 to 10 kg (11 to 22 lb.) of mass for each 20 mm^2 (0.03 $in.^2$) of weld area. Since the energy of a rotating mass is equal to the moment of inertia of the mass multiplied by velocity squared, making a slight change in velocity can be the most convenient way to bring the system into balance. The mass should be kept to a minimum for quick braking, and the flywheel type of geometry should be used where possible.

The inertia method is versatile and offers far greater weld reproducibility than the pivot method. It is recommended for parts with joint diameters greater than 20 mm.

The spin welding process takes place in five phases.

In the first phase, friction is generated by the moving part of the assembly before any melting takes place.

In the second phase, the heat is generated mostly by wear particles that flow through the melted material.

In the third phase, melted plastic reaches glass transition temperature, the highest temperature the polymer will attain during the spin welding process.

In the fourth phase, the material reaches a steady state, where the temperature is considered constant. The heat generated in the previous phases equals the heat lost through flashing or through thermal radiation in the wall stock. The penetration of the weld into the parts is proportional to the amount of time.

In the fifth or last phase, the rotation stops and an axial pressure is applied for the holding time. The parts are let to cool under pressure, and the weld is completed.

Cycle control is relatively simple, and the cycle should be only long enough to ensure complete welding. The basic cycle is similar to a drill press operation: the driving tool advances and retracts to complete the cycle. A complete cycle time lasts only as long as air pressure is applied to the pneumatic cylinder during the forward stroke.

It is very important in spin welding that nesting fixtures hold the parts securely. Ribs and recesses, where possible, provide an effective means of locating the parts when rotating.

Drag, axial spindle positions, and time are the variables that influence the welding process. Position and time are variables that can be used to control the weld. Position regulation, which ensures that a given amount of polymer is melted and displaced during each cycle, may be governed by a limit switch or stop.

It is essential that the motion be stopped once melting occurs to allow the melt to solidify under pressure. Failure to stop quickly, particularly with the fast-setting resins, will cause tearing of the joint, resulting in a weak weld.

Short weld time decreases flash and localizes heat at the rubbing surfaces for minimal residual stress and distortion. It is usual practice to spin the parts for a fraction of a second (0.1 to 0.3 seconds) at recommended surface speed. Overall cycle time for this process, including handling, is between one and two seconds, allowing a production rate as high as 60 assemblies per minute. Depending on the part size, the cycle can vary up to 15 seconds. Holding time generally requires 0.5 to 5 seconds, during which the assembly cools down to ambient temperature and the weld solidifies. Axial pressure during the friction phase is 0.07 to 0.15 MPa (10 to 20 psi). The axial pressure applied after the rotation has stopped is increased to 0.1 to 0.35 MPa (15 to 50 psi).

5.6.2 Equipment

The type and amount of equipment required for spin welding depends upon the volume of production–whether it is normal, low-volume, or sample production.

The basic spin welding equipment necessary for normal production consists of an electric motor with optional brake; speed control with V-belt and pulleys; air cylinder with solenoid air valve for stroke; driving tool; electric/electronic timer for controlling the stroke; chuck for holding the part; and a nest for the mating part (see Fig. 5.43).

A vertically mounted pneumatic cylinder contains the rotating shaft. The air cylinder, valves, and a timer can be attached to a good-quality drill press to obtain the necessary automation. The process can be further automated by adding automatic feed and setup by air or with mechanical ejection of the welded part.

For low-volume or sample production, a manually operated drill press, a driving tool, and a chuck to hold the parts can be sufficient.

5.6.3 Welding Parameters

The heat generated between surfaces is a function of relative surface velocity, contact pressure, and duration of contact. It is also a variable of material properties such as coefficient of friction and heat transfer capacity. Experimental data indicate a linear variation in the relationship between velocity and pressure.

Pressure must be applied uniformly to the surface being welded. It must be great enough to force any bubbles, contamination, or degraded material from the joint. There is, however, a definite limit to the amount of pressure that may be applied to a part. If too great a pressure is applied, the part will become deformed because pressure would be applied while the interface of the two parts is melting. In many applications, best results are obtained with an increase in pressure after initial melting. The material is thus squeezed between the joint surfaces as it sets up. Additional pressure is required at the higher temperatures to prevent degrading or bubbling of thermoplastic material. Adjustments in contact pressure should be made to suit specific applications and joint configurations.

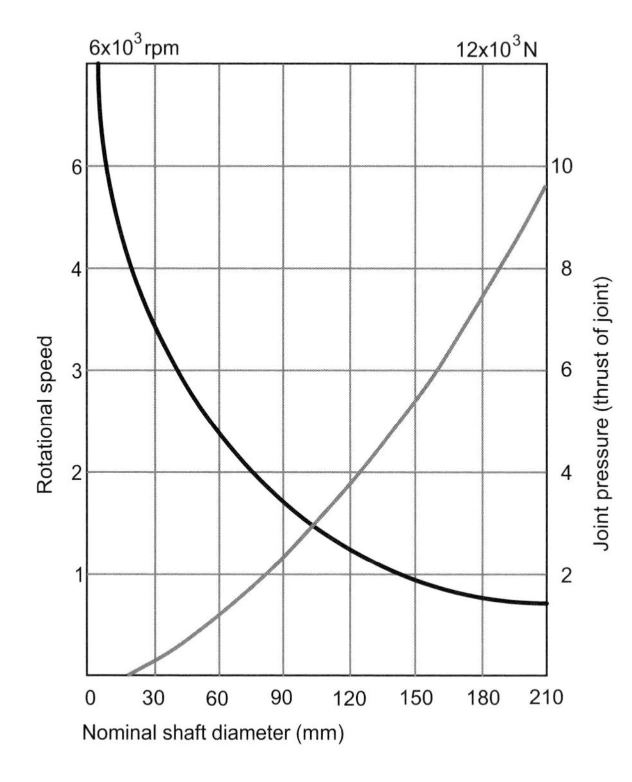

Figure 5.44
Variation of the joint pressure and of the rotational speed as functions of the nominal shaft diameter

The surface velocity of a rotating cylinder is a function of the cylinder diameter and of the number of rotations per minute. For a solid disc the average velocity of a point on the surface is 0.67 times the circumferential velocity. For a hollow part, the average velocity of a point on the surface is given by

$$V = \frac{\pi\omega\left(r_o^3 - r_i^3\right)}{9\left(r_o^2 - r_i^2\right)} \tag{5.21}$$

where

V = average point velocity (m/min)

ω = angular velocity (rpm)

r_o = outside radius (mm)

r_i = inside radius (mm)

Figure 5.44 shows a plot of suggested speed and contact pressure against joint diameter. Table 5.1 shows a suggested range of point velocity and initial contact pressures for various plastic materials.

Table 5.1 Average Velocity of a Point on the Surface and Initial Contact Pressures for Various Plastic Materials

Material	Average velocity of a point on the surface (MPS)	Initial contact pressure (MPa)
Acrylic	3 to 10	0.1 to 1
Acetal	1.5 to 10	0.2 to 1.3
Polycarbonate	2 to 12	0.1 to 1.2
Polyamide	1.5 to 15	0.2 to 1.5

Preheating the parts to be welded helps to ensure better welds. In some cases, the parts can be welded as they are removed from the molding machine. This method works especially well with polymers that have a high melting point. Some plastics cannot be melted by friction alone, and preheating makes it possible to spin weld these materials.

5.6.4 Joint Design

Joint geometry is an important factor in weld quality and aesthetics. Many design configurations will produce welds, but certain principles should be considered to obtain the optimum strength and appearance of the weld.

Desirable features of vibration welding-joint design include

- maximum weld area with a minimum material for structural economy
- self-alignment to reduce molding and welding tolerances
- progressive contact from the central area of the lip of the joint to prevent trapping of air
- rigidity to prevent bulging under high welding pressure.
- general symmetry for uniform melting and melting distribution along the contact area
- no points of stress concentration, such as sharp corners, notches, or sudden variation in the wall stock
- compatibility with other aspects of the overall design for the system or subsystem

Figure 5.45 illustrates several commonly used joint design configurations. Taper, tongue-and-groove, and step joint designs increase joint area. The limitation on weld area is the wall thickness or drag created by increased surface. Figure 5.45(j) and Fig. 5.45(l) show several joints that act as flash traps.

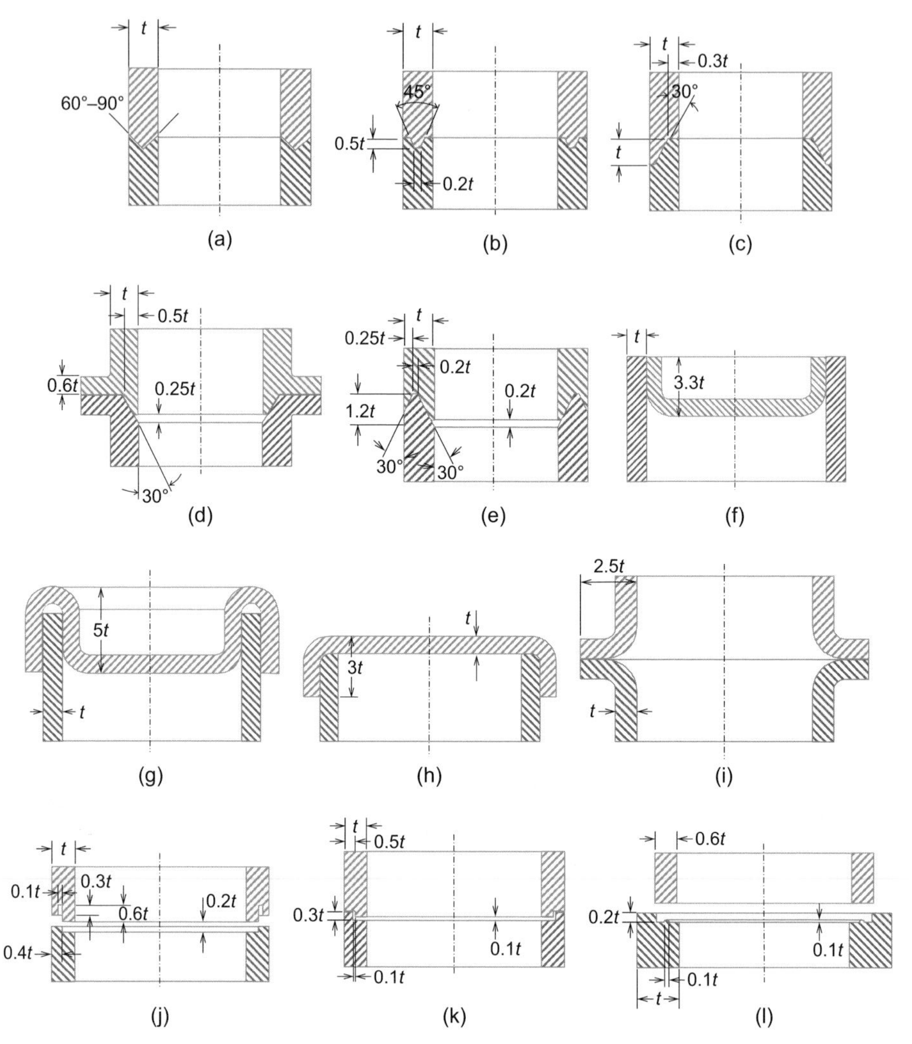

Figure 5.45 Joint design for spin welding: (a) taper; (b) tongue-and-groove; (c) shear; (d) flat shear; (e) reverse tongue-and-groove; (f) vertical; (g) double vertical; (h) vertical curve; (i) horizontal; (j) step with exterior flash trap; (k) step; (l) step with energy director

Using a taper or tongue-and-groove joint design will also improve alignment and reduce wobble.

In spin welding, excess melt that is forced out from between the mating weld surfaces is called flash. The joint itself may be designed to trap flash caused by the welding action at the surface of the joint. One method of trapping flash is to leave a clearance of about 0.25 mm (0.01 in.) between the halves of the part of the outside surface. When the part is welded, rather than run out, requiring postweld deflashing, the flash will fill this clearance.

The actual design will also depend upon the material used and the end-use requirements of the final assembly.

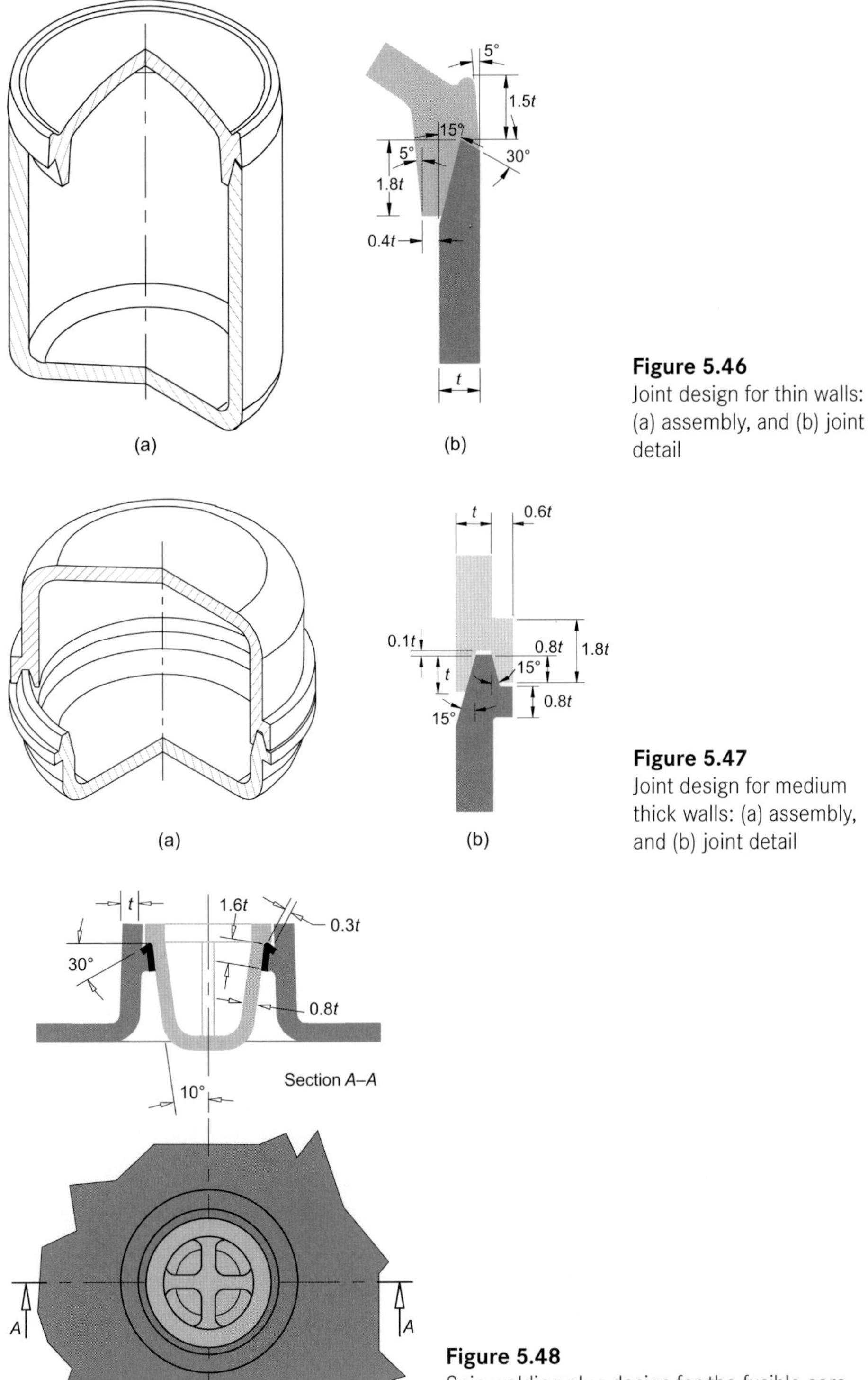

Figure 5.46
Joint design for thin walls: (a) assembly, and (b) joint detail

Figure 5.47
Joint design for medium thick walls: (a) assembly, and (b) joint detail

Figure 5.48
Spin welding plug design for the fusible core injection-molded air intake manifold of the Quad 4 engine made by General Motors LLC

5.7 Hot Plate Welding

Hot plate welding is achieved by placing a heating strip between the parts to be assembled. When the edges of the parts become soft, the strip is removed and the parts are quickly brought together. The hot plate welding method is suitable for welding two flat parts by their edges. It is commonly used for joining plastic pipes and a variety of hollow parts.

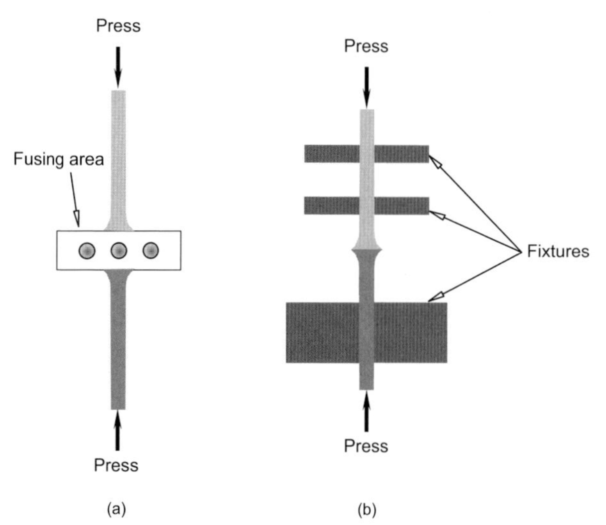

Figure 5.49
Hot plate welding

The melt depth or penetration is regulated by the amount of time the plastic parts are in contact with the heating plate. Depending on the part size, the time of contact with the heating plate usually varies between 1 and 6 seconds. The parts are held against each other under constant pressure, allowing the melted material to cool and weld molecularly.

Figure 5.50
Hot plate welder with vertical heat platen and hydraulic operation *(Courtesy of Forward Technology Industries, Inc.)*

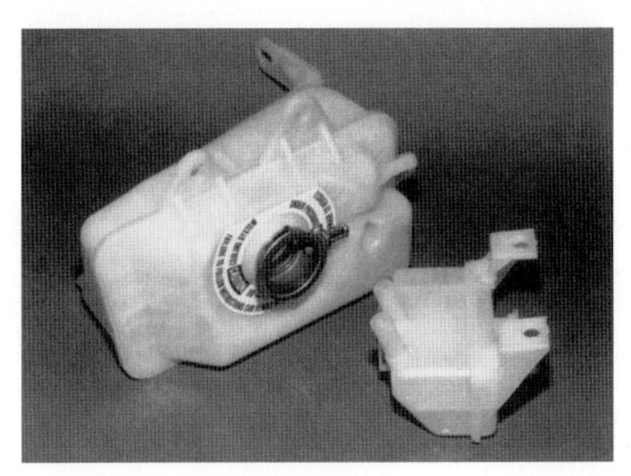

Figure 5.51 Automotive coolant system pressure reservoir (left) and brake fluid reservoir (right) made of polypropylene and assembled with hot plate welding process. The coolant reservoir sustains pressures of 0.11 MPa (16 psi) and the brake reservoir sustains internal pressures up to 0.83 MPa (120 psi) *(Courtesy of Forward Technology Industries, Inc. and Miniature Precision Components, Inc.)*

Holding time (cooling time under pressure) again depends on the part size and usually ranges from 3 to 6 seconds. The total cycle time can reach 20 seconds for large parts, although much less time is required for small parts.

Hot plate parameters are heating plate temperature, shape and coating, heating time, axial pressure, joining time, and cooling time.

The plastic parts are secured into the fixture, which precisely controls the alignment and mating of the parts during the welding process.

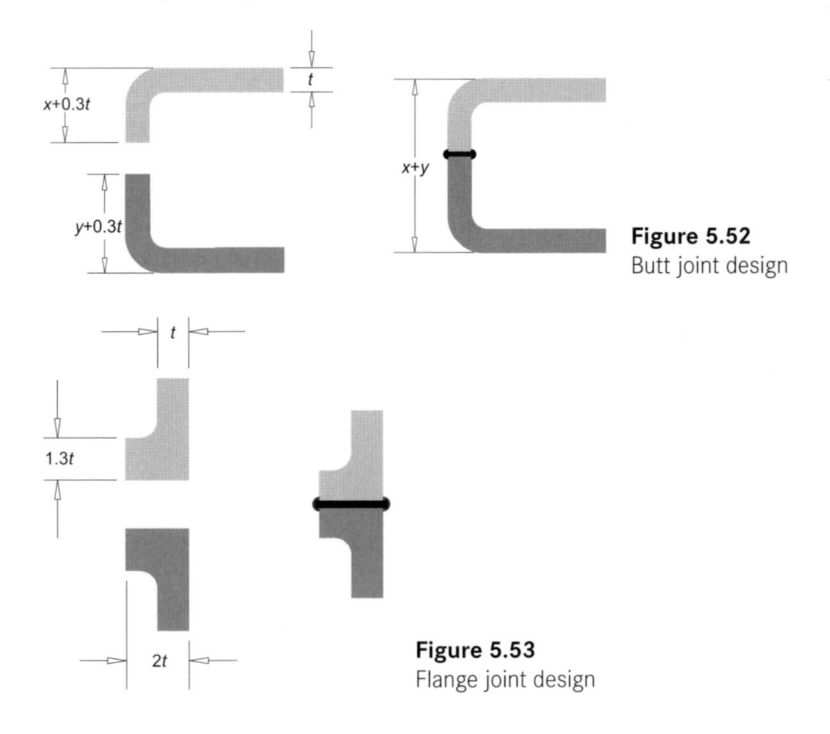

Figure 5.52
Butt joint design

Figure 5.53
Flange joint design

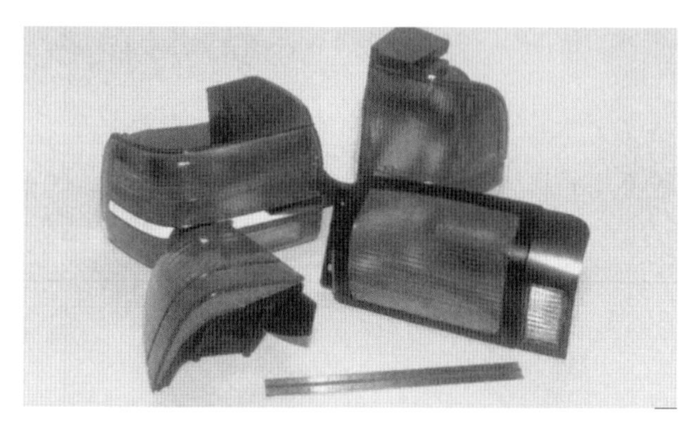

Figure 5.54 Hot plate welding of a tail lamp employing an energy director or butt joint design materials included: PMMA (acrylic) lens to an ABS lamp housing; PC (polycarbonate) lens to same lamp housing; and PC lens welded to a lamp housing made of PC/ABS alloy *(Courtesy of Branson Ultrasonics, a business of Emerson Electric Company)*

5.7.1 Process

While the plastic parts are held in the fixture by mechanical gripping fingers or pneumatic devices such as vacuum cups, a heated plate is placed between the parts. To melt the edges of the parts, the fixtures are brought into a motion normal to the welding plane, pressing the two parts against each other until the edges are molten and plastic material is displaced. The initial melting produces a smooth edge, free of any warps, imperfections, or sink marks.

Positive stops are employed in the tool design to provide an accurate control for the amount of melt as well as for the depth of the weld or seal. Typically, 0.4 mm (0.016 in.) of polymer is displaced during the melt cycle, and about 0.4 mm (0.016 in.) of material is compressed in the holding period. To produce the quality known as *regulated melt depth*, the melting and material displacement continues until the melt stops on the hot plate meet the tooling stops in the fixtures.

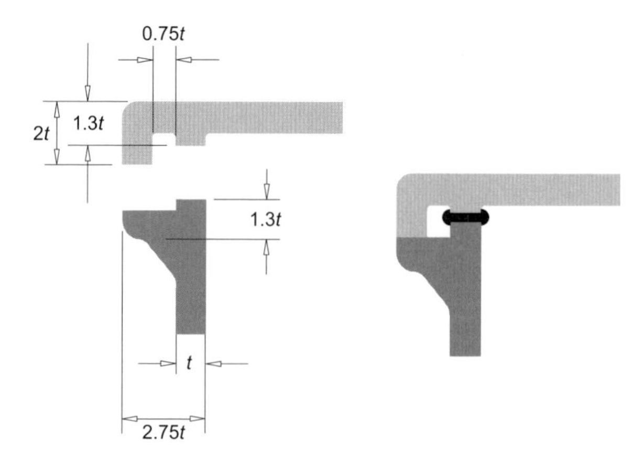

Figure 5.55
Recessed joint with flash trap

Once the parts' edges have softened, the fixture opens and the hot plate is removed. The fixture immediately closes and presses the two parts against each other. They are held together at a preset pressure for 2 to 7 seconds and the weld is completed.

The fixture opens and releases the assembly. Parts can be loaded and unloaded manually or automatically using a robot arm.

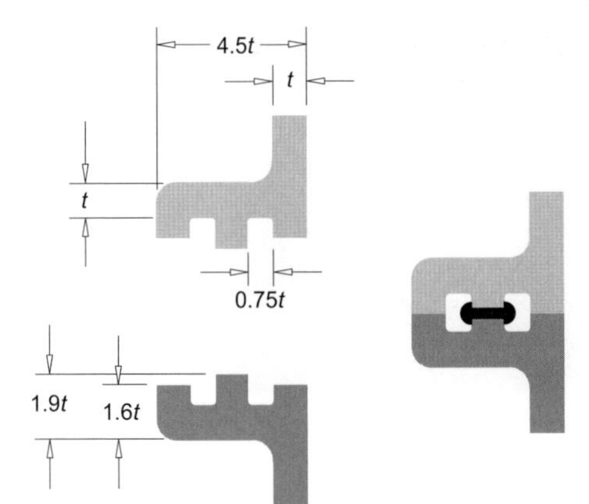

Figure 5.56
Butt joint design with flash traps

5.7.2 Joint Design

A number of joint designs can be used for the hot plate welding method. Figure 5.52 through 5.56 show some of the most common hot plate joint designs. They are all based on the butt joint.

Figure 5.52 shows a simple L-shaped butt joint that can be successfully used in many applications. The length of the base should be twice the wall thickness (t), and the base height is $1.3t$ of the same wall. Depending on the tolerances of the design and the process, approximately $0.15t$ will melt on each part, creating a joint with $2t$ final wall stock in the joint area.

Figure 5.55 shows a variation of the joint described above. Again, about $0.15t$ will melt on each part to form the joint. The vertical lip of the upper part masks the flash trap and weld area, creating a flush show surface. Figure 5.56 shows a butt joint that can be used for enclosures where visible flash is not an issue.

Where both wall structure and appearance are important, the design shown in Fig. 5.56 can be used. Both sides of this joint can serve as show surfaces.

A variation of the butt joint design, shown in Fig. 5.52, was used for the windshield washer fluid bottle of the M-Class SUV (sport utility vehicle) manufactured by Daimler Corporation under the Mercedes-Benz nameplate [179].

The windshield washer fluid bottle, shown in Fig. 5.57, is made of two polyethylene (PE) injection-molded parts that are welded using a hot plate welder. Each part has a wall stock of $t = 2 \div 2.5$ mm and a flange width of $3t$.

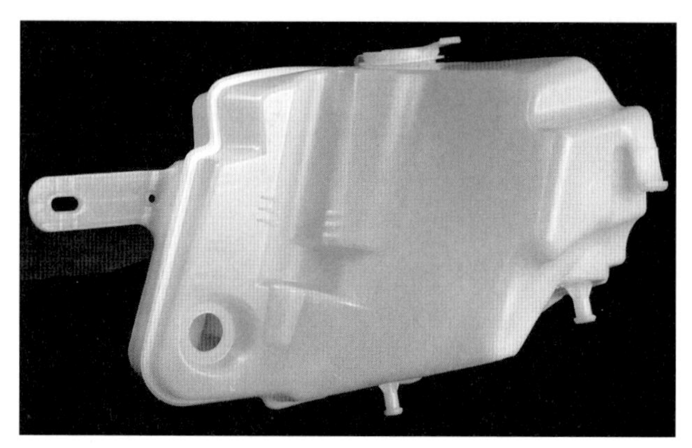

Figure 5.57 Front view of the windshield washer fluid bottle assembly used for the Mercedes-Benz M-Class Sport-Utlility Vehicle or SUV *(Courtesy of ETS, Inc.)*

Both parts incorporate a rather deep draw and show a boxy-type design. These design features combined with the large shrinkage of the semicrystalline polyethylene resin used for the application create a warpage environment. A large percentage of components do warp. During the hot plate welding operation, higher holding (cooling phase) pressures are exercised to overcome gaps in the joint area because part warpage creates a built-in stress.

Being located under the hood, the bottle is exposed to temperature gradients of almost 120°C (250°F); it varies between -40°C and +93°C (-40°F to +200°F). In time the stresses, created by temperature variation combined with the vibration loads due to the road surface/tire-suspension interrelation and the dynamic loads generated by the fluid present in the bottle during sudden acceleration or braking, initiate microcracks in the joint.

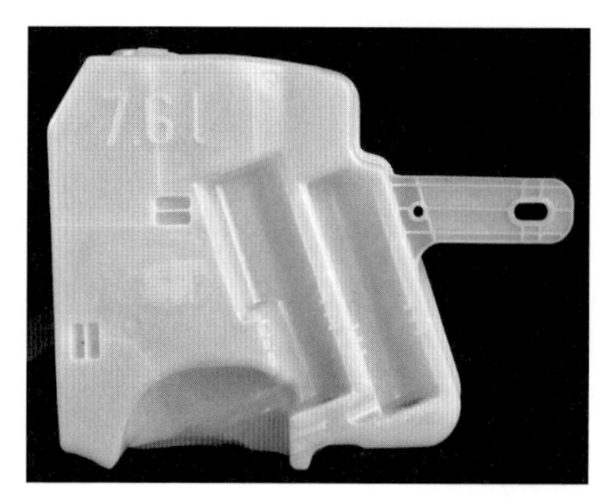

Figure 5.58
Side view of the 7.6-liter windshield washer fluid bottle assembly made of two injection-molded polyethylene components *(Courtesy of ETS, Inc.)*

Figure 5.59
Butt joint design detail *(Courtesy of ETS, Inc.)*

The overall bottle design is marginal. The weld seam is placed in a vertical plane. Soon after microcracks appear in the welded joint the washer fluid is lost. To prevent such failure the weld seam should be placed into a horizontal plane, thus not losing all of the washer fluid when the weld joint fails [179].

Table 5.2 Joint Strength Capability of Various Polymers

Polymer	Retained Strength vs. Polymer Matrix (%)
ABS	80
CA	80
CAB	80
CAP	80
EVA	80
HDPE	100
LDPE	100
PA	90
PC	80
PMMA	80
POM	90
PP	100
PS	80

Table 5.2 Joint Strength Capability of Various Polymers *(continued)*

Polymer	Retained Strength vs. Polymer Matrix (%)
PS-HI	90
PSU	60
PVC-rigid	90
PVC-flexible	90
UHMWPE	90

Table 5.2 shows the highest joint strength achievable for various unreinforced and unfilled polymers. If the polymer has fillers or reinforcements the hot plate joint strength is still based on the polymer matrix. Fillers and reinforcements do not improve the joint strength.

5.8 Vibration Welding

Vibration welding is an improved friction welding technique used for joining a wide variety of thermoplastic parts. This method offers greater flexibility in the design of plastic components; it facilitates the assembly of larger parts measuring up to 550 mm (21.5 in.) long by 400 mm (16 in.) wide and irregularly shaped parts that cannot be assembled economically by any other means.

The process principle is simple. Two parts are clamped together. One part is held stationary while the second part is vibrated against it. Friction along the common interface generates heat that welds the two parts together.

Frictional heat is generated when one part is vibrated through a reciprocating displacement of 0.5 to 4 mm (0.02 to 0.16 in.) at a frequency of 120 or 240 cycles per second (cps). When a molten state is reached at the joint interface, vibration is stopped and the parts are automatically aligned. Clamping pressure is maintained briefly while the molten polymers solidify to form a bond approaching the strength of the base material or matrix.

Almost all thermoplastics can be welded using the vibration welding technique whether injection molding, extrusion, or thermoforming methods are used to manufacture the parts.

Vibration welding does not depend on the transmission properties of the polymer to achieve a weld; if the parts can be vibrated relative to each other in the plane of the joint, this process can be used successfully.

Like spin welding, vibration welding has five phases.

In the first phase, vibration is used to generate heat through friction while the parts are still solid.

In the second phase, internal friction occurs between the plastic layers.

In the third phase, the polymer reaches its glass transition temperature.

In the fourth phase, the process reaches a steady state; the temperature of the melt still being generated equals the heat loss due to flash and losses through the wall stock.

In the fifth and last phase, the vibratory motion stops and the assembly is allowed to cool to ambient temperature under pressure.

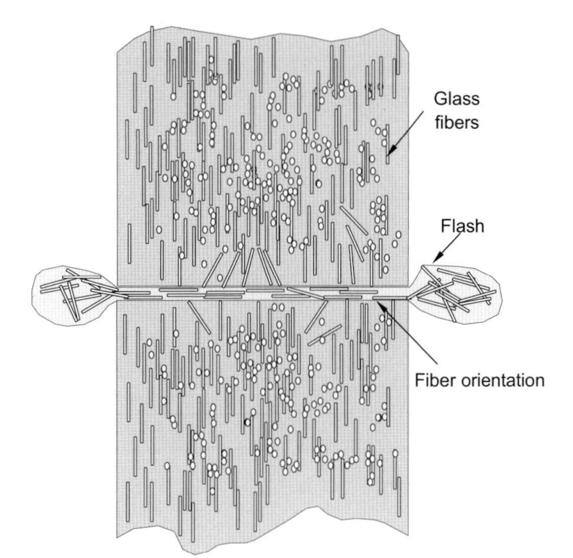

Figure 5.60
Vibration welding of glass-reinforced polymer (detail)

Mechanical properties of glass-reinforced thermoplastics are affected more by amplitude than are those of unfilled polymers. Glass fibers protrude through the surface, mostly at the center of the weld. The fibers orient themselves in the direction of the flash normal to the wall stock. The weld area will exhibit the mechanical properties of the matrix material (see Fig. 5.60).

Using higher pressure reduces the height of the molten film, resulting in a low weld thickness. Fiber orientation increases as the melt flushes outside the wall stock.

When lower pressures are used, the fibers overlap the weld area and again the weld area will exhibit the mechanical properties of the matrix material.

Generally, the strength of a joint created by vibration welding of fiber-reinforced thermoplastic parts will not exceed that of the matrix polymer. Weld strength is improved if the melt flow direction, for injection-molded parts, is parallel to the direction of the load.

Vibration welding is a fast process with cycle times between 1 and 10 seconds. Hermetic joints of approximately 1.4 MPa (100 psi) are easily achievable.

5.8.1 Process

During the vibration welding process, the distance one part moves in relation to the other is an important process variable.

In a process called cross-thickness vibration welding, two parts can be welded using a vibration motion normal to the wall stock. The moving part is exposed to room temperature in a sinusoidal fashion, and the polymer is exposed to intermittent cooling. The fraction of the joint exposed to ambient temperature is [140]

$$s = A \sin 2\pi\nu t \tag{5.22}$$

Notations:

p_0 = pressure

N = normal force

$F_{Friction}$ = friction force

ν = frequency

μ = coefficient of friction

s = distance

t = time

A = amplitude

T_{Melt} = melting temperature

$T_{23°C}$ = ambient temperature

For a part with a wall stock of 3 mm (0.12 in.) and amplitudes of 0.5 to 0.75 mm (0.02 to 0.03 in.), the portion exposed to intermittent cooling varies from 0.3 to 0.5 mm. Therefore, only 30 to 50% of the weld surface was exposed to room temperature.

The cross-thickness vibration welding method also affects the applied pressure, which is maintained during the process by the normal force N.

The variable p_0 is the pressure applied against the joint width. During welding, the contact area varies. Therefore, the applied pressure will vary as well:

$$p_0 < p < 2p_0 \tag{5.23}$$

The pressure variation could be significant for large ratios of applied amplitude versus part wall stock thickness.

In the first phase, also called the *solid friction phase,* the parts being joined are in contact with each other over their full surface area. Heating is provided by the mechanical friction generated by rubbing faces of the two parts under a normal force N.

The energy generated in the joining zone can be calculated as a product of the normal force and displacement.

Frictional force is

$$F_{Friction} = \mu N \tag{5.24}$$

The temperature variance is

$$\Delta T = T_{Melt} - T_{23^\circ C} \tag{5.25}$$

The time at which the transition to unsteady melt-film formation takes place can be obtained theoretically by employing equations determined by Carslaw and Jaeger [34].

The temperature dependence of the material data (the friction coefficient) as well as the elastic deformation, which can absorb part of the vibration, can bring about inaccuracies in the calculations. These effects will increase the solid friction period and can prevent a melt film from forming in the joint altogether.

In the second phase, also known as *unsteady melt film formation,* melting occurs as a result of the shear heat, increasing the melt film thickness. The applied pressure squeezes melt out into the weld flash.

The melting, and therefore the increase in film thickness, can be calculated through approximation using equations developed by Potente and Uebbing [118].

Allowance should be made for part melt produced during the process to be squeezed out under pressure. This significantly reduces the melt thickness.

According to Potente, Michel, and Ruthmann [117], the amount of flash can also be approximated.

Unsteady melt formation occurs when the amount of melt transported into the weld flash during a given amount of time is equal to the amount of new melt created by dissipation.

The third welding phase or *steady phase* is characterized by the equilibrium that prevails between the increase in melt thickness due to shear heating and the reduction in thickness due to melt being squeezed out into the flash. Expansion by heat and reduction by squeezing work against each other in keeping the melt thickness constant.

The fourth phase or *holding phase* is the time interval during which the melt is allowed to solidify under pressure into a firmly welded joint. The parts being joined are held together under a set pressure for the entire holding time. The pressure continues until the parts have cooled below the flow temperature (glass transition temperature for amorphous thermoplastic polymers; crystalline melting point for semicrystalline and crystalline thermoplastic polymers). A pronounced squeeze flow takes place between the parts being joined. The weld seam then reaches its definitive form and strength. This process can be compared to the joining process in hot plate butt-welding.

Tests have shown that the highest possible melt thickness must be obtained in the third, stationary phase if an optimum weld quality is to be achieved. The applied joint pressure must be relatively low. However, lowering the pressure increases the time needed to achieve the steady state. Therefore, it is advisable to use a variable pressure in practice. A high pressure is best at the beginning of the process; once the melt is achieved, the pressure should be reduced in order to attain higher melt thickness.

The unsteady process can be calculated in short time intervals. A pressure profile based on short periods can be determined. After the pressure change, the current values of the melt thickness and the temperature should once again be passed through the routine for the calculation of the unsteady state [142].

First, lower pressure means more material will melt and increase the thickness of the melt that will be eliminated through flash. The rate of change in the displacement of the movable platen is obtained from the squeeze flow of melt. This flow will increase again when the melt thickness becomes larger and will remain constant once the steady state has been reached.

5.8.2 Equipment

The vibration welder employs a simple vibrator mechanism consisting of only one moving element with no bearing surfaces. Reciprocating motion is achieved by magnetic force, alternating at 120 or 220 cps, acting directly on a mechanical suspension.

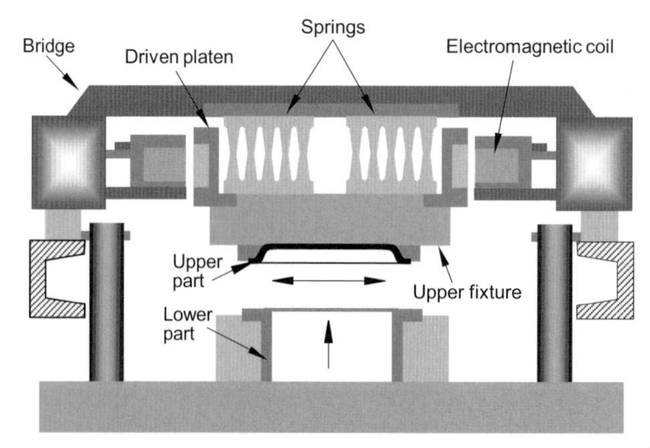

Figure 5.61
Vibration welding equipment

The major components of the vibrator are shown in Fig. 5.61. They are a set of flat springs, two electromagnets, a vibrating element (drive platen), and a clamping mechanism. The springs have three functions: to act as resonating members, to support the vibrating element against vertical welding pressures, and to return the vibrating element to the aligned position when the magnets are de-energized.

The vibrating element engages and holds the plastic part to be vibrated; the stationary element holds the other part of the assembly. Pressure is applied to the parts by a pneumatically operated clamping mechanism that engages the stationary element or tray. This locks onto the vibrator housing and pulls the part against the vibrating element during the welding cycle.

The vibrator is mounted on a frame that incorporates a power supply and a tray lift mechanism, forming a complete plastic assembly system. The modular construction of the vibration welder also allows the individual components to be used in a variety of automated systems.

Fixtures for vibration welding are usually simple and inexpensive. They generally consist of aluminum plates with cutouts that conform to the geometry of the part or countered cavities of cast urethane. Depending on the part size, two or more parts can be welded at the same time using a multicavity fixture.

5.8.3 Joint Design

Vibration welding calls for some specific design requirements. Two of the most important are that the parts be free to vibrate relative to one another in the plane of the joint and that the joint can be supported during welding.

The basic joint design for vibration welding is the simple butt joint (Fig. 5.62). A flange is generally desirable unless the wall is sufficiently rigid or supported to prevent flexure. A flange also makes it easier to grip the parts and apply uniform pressure close to the weld.

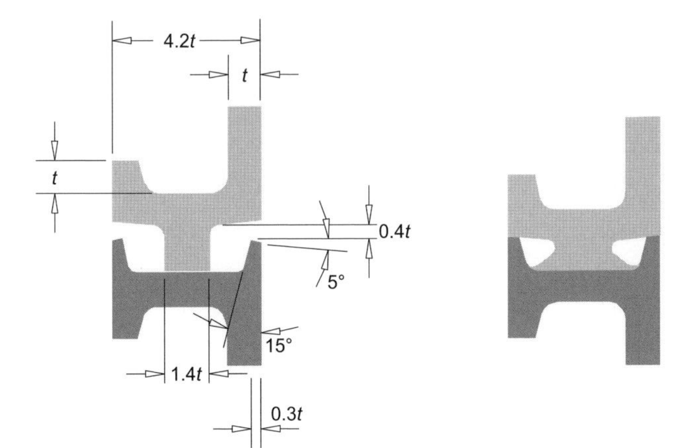

Figure 5.62
Vibration welding joint design detail: (a) before, and (b) after assembly

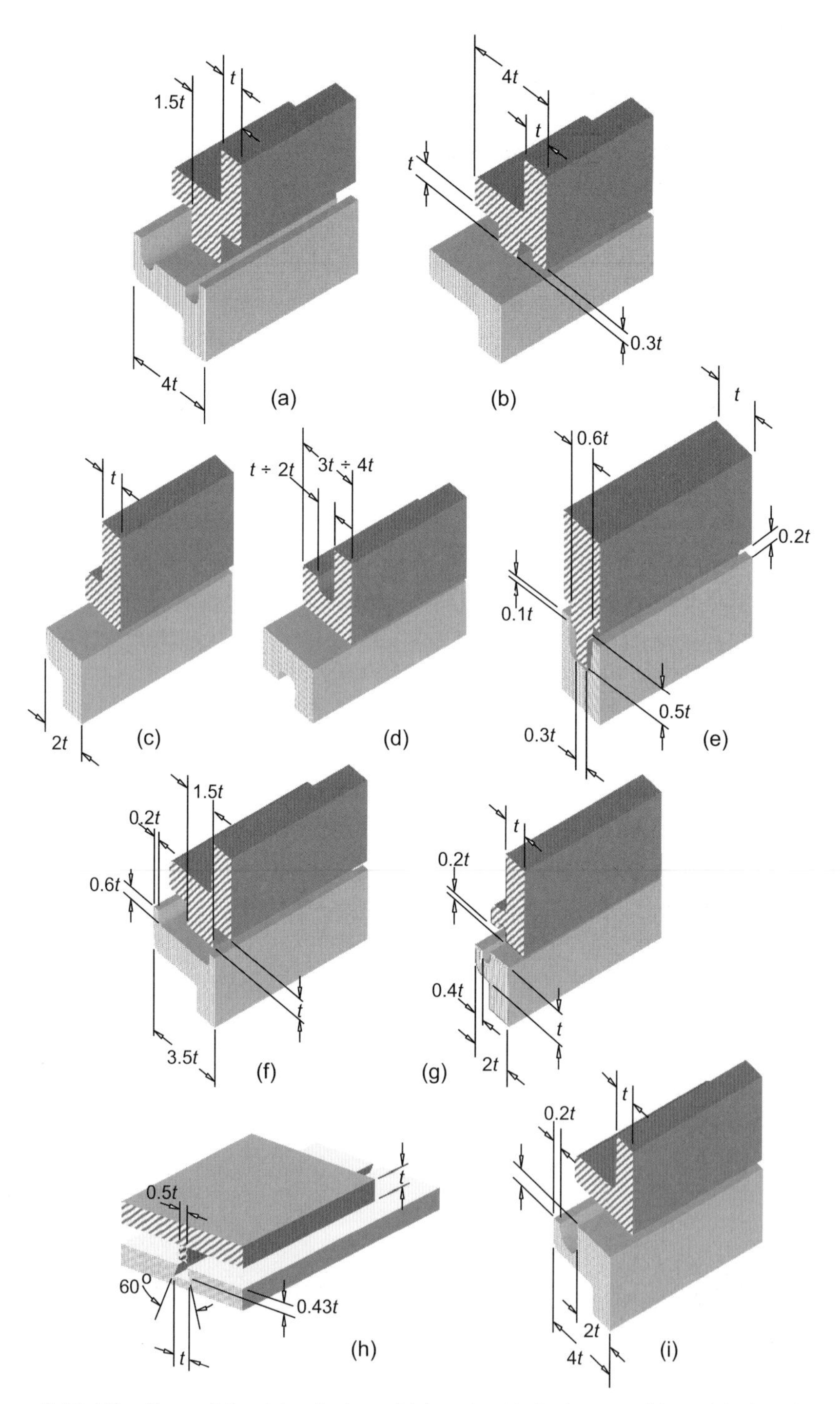

Figure 5.63 Vibration welding joint designs: (a) bench with flash traps; (b) straight bench; (c) tongue-and-groove; (d) double L; (e) ridged double L; (f) grooved bench; (g) double L with flash traps; (h) dovetail; (I) double L with single flash trap.

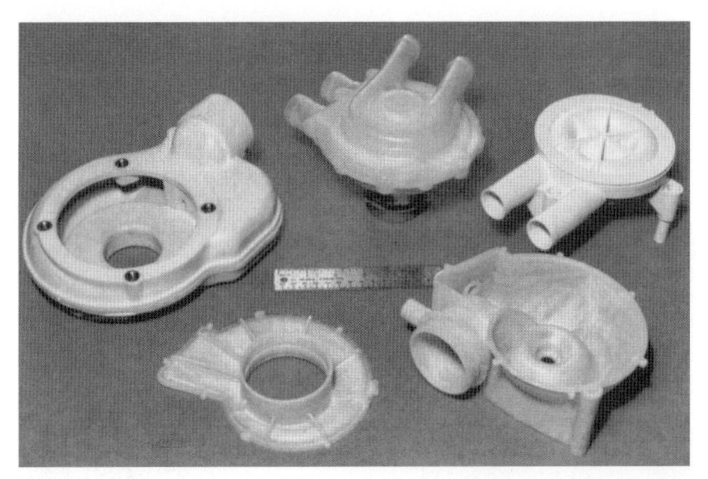

Figure 5.64 Pump for washer and dishwasher appliance assembled through a vibration welding process using a butt joint design. The welding provides a hermetic seal, and the assembly can withhold 0.52 MPa (75 psi) of internal pressure *(Courtesy of Branson Ultrasonics, a business of Emerson Electric Company)*

For thin or long supported walls, special provisions may be necessary. Walls as thin as 0.8 mm (0.03 in.) have been welded successfully.

As in any assembly method that relies on material melting, some of the melt may be displaced beyond the immediate joint area. If such displacement material or flash is unacceptable for functional or cosmetic reasons, a joint that provides flash traps must be used. Flash traps should be volumetrically sized to the amount of polymer displaced during welding.

For example, Fig. 5.65(a) shows the Volkswagen Sharan minivan air intake manifold, which was achieved by assembling two injection-molded components using a vibration welder. Each component has a wall stock of 2.5 mm. The vibration welding joint design is similar to the one shown in Fig. 5.63(e), where the flange width for the fixture is 5 mm, or twice the overall wall thickness. The main reason for such a large width for the fixture flange is to properly locate the upper and lower parts of the manifold in their respective upper and lower fixtures.

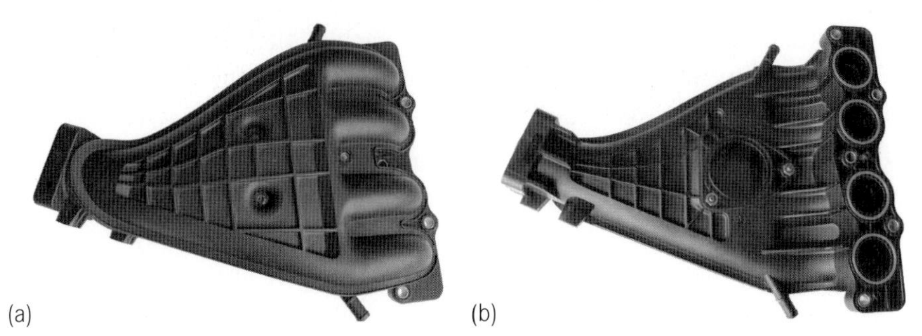

Figure 5.65 Two-piece, injection-molded air intake manifold assembled by vibration welding for Volkswagen Sharan minivan (a) from above, (b) from below *(Courtesy of ETS Inc.)*

The components are fairly large (approximately 280 by 400 mm or 11 by 16 in.) and need proper alignment in the fixtures. Any misalignment between the parts and fixtures may cause parts to break during vibration welding. The load required to move the two springs attached to the bridge of the vibration welder (see Fig. 5.61) is very high: it can easily exceed 135,000 N (approximately 30,000 lb.) per each spring.

Figure 5.66 Warpage not exceeding 0.2 mm is acceptable *(Courtesy of ETS Inc.)*

A certain amount of warpage is acceptable in the vibration welding process. In this case the maximum of allowable warpage between the two components should not exceed 0.2 mm (or about 0.008 in.). To mold complicated parts, such as the upper and lower air intake manifold, made of polyamide 6,6, 35% glass-reinforced resin, a number of tool adjustments and revisions are required during development. Appendix C, entitled "Molding Process Data Record," shows injection-molding process parameters that should be adjusted during the development process for any given component. Furthermore, using a form entitled "Tool Repair and Inspection Record" (Appendix D), during development and later in production, will assist in keeping up with all of the changes in the tool that could affect the part quality.

5.8.4 Common Issues with Vibration Welding

The following is a list of common problems and issues often encountered in vibration welding, with probable causes and possible solutions.

Overweld is usually caused by too much energy reaching the joint area. This problem can be remedied by lowering the weld pressure, decreasing the weld time, or using smaller amplitude.

Weld quality varies from part to part. This problem can have many causes. If the source of the inconsistency rests with part tolerances, then part dimensions should be checked to verify that they are within specifications. Prototype and preproduction parts should be compared to detect any changes in tooling, processes, or conditions that could cause the variance. The problem can also be remedied by compensating for the material in the tool. A heat treatment or annealing of the parts might also provide a solution.

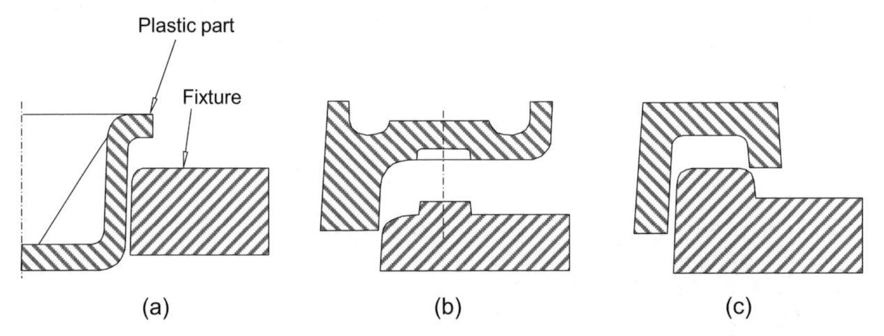

Figure 5.67 Vibration welding fixture design: (a) flat fixture; (b) recessed fixture; (c) edged fixture

Weld quality variance can be caused by the use of recycled material. Lowering the amount of regrind or verifying the quality of the regrind can correct the problem.

An uneven distribution of filler throughout the polymer is another common source of inconsistent weld quality. To fix this problem, the molding process parameters should be verified.

If the inconsistent weld quality is caused by the joint design, the existing design should be modified, or a complete redesign may be needed.

Parts become displaced after welding. This problem could be a simple matter of aligning the parts properly before welding, with the use of shims if necessary. An alignment pin–which will be destroyed during welding–can be designed into the part.

The misalignment could be caused by poor fixture design. In this case, the existing fixture design can be compensated for by realigning itself or by realigning the parts. Otherwise, the fixture will have to be redesigned.

If the part displacement is caused by warpage of the parts, ribs can be added to the parts as reinforcements.

Flash is a common welding problem that is often caused by an excessive weld. Again, lowering weld pressure, decreasing weld time, and decreasing amplitude can correct the problem.

Flash can also be caused by excessive heat due to weld areas that are too large. This can be corrected by designing relief grooves in or around the joint area.

Flash problems can be avoided by designing joints with flash traps.

Inconsistent weld in the weld area could be a case of improper support provided by the fixtures. The fixture should be redesigned to provide proper wall support. Walls of the part deflecting in the transverse direction can also cause inconsistency in the weld. Adding ribs to the part or using a fixture that provides wall support should correct the situation.

If inconsistency is caused by misalignment of the parts, the lower fixture should be realigned, and weld pressure should be checked to ensure it is not too high.

If warped parts are the cause of the inconsistency, a higher welding pressure might be the solution. Adjustments to the molding parameters of the injection-molding process can be made to prevent or minimize the warpage.

5.9 Electromagnetic Welding

Electromagnetic welding is an assembly method that uses an electromagnetic bonding agent to join two parts. The bonding agent, which has metallic powder with magnetic properties dispersed in it, is placed between the two parts. When an electric circuit is closed, a magnetic field is generated by a copper coil near the joint. The magnetic field excites the metallic powder in the bonding agent, which becomes heated through Brownian motion (random motion in a 3-D space) of the particles. Pressure is then applied to bond the two parts.

The heat is only generated in the region close to the bonding agent.

This method can be used to bond thermoplastics as well as other materials, such as fabric, paper, glass, and leather.

The bonding agent is created by extruding or laminating a matrix resin, which is chemically compatible with the materials of the parts to be assembled. The metallic powder is dispersed during the laminating process. The sheet is then die cut like a paper gasket to fit the joint. Typical metal powder content varies from 20 to 60% of the material weight, the remainder being the matrix resin. The higher the metal powder content, the shorter the cycle time; the time needed to raise the temperature varies linearly with the metal content.

Electromagnetic welding relies on the principle of inductive heating to create fusion temperature within the joint region. A thermoplastic resin is compounded with ferromagnetic particles. When this resin is exposed to an oscillating electromagnetic field, the magnetic particles become active and start oscillating in Brownian motion, generating heat. This process needs between a fraction of a second and a few seconds to reach temperatures high enough to melt the resin containing the magnetic particles. Once the resin is melted, its temperature transfers through thermal conduction to the parts being joined, making the weld possible.

One of the major advantages of this method is that no stresses are induced in the two parts. This makes electromagnetic welding suitable for assemblies that require no or low warpage due to internal stresses generated at the joint interface.

5.9.1 Equipment

Electromagnetic welding equipment consists of an induction generator, which converts the electric current of 50 to 60 Hz into 3 to 10 MHz output frequency with powers of 2 to 5 kW; copper coils that generate a high-intensity magnetic field; fixtures or tools that hold in place the parts to be assembled; and a press (a pneumatic cylinder) that applies a preset pressure once the tool is closed.

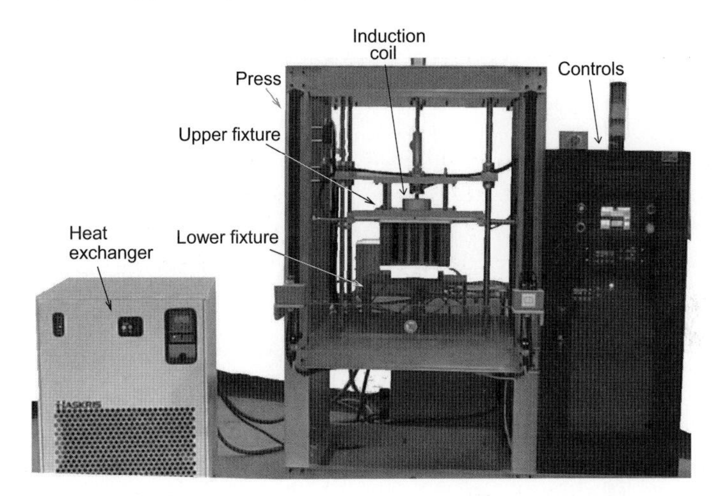

Figure 5.68 Electromagnetic welder

5.9.2 Process

The high-frequency power supply sends its output current around a closed loop of copper coil. The current in the coil creates a magnetic field that passes through the bonding resin, activating the metallic powder. The particles begin oscillating at very high frequencies, creating heat and melting the matrix of the resin where the metallic powder has been dispersed.

Induction heat is generated directly within the epoxy agent located at the joint. The heat-generating powder is incorporated in the bonding agent. The powder responds to the magnetic field generated by the induction coil, which is energized when the electric circuit closes.

5.9.3 Joint Design

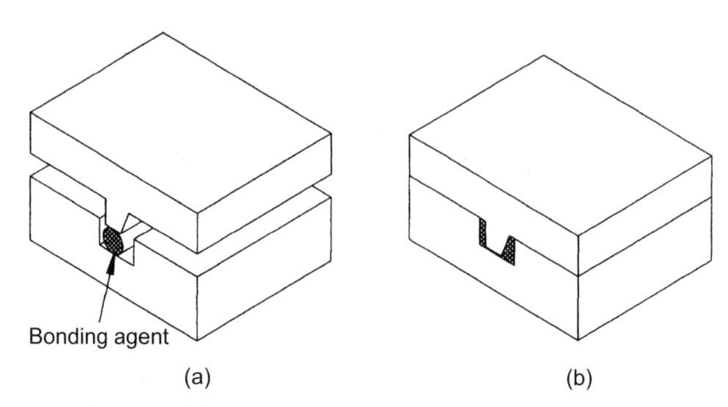

Figure 5.69 Tongue-and-groove joint design for electromagnetic welding: (a) before, and (b) after assembly

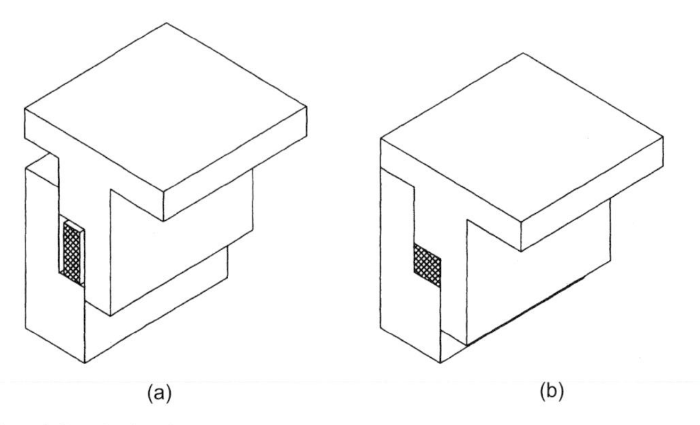

Figure 5.70 Step joint design for electromagnetic welding: (a) before, and (b) after assembly

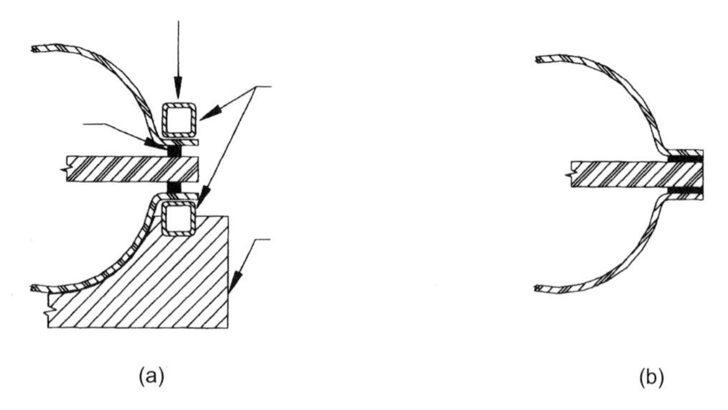

Figure 5.71 Double joint design for electromagnetic welding: (a) before, and (b) after assembly

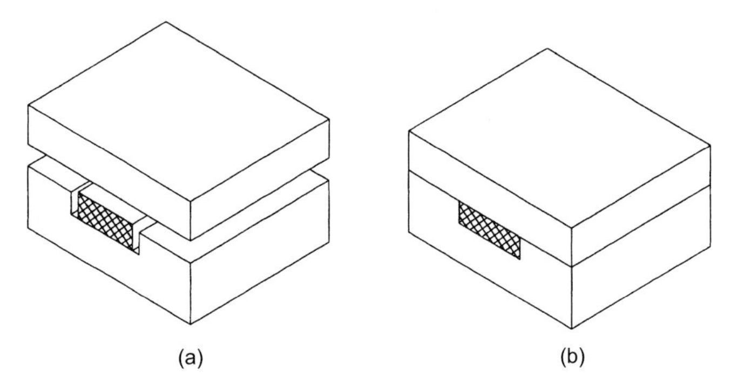

Figure 5.72 Groove joint design for electromagnetic welding: (a) before, and (b) after assembly

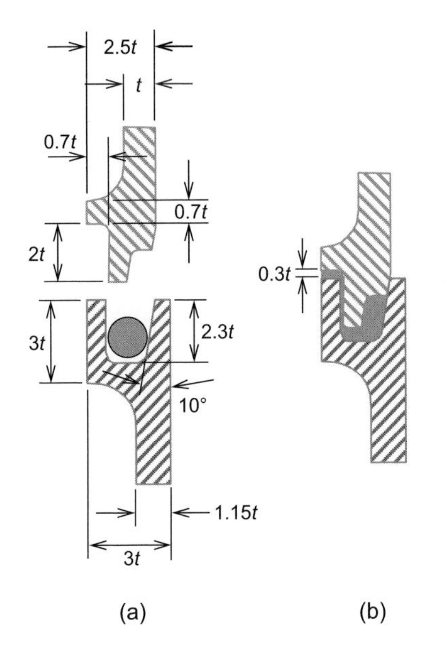

Figure 5.73
Variation of tongue-and-groove joint design for electromagnetic welding: (a) before, and (b) after assembly

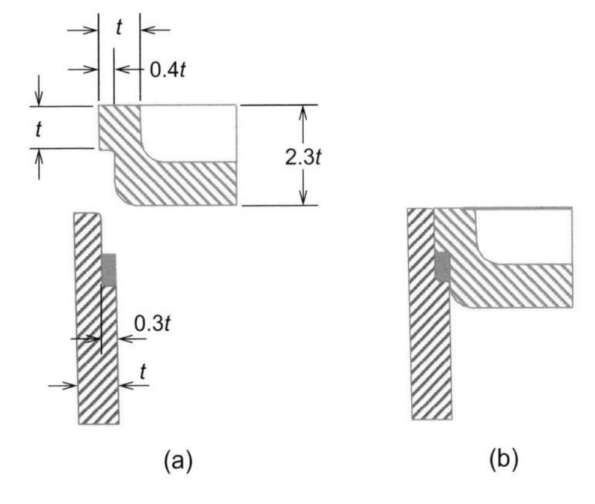

Figure 5.74
Variation for step joint design for electromagnetic welding process: (a) before, and (b) after assembly

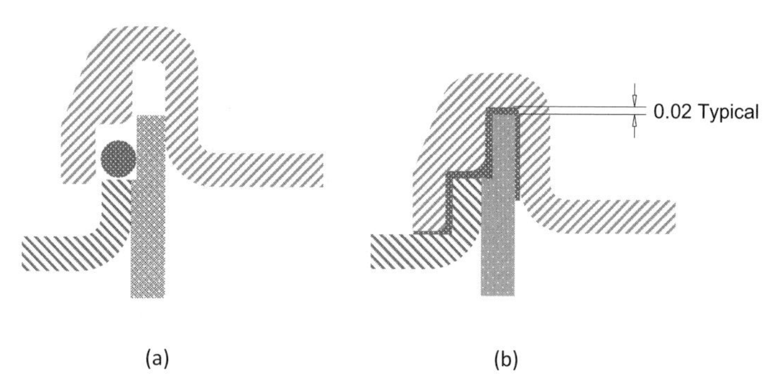

Figure 5.75 Three-piece welding: (a) before, and (b) after assembly

Figures 5.69 through 5.75 show some of the many joint designs that may be used in various electromagnetic welding applications. The tongue-and-groove joint design (Fig. 5.69) and the groove joint design (Fig. 5.72) are both good for assembling particularly large, flat parts. The step joint design is recommended for enclosure joints that fit like box lids.

The double joint design (Fig. 5.71) entails a more complex procedure, because it requires two independent electromagnetic coils, which must be synchronized to conduct both joints simultaneously during one short weld cycle. This joint also requires a special fixture designed to locate and align the parts during the process.

The variation of the tongue-and-groove design (Fig. 5.73) is recommended for assemblies that require guidance and support for the wall stock. This type of joint should be used for components that have a variance in overall dimensions due to warping.

The step joint design (Fig. 5.74) is suitable for assemblies in which the outside portion of the joint area is a show surface and no flash should be visible. A significant amount of flash will be present at the inside of the joint.

Figure 5.75 shows a design used for assembling three individual components into one.

5.10 Radio Frequency (RF) Welding

Radio frequency or *RF welding*, also known as *RF sealing*, is an assembly method that uses a high-intensity radio signal to effect bonding between two parts. The radio frequency is applied to the vicinity of the plastic joint, increasing molecular motion in the materials and causing their temperature to rise. The increase in temperature occurs first at the molecular level and then very quickly generates heat in the parts themselves.

The rise in temperature melts the polymer, the molecular chains of the two parts penetrate each other at the interface, and the weld is achieved.

Radio frequency welding or sealing is used in a variety of industries. It is particularly suitable for the medical device industry because it uses no solvents or adhesives, which would present the possibility of contamination.

5.10.1 Equipment

Radio frequency welding equipment is relatively simple. The basic requirements are electric power and compressed air.

Radio frequency welding systems are divided into two categories: high-frequency heat sealers, which provide an alternate field at 27 million cps or 27 MHz; and electronic heat sealers, which employ a radio-transmitting tube.

The RF transmitter is also called a generator because it generates radio frequency wave power rather than radio signals. The generator consists of a power supply, an oscillator, and controls. The power supply converts the electric power into high-voltage direct current (DC).

The oscillator converts the high-voltage DC into alternating current (AC) at a frequency of 27 MHz.

The equipment also consists of a press, which has interchangeable electrodes and is operated by a pneumatic cylinder. The press has two platens: a moveable one and a fixed one also called a "bed."

5.10.2 Process

In the first sequence of the process, the press lowers the moveable platen and closes the electric circuit. The RF energy flows and heats the seal or joint area. Once the material is melted, the RF energy stops. The joint is allowed to cool under the pressure exerted by the press. When the holding time is complete, the press opens and releases the assembled product. The cycle then repeats itself for the next assembly.

During the RF stage of the process, the parts to be welded are placed in a set of metal dies that are activated by an compressed air cylinder. A preset amount of pressure is applied at the joint area by the air cylinder. The radio frequency is then applied and the materials heat and melt.

There are two types of RF dies. The first type provides only nesting and location during the process. The second type is used to cut one or both of the parts to a predetermined shape. The latter type is used wherever possible because it reduces postwelding secondary operations as well as extra labor and equipment.

Polymers that have good characteristics for the RF process include ABS, cellulose butyrate, cellulose acetate, cellulose acetate butyrate, polyamide, phenol formaldehyde, pliofilm (rubber hydrochloride), polyurethane, polyvinic acetate, PVC opaque, PVC rigid, PVC semirigid or flexible, rubber, and Saran (polyvinylidine chloride).

Materials not compatible with this method are cellophane, acetal homopolymer, ethyl cellulose, polycarbonate, polyester, polyethylene, polystyrene, silicone, and polytetrafluoroethylene (PTFE), also known by the trade name of Teflon®.

A few simple calculations can determine the power required from the equipment to perform a proper weld. First, it is necessary to calculate the total joint area (length times width). A typical power requirement is 3 kilowatts (kW) per square inch (4.65 W/mm²). Or one can use the graphic method shown in Fig. 5.76.

The bond strength can be affected by various methods of sterilization used in the medical device industry, such as gamma radiation and the use of elevated temperatures. Medical devices are exposed to 5.0 to 5.5 Mrads (a unit of measure for radiation) of gamma radiation in the sterilization process. This radiation, as well as heat cycles during sterilization, can affect the strength of the joint. Leighton, Brantley, and Szabo [89] tested the effects of gamma radiation on various polymer assembly joints, with the results as shown in Table 5.3.

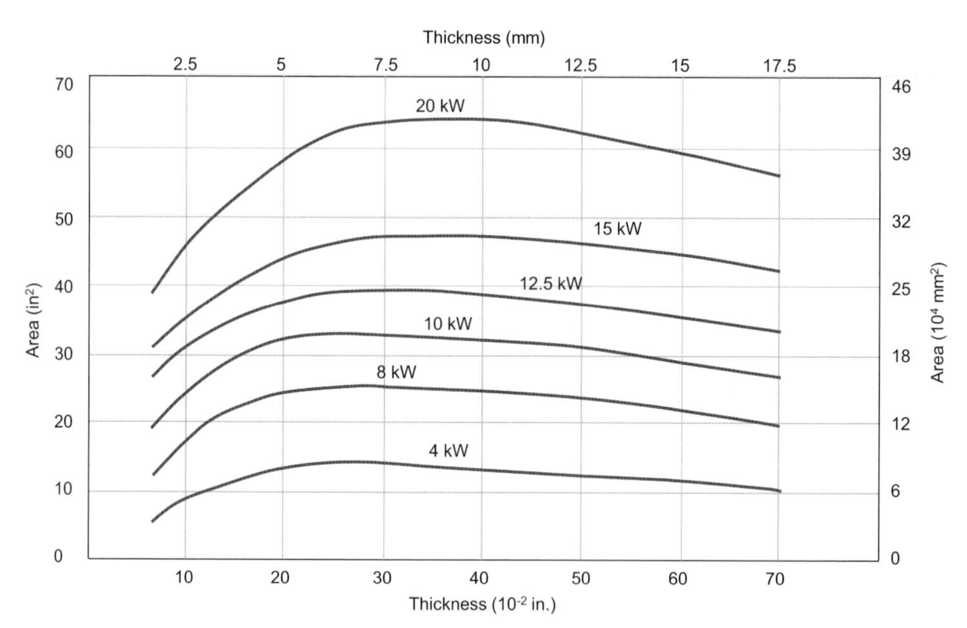

Figure 5.76 In radio frequency welding, the power required is a function of part thickness and joint area

Table 5.3 The Effects of Sterilization by Gamma Radiation on Joint Area RF Weld Strength

Polymer	Assembly joint strength variance
Part 1 Clear flexible PVC 65A Part 2 Clear rigid PVC 80D	+0% (same as for assemblies not exposed to radiation)
Part 1 Clear flexible PVC 65A Part 2 Clear flexible PVC 65A	+1% (better than assemblies not exposed to radiation)
Part 1 Clear flexible PVC 80A Part 2 Clear rigid PVC 80D	+3% (better than assemblies not exposed to radiation)
Part 1 Clear flexible PVC 80A Part 2 Clear flexible PVC 80A	+5% (better than assemblies not exposed to radiation)
Part 1 Clear rigid PVC 80D Part 2 Clear rigid PVC 80D	−5% (lower than assemblies not exposed to radiation)

Radio frequency sealing can be used to bond parts made from similar or dissimilar materials. Testing should determine which material combinations are suitable for this welding method.

5.11 Laser Welding

Laser is an acronym for light amplification by stimulated emission of radiation, which describes the physical process leading to the production of the laser beam.

Employing laser light energy in welding thermoplastic polymer components is a noncontact method that has grown in importance over the past few years. Initially, the lasers used for welding thermoplastics were based on *carbon dioxide* (CO_2), which have, as lasing medium, a gas mixture and *neodymium-doped yttrium aluminum garnet (Nd:YAG)*, using a crystal as a lasing medium for solid-state lasers.

They were followed by diode (also called semiconductor) with pulsating or *continuous-beam lasers*. New developments have created diode lasers with longer wavelengths in ranges between 1,470 and 1,550 nm. Lately, *fiber lasers* were developed with an optical fiber doped with rare earth elements as lasing medium. For example the *thulium fiber laser* systems operate in the spectral range of 1,800 to 2,100 nm.

5.11.1 Equipment

Laser welding systems vary greatly in design and complexity, depending on their mode of operation. The most commonly used welding systems are *diode lasers* and fiber lasers.

Typical laser welding systems have an energy source (usually referred to as the pump or *pump source*), a gain medium or laser medium, two or more mirrors that form an optical resonator, and the fixtures.

Laser pumping is a process by which the atoms are excited and their energy level is increased from low to high energy in order to achieve population inversion. When all the atoms jump into an excited state, the population inversion is reached.

Gain or *lasing medium* could be a liquid (for example, ethylene glycol), a gas (for example, carbon dioxide), a solid (for example, YAG or *yttrium aluminum garnet*), or even semiconductors (for example, a crystal with uniform dopant distribution). In the lasing medium, spontaneous and stimulated emission of photons takes place, leading to the phenomenon of optical gain, also known as "amplification."

Light from the medium, produced by spontaneous emission, is reflected by the mirrors back into the medium, where it may be amplified by *stimulated emission*. The light may reflect from the mirrors and thus pass through the lasing medium many times before exiting the cavity as the laser beam.

An interface between the machine and the operator is established by the *controls* that allow proper monitoring of the laser welding system. The controls, via logic circuits, provide feedback to the operator regarding the condition of the machine, welding parameters, and machine status.

Activated electrically or pneumatically, the press, also called the *actuator*, provides movement to the upper fixture or platen. It moves the part located in the upper fixture in a vertical direction until it makes contact with the mating part in the lower fixture. Then, it further applies a small pressure during the weld-and-hold cycle. The closed-loop system used by the actuator applies the predetermined load to both components. Certain systems use accurate displacement control to monitor proper movement and force.

The jig that allows proper positioning of the lower component being welded is called the *lower fixture*. Proper location and alignment of the part is required, especially for components that have tight tolerances.

One of the most critical and complex features of the welding systems is represented by the *upper fixture*. It is within the upper fixture that the laser beam is delivered to the assembly being welded. This component varies between laser welding systems, depending on the actual design concept and heating configuration employed.

All manufacturers of laser welding systems require that the operator be protected from the radiation by some type of *enclosure* or *eyewear*. The machine enclosure must be certified by the U.S. Federal Drug Administration (FDA) to ensure adequate operator protection.

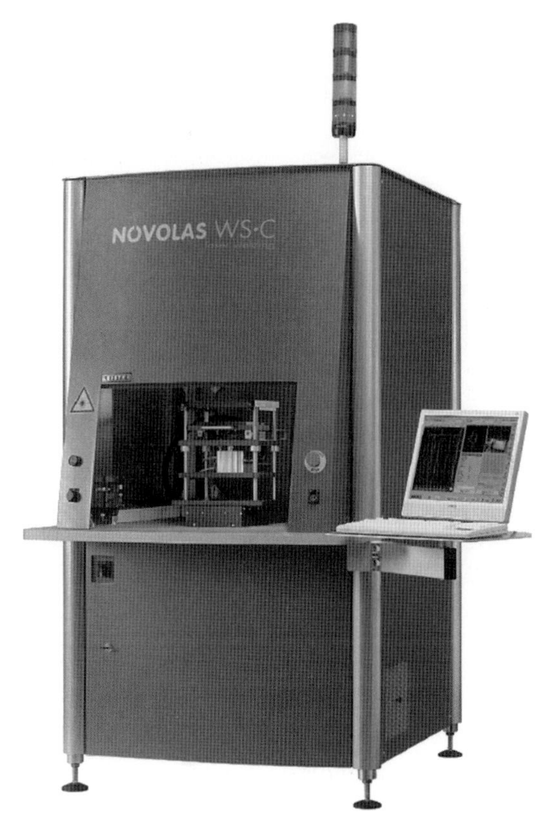

Figure 5.77
Laser welding system *(Courtesy of Leister Technologies AG)*

Figure 5.77 shows an infrared laser machine that works by illuminating the entire surface simultaneously. It is important to note that, for safety reasons, there are no portholes for viewing the operation. However, in this particular model, there is a monitoring system that allows the operator to watch the process via a special infrared radiation-sensitive video camera.

5.11.2 Process

There are laser welding techniques available, such as noncontact welding (similar to hot plate welding) and transmission welding. These techniques are detailed next.

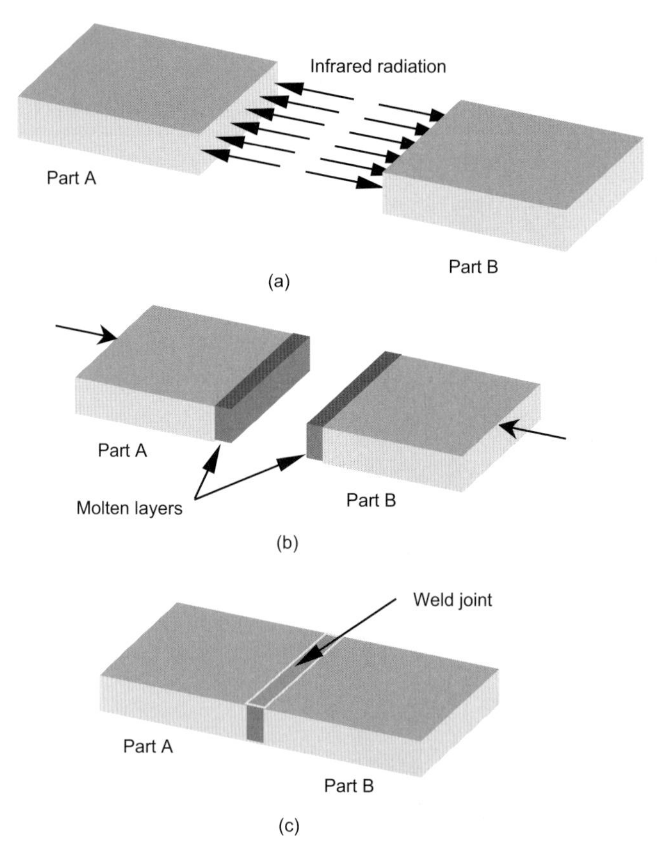

Figure 5.78 The three phases of the noncontact laser welding method: (a) the heating phase, (b) parts being brought together, and (c) the welding phase

5.11.3 Noncontact Welding

Noncontact welding (see Fig. 5.78) is similar to hot plate welding. It is also called *contour welding*. This technique makes laser welds similar to the way a pen makes lines on a sheet of paper. Parts can be moved relative to a laser with fixed optics, or the laser head can be mounted at the end of a robot arm or other positioning system.

Various system concepts are used, from rectangular coordinate systems in 2-D and 3-D and six- and seven-axis robotic cells, to simple two-axis systems for welding lids to container parts. Tooling costs are fairly low, so these systems lend themselves well to low-volume or job-shop production. Cycle times are longer than for other approaches, but this disadvantage can be overcome to a degree by either increasing laser power or mounting multiple lasers.

The thermoplastic surfaces of the two or more components to be joined are heated by direct laser exposure for a sufficient length of time to produce a molten layer, usually for two to ten seconds. Once the surface is fully melted, the laser head is withdrawn from between the parts, the parts are pressed together, and the melt is allowed to solidify and form a joint. The steps of noncontact welding with laser welding are depicted in Fig. 5.78.

5.11.4 Transmission Welding

The process for the transmission welding (TW) technique requires having at least two parts: the first component made from thermoplastic polymer transparent to the laser beam, and the second component made of laser-absorbing or opaque resin.

The TW method is also called *laser penetration* (LP) welding, which operates contact-free and enables direct process control. From the laser head (see Fig. 5.79) the light passes through the upper component made of laser-transparent polymer with almost no absorption and melts the surface of the underlying component made of laser-absorbing polymer (see Fig. 5.80). Applying gently compressive loads on the two components guarantees good thermal conduction so that the underside of the upper component also melts.

Figure 5.79
Laser head *(Courtesy of Leister Technologies AG)*

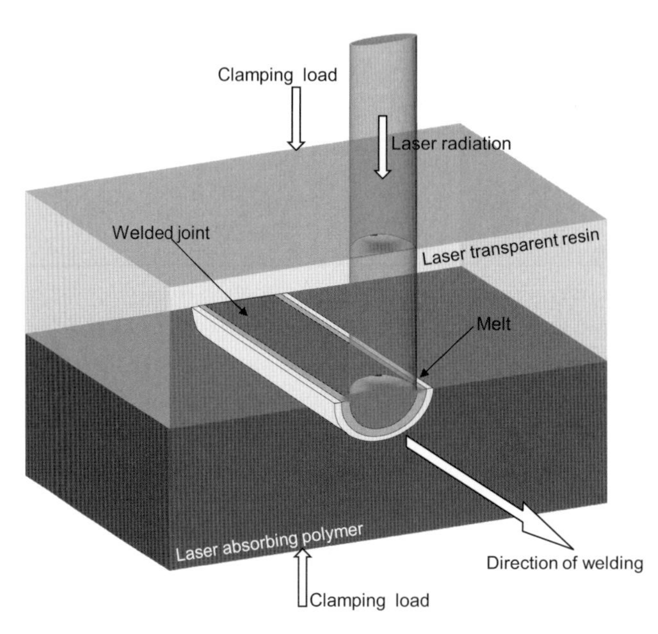

Figure 5.80 Transmission welding schematic

The use of absorbing pigments, such as carbon black, causes the laser light, after passing through the upper thermoplastic transparent part almost *unscattered*, to be absorbed by the lower thermoplastic component, which is colored with carbon black, a laser-absorbent material.

When the light radiation is forced to deviate from a straight trajectory by one or more paths due to localized nonuniformities in the polymer through which it passes, it is called *scattering* (see Fig. 5.81). It represents the deviation of reflected radiation from the angle predicted by the law of reflection. *Diffuse reflections* are reflections that undergo scattering, while mirror-like reflections, also called specular reflections, are known as *unscattered* reflections.

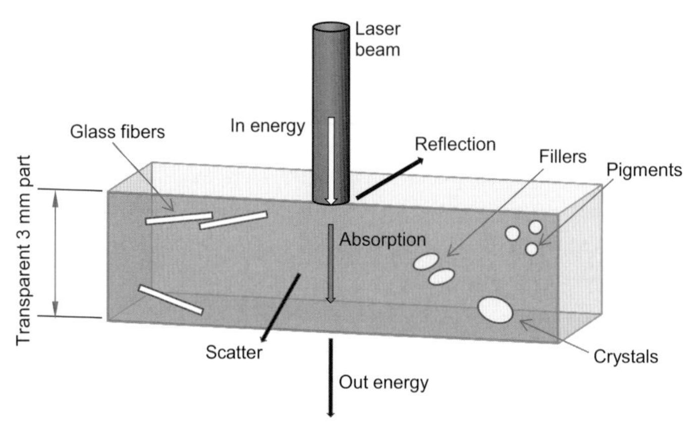

Figure 5.81 Scattering

In polymers, the scattering behavior is encountered mostly in crystalline resins, and in both amorphous and crystalline resins reinforced with glass fibers, certain color pigments, additives used to create flame-retardant polymers and, probably most important, *titanium dioxide* TiO_2 pigments used to create a deep white color in polymers.

Components made of polymers transparent to the laser radiation transmit the laser beam without any considerable heating within the resin. In the second component, made of opaque resin, the laser radiation beam is absorbed almost entirely in a thin layer near its surface; the laser energy is transformed into heat and the plastic is melted. By heat conduction, the upper, transparent joining component becomes plasticized. Once the melt layer solidifies, a solid, invisible joint is formed. A certain percentage of the light incident of the laser on the transparent part is lost due to reflection from the polymer surface (see Fig. 5.82). The laser light that penetrates the top surface of the laser-transparent part is further reduced as a result of absorption as the light passes through the material thickness. To achieve high-strength joints, the thickness of the transparent component is limited to a range of 3 mm to 6 mm to prevent laser-energy absorption.

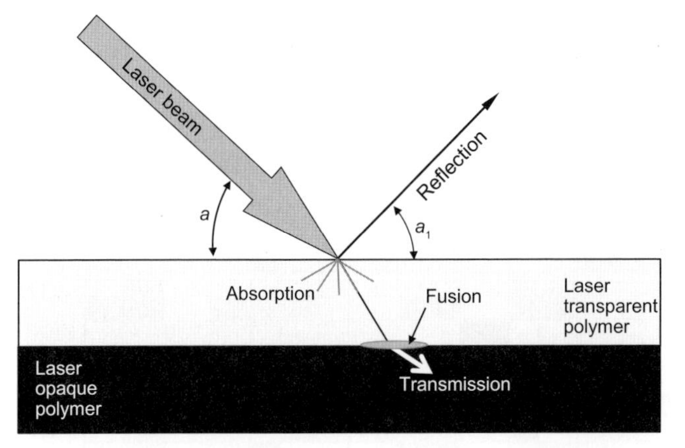

Figure 5.82
A certain percentage of the laser beam is being reflected, some is absorbed by the transparent polymer, while the remaining portion is transmitted to the laser-opaque polymer

A transparent material that scatters the light to a large extent (often because of reinforcing agents such as glass fibers or fillers such as talc and $CaCO_3$) results in increased absorption because of the longer path that the light must travel before leaving the material. The presence of spherulites or reinforcing agents may cause laser energy reflection originating in the bulk material. The combined effect of surface and bulk reflection and absorption of the incident laser light is lower transmission of the laser beam through the transparent part. Less energy is available to create a proper weld.

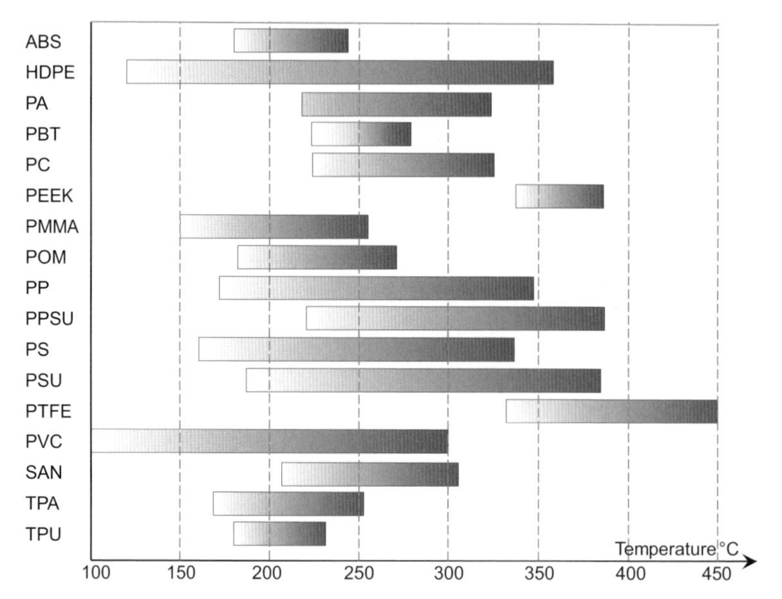

Figure 5.83 Temperature gradients in degrees Celsius for polymers softening and fusion when laser welded

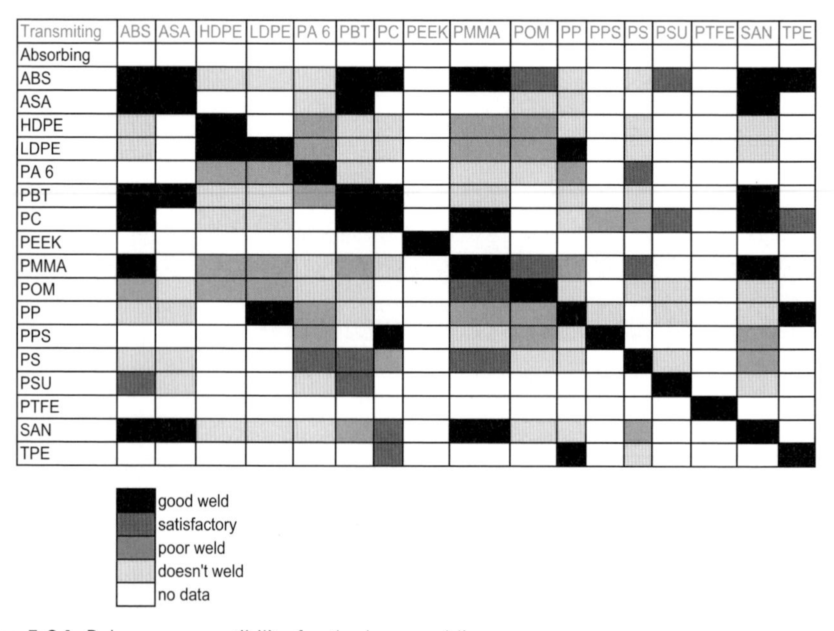

Figure 5.84 Polymer compatibility for the laser welding process

Amorphous thermoplastic polymers are generally more suitable for the laser welding process. Crystalline resins, such as polyethylene (PE) and polypropylene (PP), tend to promote internal scattering of the radiation, thus reducing the strength of the weld line. Therefore, scattering materials often limit the clear component of the final assembly to a thickness of 3 to 6 mm (0.125 to 0.25 in.).

Transmission welding enables the visual integration of the joining process with online monitoring. An example would be monitoring the power of the diode laser and monitoring or controlling the temperature gradient of the welding process. These types of control and monitoring are not possible with other welding processes in which heat is generated in the joint by friction.

Transmission welding is simple to use when one component is made of a thermoplastic polymer pigmented with carbon black to create opacity for the laser beam, while the other part is made of a thermoplastic transparent to the laser beam (Fig. 5.85(a)). The same figure also shows that it is also quite simple to TW black pigmented thermoplastic components. However, in order to effect the required melt, the component through which the laser beam passes should have a lower amount of pigment than the component that should attract the laser beam.

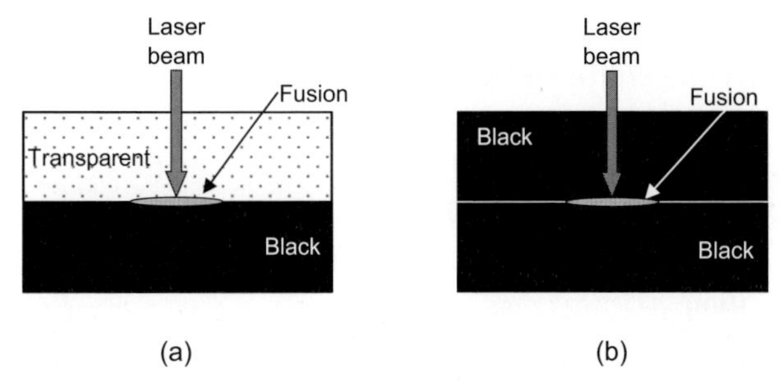

Figure 5.85 Transmission welding (TW): (a) transparent and black polymers, (b) nigrosin black and carbon black polymers

When the pigment used to color the thermoplastic resin is not black, achieving a proper joint strength becomes more challenging, as shown in Fig. 5.86. The most challenging joints to weld involve both components being transparent to the laser beam or both being pigmented with titanium dioxide (see Fig. 5.87). TiO_2 pigment permits the most scatter within the polymer; thus the laser beam loses its energy, and welding the multicomponent assembly becomes very difficult–sometimes even impossible.

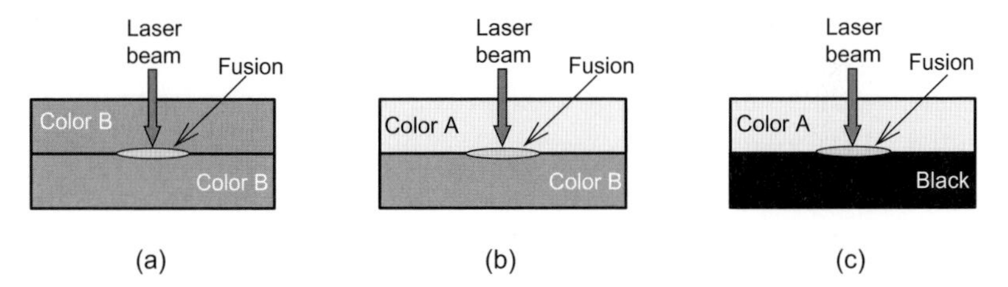

Figure 5.86 TW: (a) color B and color B polymers, (b) color A and color B polymers, (c) color and black polymers

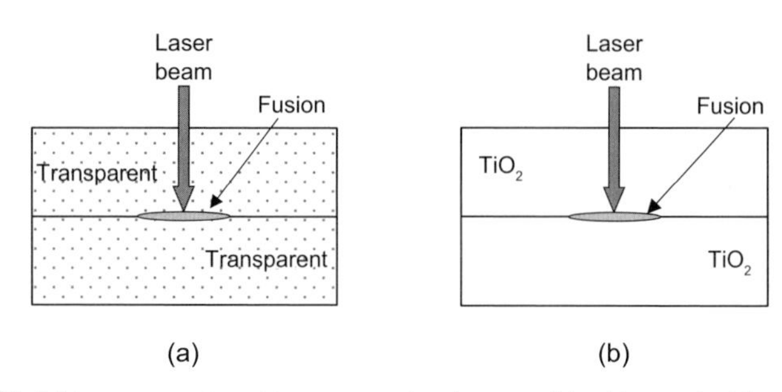

Figure 5.87 TW: (a) transparent and transparent polymers, (b) white and white polymers

One of the first applications to reach the marketplace using the TW method was the ignition key used for the E-Class and S-Class Mercedes-Benz vehicles starting in the late 1990s (see Fig. 5.88). While carbon black is an effective additive for making the thermoplastic polymer absorb infrared radiation, it represents a black body and absorbs nearly the entire electromagnetic spectrum. Thus, the component will appear black and may even have some electrical properties, depending on the amount of carbon black that was added. In some applications it is important that both components appear black. In order to weld such an application with TW, it is possible to add an infrared dye to one component to make it appear black but transmit infrared radiation.

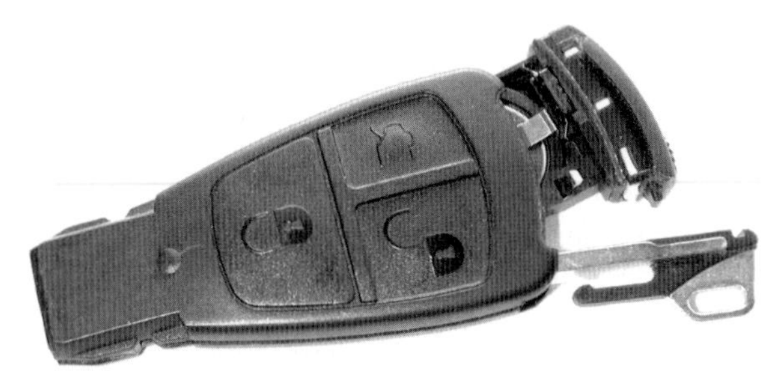

Figure 5.88 Laser welded ignition key housing used for Mercedes-Benz automobiles *(Courtesy of ETS, Inc.)*

5.11.5 Intermediate Film & ClearWeld™ Welding

The intermediate film laser welding method is a technique similar to transmission welding that allows assembly of nonweldable combinations of thermoplastics.

This process was developed to weld thermoplastic components that are transparent to the laser beam. However, this process employs consumables represented by the intermediate thermoplastic film. Absorption of the laser beam is achieved by the intermediate multilayered absorbing film placed in the joint area.

The laser beam melts the thermoplastic polymeric film, and through thermal conduction the actual components on either side of the film also are melted. Intermediate thermoplastic layers in combination with adhesion promoters can be applied to weld otherwise incompatible thermoplastics.

For example, if an assembly of two components made of PP, one pigmented with carbon black and the other in natural color, is welded with the transmission laser welding method, the joint can withstand separation loads exceeding 1,200 N. Both parts made of PP do melt around 165°C. If the two parts are instead made of polyamide 6 (PA 6), one black and the other natural color, and are welded the same way, the joint strength capability exceeds 1,900 N. The melt temperature for PA 6 in this case is about 220°C.

Transmission laser welding of components made of dissimilar thermoplastics like PP and PA is restricted because they exhibit large gradients of melt temperatures. For such cases, employing a thermoplastic intermediate multilayer film (also called consumable film) allows welding of dissimilar polymers.

For example, if the laser-trasparent thermoplastic part is PP and the carbon black pigmented component is polyamide using 1-mm-thick multilayer transparent film, the joint strength exceeds 450 N pulling load. In this case, the first layer of the film is made of PP, followed by a layer of adhesive promoter (such as Admer made by Mitsui Chemicals), and the last layer is made of PA.

However, if the laser-transparent polymer is PA and the carbon black pigmented polymer is PP using 1-mm-thick multilayer pigmented film, the joint strength exceeds 500 N pulling load. In this case, the pigmented film is the first layer made of PA, followed by a layer of adhesive promoter, followed by a layer of PP.

A few years ago, a new approach to TW welding was developed by TWI, a British R&D organization specializing in polymer joining, and Gentex Corporation of Pennsylvania. They developed and patented a series of unique coatings and resins capable of absorbing the laser beam energy while remaining virtually colorless. When these coatings are applied at the interface of two pieces made of thermoplastic polymer, the energy of the laser is harnessed and converted directly into heat necessary to create a weld directly at the interface between the thermoplastic components. An extremely thin layer of this weld-enabling material absorbs the laser energy very efficiently, acting as focal point for the laser. A localized melt of the thermoplastic polymers occurs, resulting in an instant weld. No particulate is generated, no cure time is required, and no color is visible. The trade name of the process is Clear-Weld™. There are many ways to apply these materials: film, solutions, inks–they can even be precompounded into the polymer(s) used to make the components.

5.11.6 Polymers

Laser welding is a relatively new technique for assembling thermoplastic components. It has often been difficult to determine which thermoplastic polymers are weldable using this process. Table 5.4 shows a number of polymers that have been tested and work well with laser welding techniques.

Table 5.4 Polymers Successfully Welded

PC (polycarbonate)	PS (polystyrene)	PP (polypropylene)
PMMA (polymethylmethacrylate)	ABS (acrylonitrile butadiene styrene)	PK (polyketone)
EVOH (ethylene-vinyl alcohol copolymer) [193]	PVC (polyvinyl chloride)	Elastomers
Acetyl	PE (polyethylene)	PA (polyamide)
PTFE (polytetrafluoroethylene) [63]		

As explained earlier, various dyes and pigments interact with infrared radiation in quite different ways. Dyes tend to dissolve within the thermoplastic polymer, and thus their particle size approaches the size of the molecular structure of the resin. Because the particle size is small, dyes in general do not promote internal scattering, as compared with pigments that tend to be inorganic and do not dissolve within the thermoplastic polymer. The pigment particle size depends on the manufacturing technique employed. So, pigments tend to act as very small, randomly oriented mirrors, dispersed throughout the resin, promoting internal scattering.

5.11.7 Applications

There are multiple components in an automotive fuel system, which supplies the engine with gasoline. Unfortunately, gasoline and gasoline vapors can escape from the fuel system during combustion. The liquid vapor separator or LVS for short is one component designed to prevent that escape. Located between the fuel tank and an activated carbon filter, the LVS separates liquid gasoline from gasoline vapors. While the gasoline is returned to the tank, the gases are released into the open air after having been filtered with the activated carbon filter. In addition to protecting the environment, this approach also saves fuel.

Figure 5.89 Liquid vapor separator assembly used in the VW Polo *(Courtesy of Leister Technologies AG)*

The Volkswagen Polo, manufactured for the Indian market, uses this type of LVS, made of two injection-molded components made of thermoplastic acetal polymer (POM) that have been laser welded with a 940 nm wavelength diode laser. Joint strength exceeds 44 psi or 3 bars (0.3 MPa) burst pressure. The parts were molded and assembled in China by Shenzhen Yuanwang Industry Automation Equipment. A pneumatic cylinder mounted on the side applies a small pressure on the joint area. Short cycle times necessary to laser weld the two components are achieved with an integral two-cycle rotary indexing table with four holders.

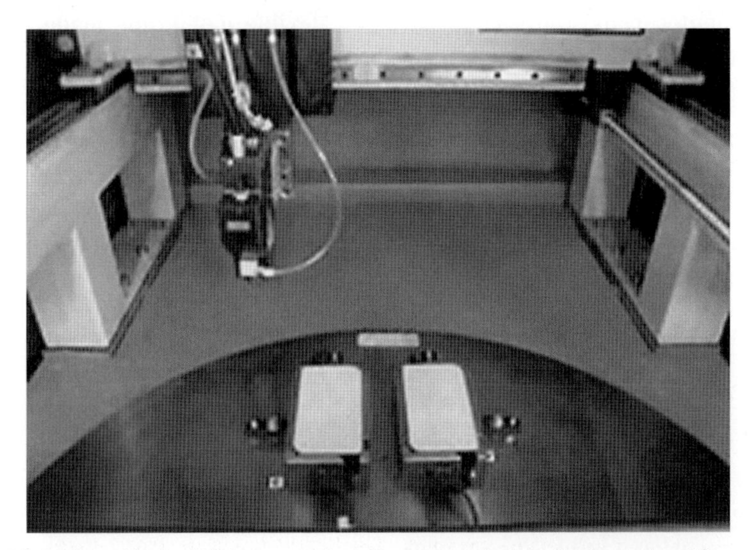

Figure 5.90 Index table for LVS laser welding *(Courtesy of Leister Technologies AG)*

The pressure required to hold the two components together during welding is applied to the joint area using a clamping device that allows the laser beam to pass through. A welding speed of 50 mm per second (approximately 2 in. per second) allows for a very short cycle time to assemble the component. The lower part of the LVS, which is opaque to the laser beam, is made of POM (acetal) thermoplastic colored with 1% by weight of carbon black pigment. The upper component, which is the laser-transparent component, is also made of the same polymer but is uncolored—of only natural color (NC). The contour welding technique employed here allows flexibility regarding component dimensions and part design. It also permits the welding of components made of different wall thicknesses.

Figure 5.91
Flexible tap for beverage industry *(Courtesy of DILAS Diodenlaser GmbH and Scholle Europe France SAS)*

Another laser-welded application is the flexible tap for the beverage industry, trade named FlexTrap®, created by Scholle Europe France SAS. This tap gives the user specific control of the beverage flow rate and features gravity dispensing and automatic shutoff. Figure 5.91 shows the flexible tap assembly made of two separate laser-welded components. The first component of the assembly is the PP carbon black pigmented body; the second part is a flexible membrane made of thermoplastic elastomer (TPE) colored with red dye, representing the tap button.

To manufacture the flexible tap assembly, the initial approach was to use a scanning head and to perform a circular contour laser weld. The scanning head has the advantage of flexibility for welding contour, but when millions of parts are produced with the same geometry, it falls short of productivity requirements. Programming the software to guide the scanning head and overall machine integration are further complications compared to a fixed beam approach.

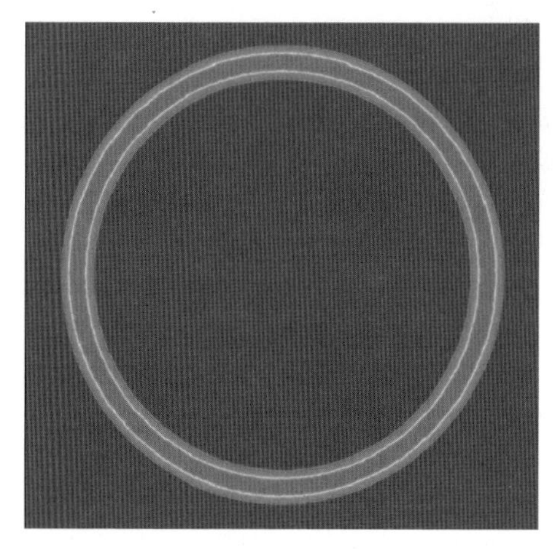

Figure 5.92
Optical imaging unit generating a ring focus beam *(Courtesy of DILAS Diodenlaser GmbH)*

Using an optical imaging unit generating a ring-focus right away (see Fig. 5.92), without using moving parts such as scanning mirrors, cuts the welding cycle time almost in half.

Figure 5.93
The axicon is properly modifying the laser beam into a ring shape *(Courtesy of DILAS Diodenlaser GmbH)*

To realize a ring focus beam layout a fiber-coupled diode laser source was used. Once the laser beam reaches the optical imaging unit, it splits from the fiber with the help of an axicon, a conical prism lens that transforms a laser beam into a ring-shaped image. Then, by employing additional spherical lenses, the size of the ring image is scaled properly and then focused onto the joint area (see Fig. 5.93). The working distance from the laser head to the welding joint should be less than 142 mm (approximately 5.5 in.).

Figure 5.94
Air-cooled, fiber-coupled diode laser welder *(Courtesy of DILAS Diodenlaser GmbH)*

The equipment used for this application is a maximum 50-watt air-cooled compact fiber-coupled diode laser system from Dilas. The entire cycle time for welding the flexible membrane to the rigid tap body is only 0.4 seconds. Using a contour welding approach would increase the cycle time to 0.7 seconds (a 75% increase) because the scanner would need to pass twice (720°) to achieve a proper weld strength. Because the heat is supplied by the axicon lens simultaneously due to the ring-shaped image of the beam, this method results in a better, more homogeneous weld joint.

Figure 5.95
The transparent glass fixture of the clamping tool is shown and the exit of the imaging optics (upper right). The parts being assembled are pressed from below against the glass fixture. The pink illumination is used for the camera inspection *(Courtesy of DILAS Diodenlaser GmbH)*

Both parts–the flexible membrane and the rigid body–are clamped around 5 psi to 10 psi during the 0.4 seconds of cycle time. The clamping load is provided from below by a transparent piece of glass fixture. The laser ring-focus is imaged through the glass onto the joint area. The laser beam passes through the upper component, the red-dye-colored flexible membrane. Then it is absorbed by the carbon black pigment used to color the rigid PP body placed below. The lower part to be joined absorbs laser energy and is heated and melted.

Figure 5.96
Flexible tap assemblies *(Courtesy of DILAS Diodenlaser GmbH and Scholle Europe France SAS)*

The flexible tap is used in beverages packaged into cardboard boxes (see Fig. 5.97)

Figure 5.97
Cardboard box for beverage industry *(Courtesy of DILAS Diodenlaser GmbH and Scholle Europe France SAS)*

5.12 Conclusion

As discussed in this chapter, there are many different ways of welding thermoplastic components. Depending on the application, materials, design, and conditions under which the product will be used, it is possible to weld plastic parts effectively.

Ultrasonic welding should be used for large-volume production and in cases involving small parts with tight tolerances—for example, when the joint for crystalline polymers is less than 70 mm in length.

Spin welding, which occurs in one plane, requires at least the rotating part to be fully round. This method will have the most flash at the joint. Very large parts can be welded by spinning.

Hot plate welding allows welding in 3-D space. This technique has been used worldwide for more than 50 years for parts that can be as large as 2,000 mm. The joint strength is lower than that of other techniques because the heat is generated electrically and not by friction.

Vibration welding allows welding 15° off plane rather large parts exceeding 1,500 mm. Both the upper and lower nests must keep components properly located during the welding process.

Electromagnetic welding is using consumables—bonding agents—allowing assembling dissimilar materials, such as amorphous and crystalline thermoplastics, where the joint length can exceed 6,000 mm. It allows welding in 3-D space.

Radio frequency welding is used mostly in the packaging industry, and the 2-D joints could be up to an overall size of 1,000 × 2,000 mm.

Laser welding provides an invisible joint to the end user. While welding clear polymer to clear polymer and white polymer to white polymer is challenging, it also allows welding in 3-D space when the laser head is placed on a robot hand or the joint area is moved in relation to the stationary laser head.

The welding methods presented here require, in most cases, that the components being assembled be made of the same polymer or the same family of polymers.

Other considerations should be taken into account when selecting an assembly procedure, such as cost and experience with any given welding method.

6 Press Fitting

6.1 Introduction

Press fitting is a very simple assembly method that uses no additional components or fasteners. One part is force fitted into a mating part. Press fitting can be used to join components made from the same or different materials, and all material properties must be considered when the joint is designed.

An important characteristic of press fits is that the male part or "boss" must be larger than the highest tolerance of the hole in the mating part in order for the fit to succeed. If these dimensions are equal, they create a "slide fit," which will not be secure. This chapter will outline the theory of press fits before presenting a couple of actual case histories.

Press fits take many different forms. We will be looking specifically at the *hub shaft* type of press fit. The goal is to evaluate polymer use, geometric options, or the torque and force that can be transmitted through a hub shaft assembled by interference fit.

The thermoplastic components used in the example are a glass-reinforced resin shaft and an unreinforced thermoplastic hub. Working temperatures applied to the example range between 23°C (73°F) and 93°C (200°F).

The main consideration in this example is the product's life expectancy, which can be estimated by understanding the creep safety factors of the thermoplastic materials. Step-by-step calculation procedures and multiple input options are presented to determine the general parameters for the assembly. Temperature variance, transmitted torque, and material conditions are all factored in.

6.2 Definitions and Notations

Notations:

R = Radius shaft/hub

E = Young's modulus shaft/hub

E_S = Secant modulus shaft/hub

v = Poisson's ratio shaft/hub

h = Joint depth

σ_{Design} = Design (allowable or permissible) strength shaft/hub

F_{In} = Assembly force

F_{Out} = Pull-out force

μ = Coefficient of friction

p_c = Pressure at contact surface between hub and shaft

α = Coefficient of thermal expansion/contraction shaft/hub

β = Geometric factor

$\sigma_{Tangential}$ = Tangential strength

M = Transmitted torque

T = Temperature

n = Safety factor

i = Total interference

6.3 Geometric Definitions

To assemble the two parts, an insertion force F_{In} is required. The insertion force is influenced by the hub radius and joint depth. A geometric factor, a function of the inside and outside hub diameters, is

$$\beta = \frac{R_o^2 + R_i^2}{R_o^2 - R_i^2} \tag{6.1}$$

where R_o = hub outside radius

R_i = hub inside radius

6.4 Safety Factors

Two safety factors are required relative to the yield strength of the thermoplastic materials considered. The first safety factor is a recommended value for polymers: 1.5 for unreinforced thermoplastic resins, and 3 for reinforced polymers. The latter safety factor covers imperfections in the injection-molding process, such as the *weld (knit) line* in the hub due to the gate location. Weld lines may have varying locations based on injection-molding parameters such as mold temperature and cycle time, and tool design such as gate location, gate size, and venting location. It can be said that the component strength cannot exceed the strength of the weld line.

Employing these two safety factors, one can determine the polymer design strength, σ_{Design}, and the polymer final (weld line) strength, σ_{Final}:

$$\sigma_{\text{Design}} = \frac{\sigma_{\text{Yield}}}{n_{\text{Design}}} \tag{6.2}$$

$$\sigma_{\text{Final}} = \frac{\sigma_{\text{Design}}}{n_{\text{Final}}} \tag{6.3}$$

6.5 Creep

The increase in strain value with time under a constant stress is referred to as creep (see Section 1.9). Different thermoplastics or thermoset materials have different rates of creep. Usually the rate of creep varies with the polymer composition (fillers, additives, reinforcements, pigments, lubricants, and so on), the temperature to which the assembly is exposed, the stress level the parts are subjected to, the moisture content, and other factors.

A theory of safety factors states: "Initial design safety factor decreases with time" (see Chapter 2). The safety factor equals 1 (or less) when the useful lifetime of the component or assembly has ended. The *design/fail* strain value in the end equals the yield design/fail strain at time end (t = end). In other words,

time = 0 at initial (design inception) time

time = end at the time when $n = 1$

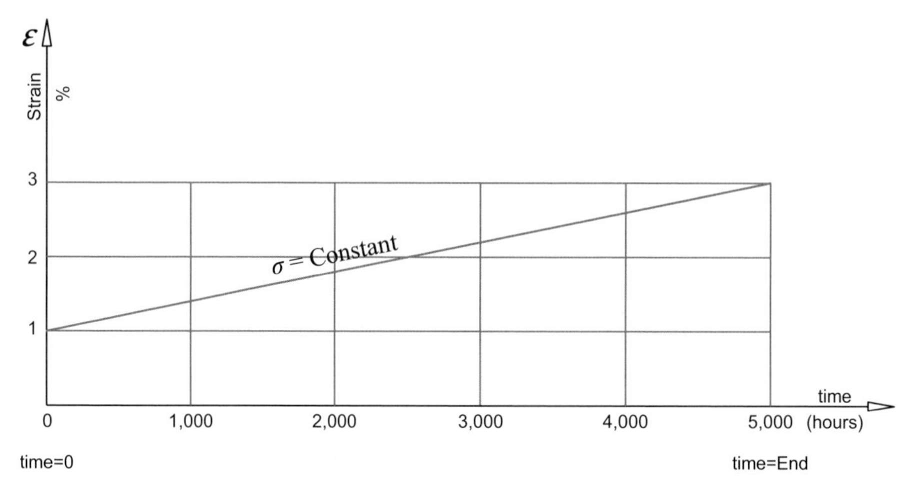

Figure 6.1 Safety factor (a) time = 0; and (b) time = end

6.6 Loads

There are three distinct loads that should be examined. As was stated earlier, F_{In} represents the axial assembly force necessary to press the shaft into the hub. The assembly operation takes place at room temperature, 23°C (73°F). The insertion force is dependent on the interference, *i*, considered.

The second force to be evaluated is the F_{Out} (pull-out force). This is also an axial force, dependent on the amount of interference available and the contact pressure at the joint surface of the two components.

The third load is the transmitted torque, *M*. This represents the ability of the assembly (shaft and hub) to carry a specific prerequisite load. The torque varies as a function of the interference between the two components and the amount of contact pressure.

6.7 Press Fit Theory

Equations developed by the French mathematician Lamé are used to define the contact pressure at the interface between the assembled parts. These equations apply specifically to thick-wall cylinders.

A. In a room temperature 23°C (73°F) case, the surface contact pressure at the interface is

$$p_c = \frac{i}{\frac{R_i}{E_{Hub}}(\beta + \nu_{Hub}) + \frac{R_i}{E_{Shaft}}(1 - \nu_{Shaft})} \tag{6.4}$$

The insertion force required to assemble the components is

$$F_{In} = 2\pi\mu h R_{Shaft} p_c = F_{Out} \tag{6.5}$$

Transmitted torque is determined as a function of the insertion force (6.5) as

$$M = F_{In} R_{Shaft} = 2\pi\mu h R_{Shaft}^2 p_c \tag{6.6}$$

It is important to recall that p_c is normal to the shaft and hub surfaces. Therefore,

$$\sigma_{Normal} = p_c \tag{6.7}$$

The maximum stress, however, is tangential to the surface:

$$\sigma_{Tangential} = \beta p_c \tag{6.8}$$

where β is the geometric factor described in 6.1.

The condition required to have a proper press fit assembly is that the yield strength of the polymer should be greater than the tangential stress present at the common surface between the two parts, i.e.:

$$\sigma_{Yield} > \sigma_{Tangential} \tag{6.9}$$

In 6.9 when the *greater than* sign is replaced with the *equal* sign, the maximum interference is determined:

$$i_{Maximum} = 2R_{Shaft} \frac{\sigma_{Yield}}{\beta} \left[\frac{\beta + \nu_{Hub}}{E_{Hub}} + \frac{1 - \nu_{Shaft}}{E_{Shaft}} \right] \tag{6.10}$$

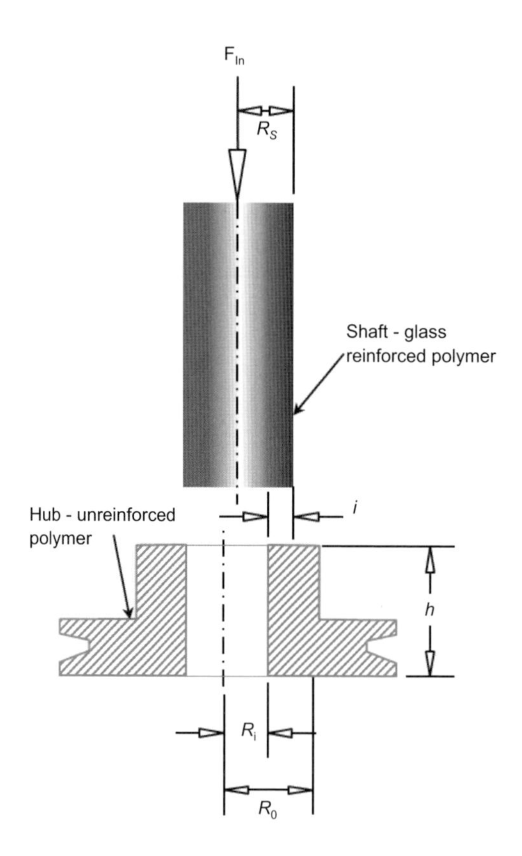

Figure 6.2
Shaft and hub before assembly

B. During operation, the temperature increases due mostly to friction. The interference will vary with the operating temperature. Above room temperature the total interference is

$$i^T = 2\left(R_o^T + R_i^T\right) \tag{6.11}$$

The shaft radius will change with temperature. So R_{Shaft}^T becomes

$$R_{\text{Shaft}}^T = R_{\text{Shaft}}\left(1 + \alpha_{\text{Shaft}} \Delta T\right) \tag{6.12}$$

The hub radius will experience similar change:

$$R_{\text{Hub}}^T = R_{\text{Hub}}\left(1 + \alpha_{\text{Hub}} \Delta T\right) \tag{6.13}$$

The initial interference was

$$i = 2\left(R_s - R_i\right) \tag{6.14}$$

Using 6.12, 6.13, and 6.14, the total interference required at a given temperature T (6.11) becomes:

$$i^T = i - 2R_{\text{Shaft}} \Delta T\left(\alpha_{\text{hub}} - \alpha_{\text{Shaft}}\right) \tag{6.15}$$

6.8 Design Algorithm

A function of unknown variables can describe the design of the assembly: F (time, temperature, material, geometry, loads) = 0.

In the above function, time represents the useful life of the product. In order to shape the function to a manageable form, the number of variables must be reduced or simplified. Therefore, the time variable will be considered for only two positions. The first value is the initial time (time = 0), when the product is first assembled. At the end of the useful product life cycle, (usually well defined in the product warranty) the assembly may fail. Time = end will represent the second time position.

Temperature will also be considered for two values: room temperature, the condition under which the product is assembled; and operating or nominal temperature.

The remaining three variables–material, geometry, and loads–can be grouped together. If any two are considered known, the value of the third can be determined. For example, if material and geometry variables are known, then loads (torque and assembly force) can be calculated. When material and loads are known, geometry can be calculated. And finally, knowing the geometry and loads, the material properties will be determined.

Subsequently, there are four cases (A, B, C, and D) that are dependent on time and temperature for each of the three options.

Table 6.1 Suggested Design Algorithm for Press Fit Assemblies

Option case	Loads (material, geometry)	Geometry (material, loads)	Material (geometry, loads)
Case A $t = 0$; $T = 23°C$	Defines F_{In} and M for initial assembly	Defines i, h, and β for transmitted torque	Defines σ_{Yield}
Case B $t = 0$; $T = 93°C$	Defines the variation for F_{Out} and M with temperature	Variation of β and h relative to temperature	Defines E, Young's modulus variation with temperature
Case C t = end; $T = 23°C$	Defines F_{Out} and M subjected to stress relaxation	Variation of h and β due to stress relaxation	Defines secant modulus due to stress relaxation
Case D t = end; $T = 93°C$	Defines F_{Out} and M due to isothermic stress relaxation	Variation of h and β due to isothermic stress relaxation	Defines secant modulus due to isothermic stress relaxation
Remarks	Further iteration required if the transmitted torque remaining after stress relaxation is smaller than the imposed value.	A couple of iterations will determine the best values for h and β.	Based on the material properties calculated, one can easily select a desirable polymer.

6.9 Case History: Plastic Shaft and Plastic Hub

It is not our intention to detail all option modes possible through all four cases mentioned. However, for demonstration purposes, the first option mode is presented in detail.

For the first option mode, the loads (insertion force, pull-out force, and transmitted torque) are considered unknown. They will be determined based on the given geometry and available material data at two distinct temperatures and the two time limits. They are initial time ($t = 0$), when the parts are assembled, and at time end when the product reaches the end of its life cycle.

The example to be examined is a pulley for a cassette tape player. This pulley's function is to transmit a certain torque through the belt.

6.9.1 Shaft and Hub Made of Different Polymers

The actual assembly for this example consists of two parts: a pulley and a shaft. Because the shaft is a structural part and has to sustain the bending moment created by the belt tension, a glass-reinforced polymer was selected. The pulley, on the other hand, needs only to be able to resist the torque transmitted by the belt without spinning on the shaft, so an unreinforced plastic material was used (see Fig. 6.2).

6.9.2 Safety Factor Selection

Two gate locations were considered for the pulley. A diaphragm gate requires degating either in the tool itself (typically achieved in the three-plate tool). De-gating can also take place after the molding cycle has been completed, by using a machining procedure as a secondary operation. The diaphragm gate was not selected for the pulley because a requirement of the part was a hub surface free of flash and microcracks that would otherwise be created by the machining operation.

The other gate design was a submarine (tunnel) gate. The gate location (Fig. 6.3) resulted in a weld (knit) line created on the diametrically opposite side.

A weld (knit) line does not have the same strength as the basic thermoplastic material. Weld lines seldom exceed 60% of the initial strength of the base polymer. As mentioned earlier in this chapter, a part can only be as strong as its weakest point. Therefore, an additional safety factor, called n_{Final}, had to be considered.

$$n_{\text{Design}} = \frac{\sigma_{\text{Yield}}}{\sigma_{\text{Design}}} \tag{6.16}$$

$$n_{\text{Final}} = \frac{\sigma_{\text{Design}}}{\sigma_{\text{Final}}} \quad (6.17)$$

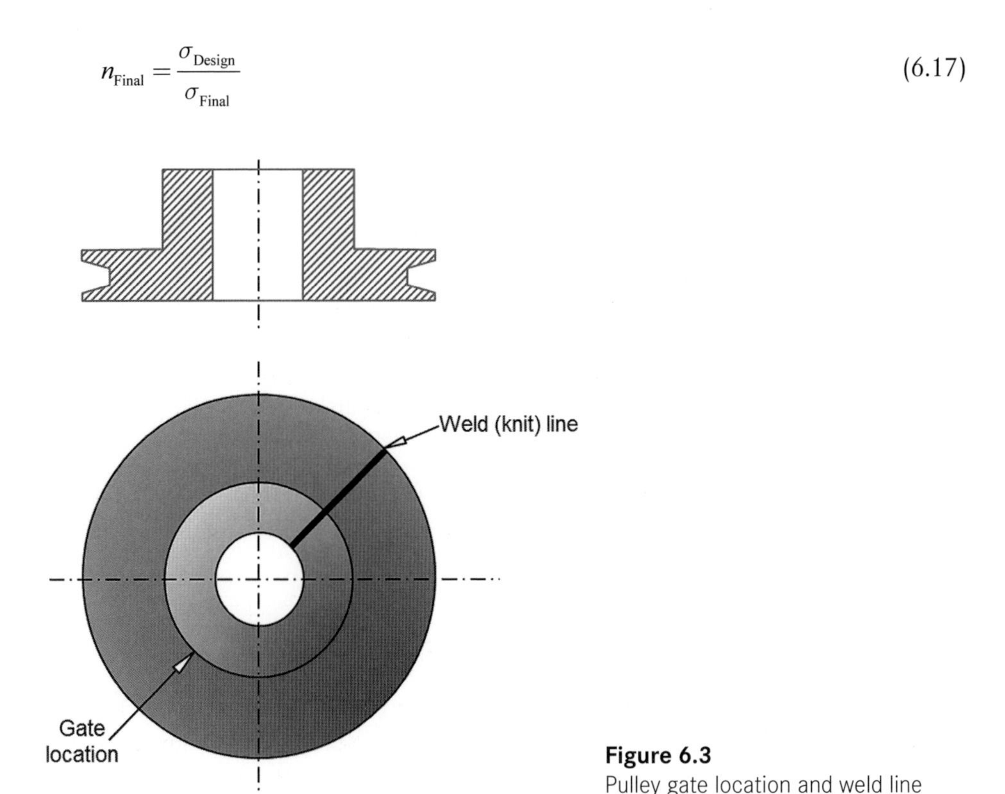

Figure 6.3
Pulley gate location and weld line

6.9.3 Material Properties

First we look at the material properties of the shaft.

The shaft material for this application is a PET glass-reinforced thermoplastic.

The stress/strain curves shown in Fig. 6.4 are typical curves that a material supplier makes available. Throughout the next few pages, we will work with the above data to put it in a useful form in order to carry on with the design calculations.

From Fig. 6.4 two curves for two temperatures of interest for these examples are selected. They are the room temperature of 23°C (73°F) and the elevated operating temperature of 93°C (200°F), as shown in Fig. 6.5.

We further isolate the stress/strain curve corresponding to room temperature. The Young's modulus for this curve is shown in Fig. 6.6.

Selecting Young's modulus as the modulus of elasticity for this polymer will result in a significant error. The shaded area between the Young's modulus line and the stress/strain curve as shown in Fig. 6.6 represents the amount of error this would induce in calculations.

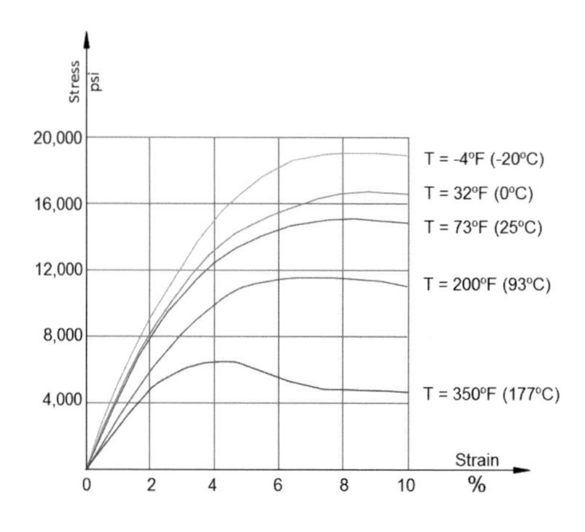

Figure 6.4
Stress/strain curves for PET glass-reinforced thermoplastic resin

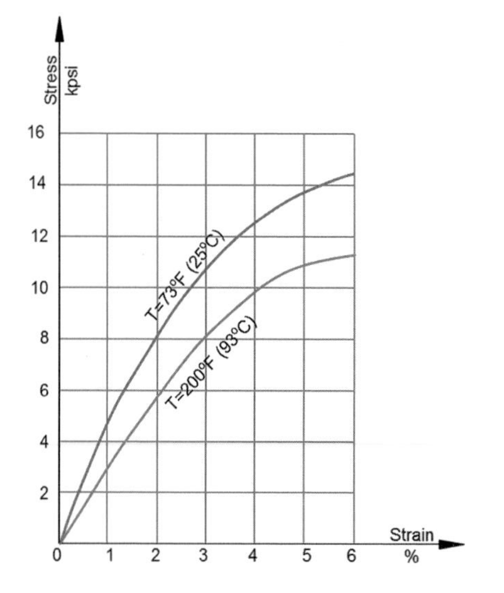

Figure 6.5
Stress/strain curves for PET glass-reinforced thermoplastic at two temperatures: 23 °C (73 °F) and 93 °C (200 °F)

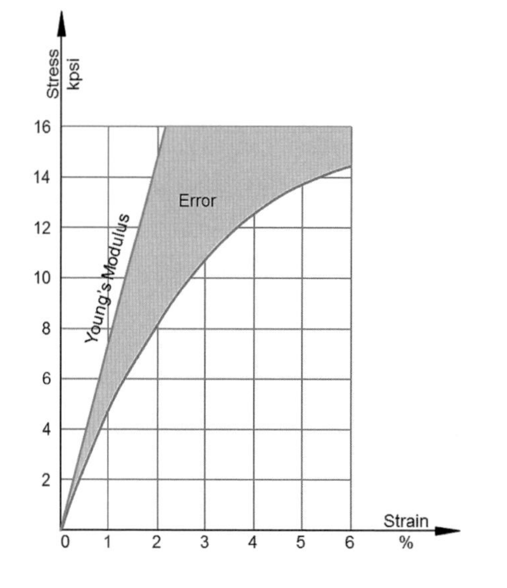

Figure 6.6
Stress/strain curve for PET glass-reinforced thermoplastic at 23 °C (73 °F) and its Young's modulus

6.9.4 Shaft Material Properties at 23°C

To minimize the error, a secant modulus is used instead of Young's modulus. When the shaded area of Fig. 6.6 is compared with that of Fig. 6.7, it becomes obvious which of the two approximations is more accurate.

The yield point, σ_{Final} as shown in Fig. 6.7, is usually specified in the technical literature. Glass-reinforced polymers do not have a distinct yield point. The region where the polymer yields under load depends on the amount of reinforcement and how well its percentage content is controlled during the final polymer compounding. In Fig. 6.7 the lowest recorded yield point within the yield region has been selected. This approach is recommended for reinforced polymers.

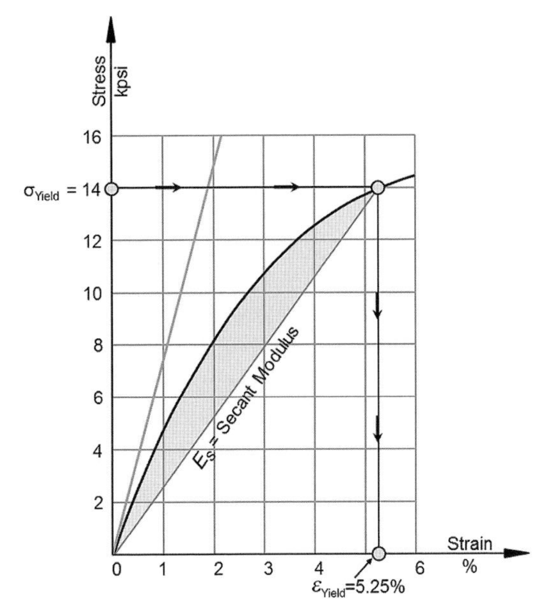

Figure 6.7
Stress/strain curve for PET glass-reinforced thermoplastic and its secant modulus

The reinforcement could be glass, metal, carbon, mineral, aramid, and so on. Therefore, the lowest recorded yield point for PET glass-reinforced resin is $\sigma_{\text{Yield},\,t=0,23^\circ\text{C}}$ = 96.5 MPa (14 kpsi). Locating this value on the sigma axis (Fig. 6.7), and by using a straight line parallel to the strain axis, an intersection with the stress/strain curve will occur. The yield strain location is obtained by projecting the point found on the curve on the strain axis, $\varepsilon_{\text{Yield},t=0,23^\circ\text{C}} = 5.25\%$. Therefore the secant modulus at yield can be determined:

$$E_{S_{\text{Yield},t=0,23^\circ\text{C}}} = \frac{\sigma_{\text{Yield},t=0,23^\circ\text{C}}}{\varepsilon_{\text{Yield},t=0,23^\circ\text{C}}} = 266\,\text{kpsi} \tag{6.18}$$

Let's review what we have accomplished so far. The stress/strain curves, as provided by a material supplier, were considered for the shaft material at different temperatures.

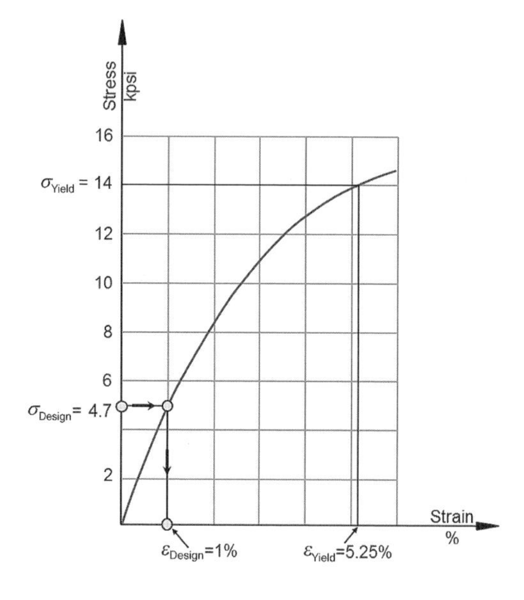

Figure 6.8
Design stress, strain, and secant modulus for PET glass-reinforced polymer at 23°C (73°F)

From the five curves present in the graph, those needed for extreme conditions were selected. The two specific curves at two different temperatures were isolated and enlarged, and the Young's modulus for each curve was identified. After the polymer yield strength was identified, the yield strain was determined. By dividing the yield stress by the yield strain, the polymer secant modulus was obtained.

$$\sigma_{\text{Design},\, t=0,23^\circ\text{C}} = \frac{\sigma_{\text{Yield},t=0,23^\circ\text{C}}}{n_{\text{Design}}} = 4.7\,\text{kpsi} \tag{6.19}$$

The safety factor considered for the shaft is n_{Design} = 3. The design stress level is:

$$\varepsilon_{\text{Design},t=0,23^\circ\text{C}} = 0.01 = 1\% \tag{6.20}$$

Following the same procedure, the value of $\sigma_{\text{Design},\, t=0,23^\circ\text{C}}$ is located on the sigma axis by moving horizontally until the stress/strain curve is intersected. A vertical line is drawn through the design point intersecting the stress/strain axis. This provides the design strain value: $\varepsilon_{\text{Design},t=0,23^\circ\text{C}}$.

The secant modulus for the design point is obtained by dividing the stress by strain. Therefore,

$$E_{S_{\text{Design},t=0,23^\circ\text{C}}} = \frac{\sigma_{\text{Design},\, t=0,23^\circ\text{C}}}{\varepsilon_{\text{Design},\, t=0,23^\circ\text{C}}} = 470\,\text{kpsi} \tag{6.21}$$

As can be observed from Fig. 6.8, the amount of error induced by $E_{S_{\text{Design}}}$ is small.

6.9.4.1 Shaft Material Properties at 93°C

In a similar manner, the polymer mechanical properties are determined for the case of elevated temperature at 93°C (200°F). Figure 6.9 shows how $E_{S_{Yield}}$ and ε_{Yield} are determined. The design values for stress, strain, and secant modulus are also determined in a similar way.

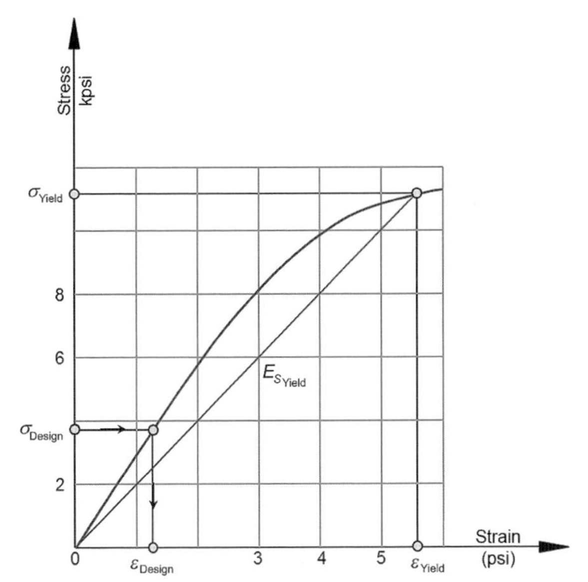

Figure 6.9
Stress/strain curve for PET glass-reinforced thermoplastic at 93°C (200°F); shaft design stress and strain and secant modulus

6.9.4.2 Creep Curves at 23°C

This assembly (shaft and pulley) will experience stress relaxation over time.

In Fig. 1.40 from Chapter 1, a typical stress-relaxation test was shown. It should be observed that for this test the load decreases with time.

Sophisticated computerized equipment is required for tests that accurately determine stress relaxation curves of thermoplastic or thermoset polymers. These tests are expensive, partly because the load has to vary continuously. It may not be considered economically feasible for companies to spend large amounts of money for such tests. In the absence of this data, creep curves were used for the example described.

Creep curves are a good tool for predicting how a material will behave in the future. It is important to note that tests of actual parts must be conducted to validate the design theory and approach.

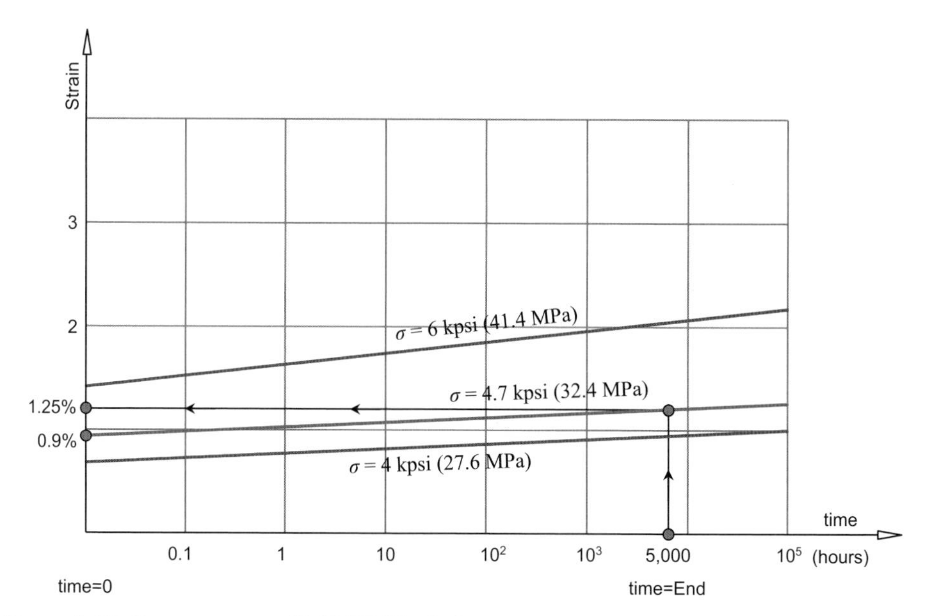

Figure 6.10 Creep curves at 23°C (73°F); shaft

In Fig. 6.10, the creep curves for the shaft material, a PET glass-reinforced thermoplastic polymer, are shown for a temperature gradient of 23°C. There are three curves. The values of $\sigma = 6$ kpsi (41.4 MPa) and of $\sigma = 4$ kpsi (27.6 MPa) are obtained through actual tests. The curve situated between these two values was obtained graphically.

The two known curves are for 4 kpsi and 6 kpsi. We want to find the creep curve corresponding to a design stress level of 4.7 kpsi. This curve will be located somewhere between those of 4 and 6 kpsi on the graph. We measure up from point zero to the beginning point of the 4 kpsi curve on the strain axis. We can say, for example, that this distance is 20 mm (0.78 in.). Therefore, 20 mm up the strain axis represents 0.7% strain in a tensile specimen loaded in tension at time = 0. By cross-multiplying, 23.5 mm will correspond to a stress level of 4.7 kpsi. Reading the strain corresponding to this distance on the strain axis, we obtain a 0.9% value at time = 0. So far, the first point of the 4.7 kpsi has been determined.

Using the same procedure, the second point of the curve situated at time = 10^5 hours is determined. Then a line is drawn connecting the two points. This represents the creep curve for our component for a stress level of 4.7 kpsi.

The curve shown in Fig. 6.10 is drawn on a logarithmic scale. After the point time = end at the 5,000 hour is located, the $\sigma = 4.7$ kpsi curve is then intersected with a vertical line; moving horizontally, the strain value for t = 5,000 hours is found.

The polymer used to manufacture the shaft component has an initial strain of 0.9% under the design load of $\sigma = 4.7$ kpsi. After 5,000 hours, the shaft will further relax under the static load by another 0.35%, 1.25% total strain deformation at time = end.

Note: Figure 1.40 from Chapter 1 shows a stress-relaxation test for tension. The material data for the shaft is in compression and not tension. The compressive material data was not available. An assumption was made that the material has a very similar behavior in tension as it has in compression.

6.9.4.3 Creep at 93°C

A similar procedure is used for determining the shaft material's creep properties for the high-temperature conditions (93°C or 200°F).

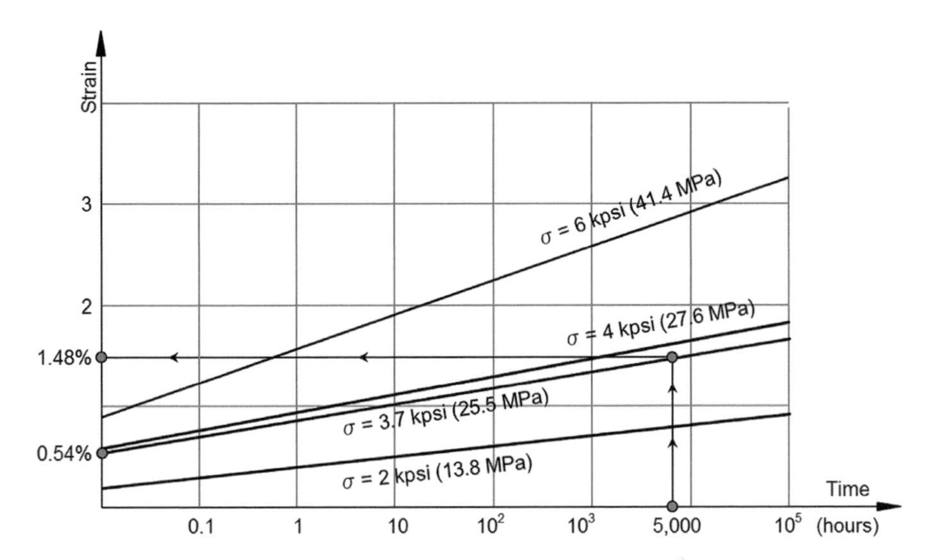

Figure 6.11 Creep curves at 93°C (200°F) for the shaft

In Fig. 6.11 four curves for constant stress are shown ($\sigma = 2, 3.7, 4, \text{and} \, 6$ kpsi). Three of the four curves were obtained from the material data. The fourth ($\sigma = 3.7$ kpsi) was determined through the graphic interpolation method explained above.

Therefore, the design strain at initial time is 0.54%, and the design strain for time = end condition is 1.48%.

Now we can determine the secant apparent modulus at time = end:

$$E_{S_{\text{Design},\,t=\text{end},93^\circ\text{C}}} = \frac{\sigma_{\text{Design},t=\text{end},93^\circ\text{C}}}{\varepsilon_{\text{Design},\,t=\text{end},93^\circ\text{C}}} \tag{6.22}$$

or

$$E_{S_{\text{Design},\,t=\text{end},93^\circ\text{C}}} = 251.98\,\text{kpsi} \tag{6.23}$$

The mechanical properties at two distinct time positions (0 hours and 5,000 hours) and two distinct temperature gradients (23°C and 93°C) for the thermoplastic polymer used for the shaft were established.

The material data calculated now can be listed:

$$\sigma_{\text{Design},t=0,23^\circ\text{C}} = 4.7\,\text{kpsi} \tag{6.24}$$

$$\varepsilon_{\text{Design},t=0,23^\circ\text{C}} = 1\% \tag{6.25}$$

$$E_{S_{\text{Design},t=0,23^\circ\text{C}}} = 470\,\text{kpsi} \tag{6.26}$$

$$\varepsilon_{\text{Design},t=\text{end},23^\circ\text{C}} = 1.25\% \tag{6.27}$$

$$E_{S_{\text{Design},t=\text{end},23^\circ\text{C}}} = 376\,\text{kpsi} \tag{6.28}$$

$$\sigma_{\text{Design},t=0,93^\circ\text{C}} = 3.7\,\text{kpsi} \tag{6.29}$$

$$\varepsilon_{\text{Design},t=0,93^\circ\text{C}} = 1.27\% \tag{6.30}$$

$$E_{S_{\text{Design},t=0,93^\circ\text{C}}} = 291\,\text{kpsi} \tag{6.31}$$

$$\varepsilon_{\text{Design},t=\text{end},93^\circ\text{C}} = 1.48\% \tag{6.32}$$

$$E_{S_{\text{Design},t=\text{end},93^\circ\text{C}}} = 250\,\text{kpsi} \tag{6.33}$$

6.9.4.4 Pulley at 23°C

A similar procedure is used to obtain the pulley's material data.

Figure 6.12 shows a typical stress/strain curve at four temperatures as provided. Two of the four curves, 73°F and 200°F, are of interest to us. We'll separate them from the rest.

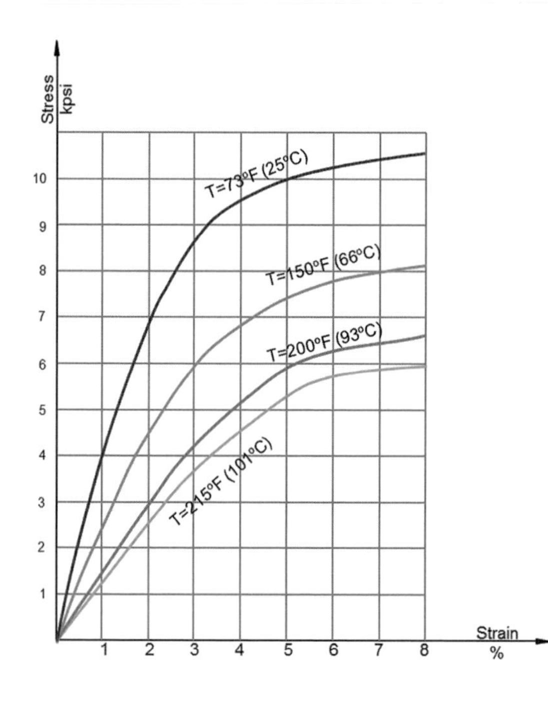

Fig. 6.12
Stress/strain curves for thermoplastic resin at various temperatures for the hub

Because the way the gate is located in the tool, a weld (knit) line will be present on the opposite side of the gate. The strength properties of the thermoplastic polymer, used for the pulley or hub, in a weld (knit) area are at best 60% of the original material properties. Therefore, we need to account for the loss of mechanical properties at the weld line. In other words, we have to establish another safety factor that will account for the manufacturing process employed to produce the hub.

If the polymer yield strength is σ_{Yield}, then the design stress level will be

$$\sigma_{\text{Design}} = \frac{\sigma_{\text{Yield}}}{n_{\text{Design}}} \tag{6.34}$$

The safety factor, n_{Design}, for an unreinforced thermoplastic polymer that has a distinct yield point is 1.5. Therefore,

$$\sigma_{\text{Design}} = \frac{\sigma_{\text{Yield}}}{1.5} \tag{6.35}$$

Accounting for processing, in this case the injection-molding process, we need another safety factor to account for the weld line weakness. Because we said that the material properties are reduced by 40% in a weld area, let's select

$$n_{\text{Final}} = 2 \tag{6.36}$$

Therefore, the stress level, taking into account the decreased strength at the weld line, σ_{Final}, is

$$\sigma_{\text{Final}} = \frac{\sigma_{\text{Design}}}{n_{\text{Final}}} = \frac{\sigma_{\text{Design}}}{2} \tag{6.37}$$

Based on the above, the secant modulus for design and final conditions will have the following formulas:

$$E_{S_{\text{Design}}} = \frac{\sigma_{\text{Design}}}{\varepsilon_{\text{Design}}} \tag{6.38}$$

and

$$E_{S_{\text{Final}}} = \frac{\sigma_{\text{Final}}}{\varepsilon_{\text{Final}}} \tag{6.39}$$

Figure 6.13 shows the stress/strain curve at 73°F for the pulley material. To obtain the stress and strain levels specified, the following procedure should be followed.

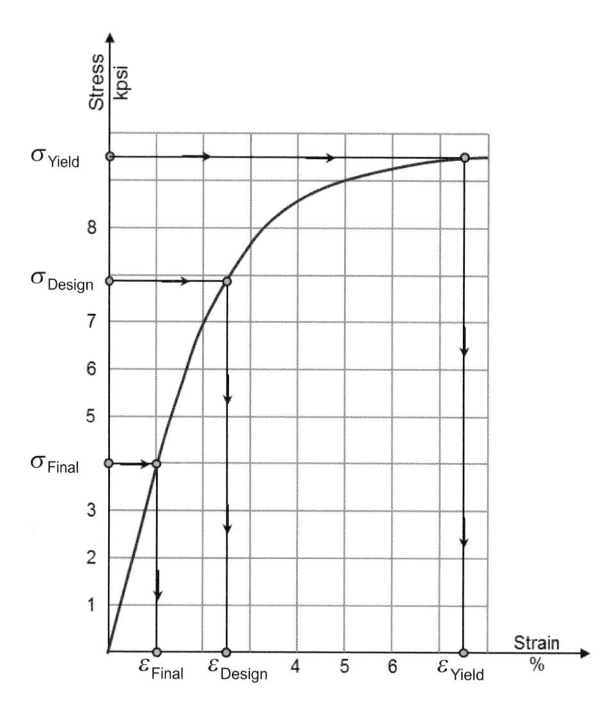

Figure 6.13
Stress/strain curve for polymer at 23°C (73°F) and its design stress, strain; final stress, strain, for hub

First, we identify on the σ (vertical or stress) axis the yield strength of the material. From that point, we move to the right, parallel to the strain axis, until the stress/strain curve is intersected. This point represents the yield point for both stress and strain for the material. Then, by moving down from that point until the ε (horizontal or strain) axis is intersected, we find the corresponding strain level at yield.

The yield stress is

$$\sigma_{\text{Yield},t=0,23^{\circ}\text{C}} = 10.5\,\text{kpsi} \tag{6.40}$$

Employing the design safety factor, we now can determine the design stress level:

$$\sigma_{\text{Design},t=0,23^{\circ}\text{C}} = \frac{\sigma_{\text{Yield},t=0,23^{\circ}\text{C}}}{n_{\text{Design}}} = 7.9\,\text{kpsi} \tag{6.41}$$

By taking into account the weld line we have

$$\sigma_{\text{Final},t=0,23^{\circ}\text{C}} = \frac{\sigma_{\text{Design},t=0,23^{\circ}\text{C}}}{n_{\text{Final}}} = 4\,\text{kpsi} \tag{6.42}$$

Corresponding to the stress levels observed, there are two strains associated with them. They are

$$\varepsilon_{\text{Design},t=0,23^{\circ}\text{C}} = 2.5\%$$

$$\varepsilon_{\text{Final},t=0,23^{\circ}\text{C}} = 1.0\% \tag{6.43}$$

Now we can determine the secant modulus corresponding to the distinct points mentioned above:

$$E_{S_{\text{Design},\,t=0,23^\circ\text{C}}} = 315\,\text{kpsi}\ (2.18\,\text{MPa})$$

$$E_{S_{\text{Final},\,t=0,23^\circ\text{C}}} = 400\,\text{kpsi}\ (2.77\,\text{MPa}) \tag{6.44}$$

6.9.4.5 Pulley at 93°C

Figure 6.14 shows the stress/strain curve for the pulley material at the high-temperature condition.

$$\sigma_{\text{Design},\,t=0,93^\circ\text{C}} = \frac{\sigma_{\text{Yield},\,t=0,93^\circ\text{C}}}{n_{\text{Design}}} = 4.2\,\text{kpsi} \tag{6.45}$$

$$\sigma_{\text{Final},\,t=0,93^\circ\text{C}} = \frac{\sigma_{\text{Design},\,t=0,93^\circ\text{C}}}{n_{\text{Final}}} = 2.1\,\text{kpsi} \tag{6.46}$$

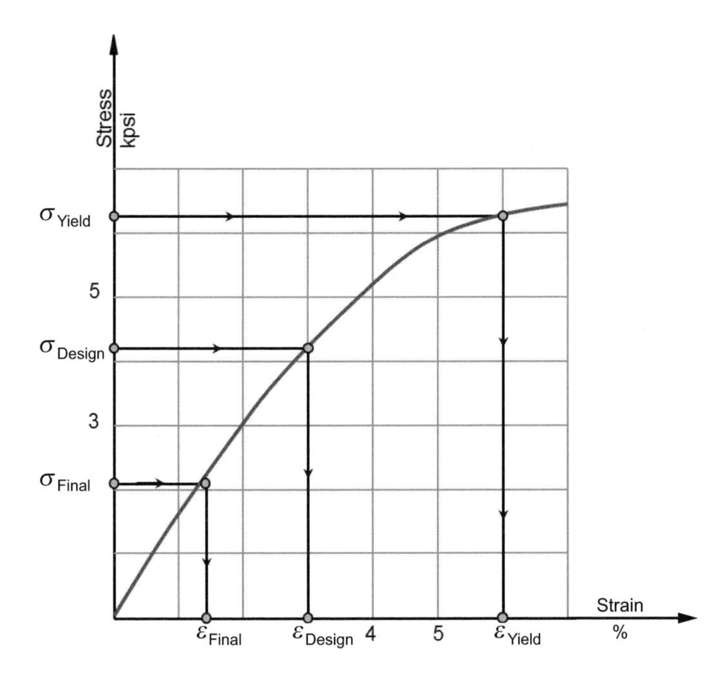

Figure 6.14
Stress/strain curve for polymer at 93°C (200°F) and its design stress, strain; final stress, strain, for hub

Corresponding to these stress levels we have the following strains:

$$\varepsilon_{\text{Design},\,t=0,93^\circ\text{C}} = 3\%$$

$$\varepsilon_{\text{Final},\,t=0,93^\circ\text{C}} = 1.42\% \tag{6.47}$$

and the following secant moduli:

$$E_{S_{\text{Design},\,t=0,93^\circ\text{C}}} = 140\,\text{kpsi (0.97 MPa)}$$

$$E_{S_{\text{Final},\,t=0,93^\circ\text{C}}} = 148\,\text{kpsi (1.02 MPa)} \tag{6.48}$$

6.9.4.6 Creep, Pulley at 23°C

Figure 6.15 shows the creep curve for the hub material at 23°C. The strain at time = 0 is 0.95%, corresponding to a continuous, constant static stress of 4 kpsi (27.67 MPa). By selecting time = 5,000 hours on the longitudinal axis and moving up until we intersect the creep curve, we can read the strain at time = end as 1.96%.

Similarly, we have

$$\varepsilon_{\text{Final},\,t=\text{end},23^\circ\text{C}} = 1.96\% \tag{6.49}$$

and

$$E_{S_{\text{Final},\,t=\text{end},\,23^\circ\text{C}}} = 205\,\text{kpsi (1.42 MPa)} \tag{6.50}$$

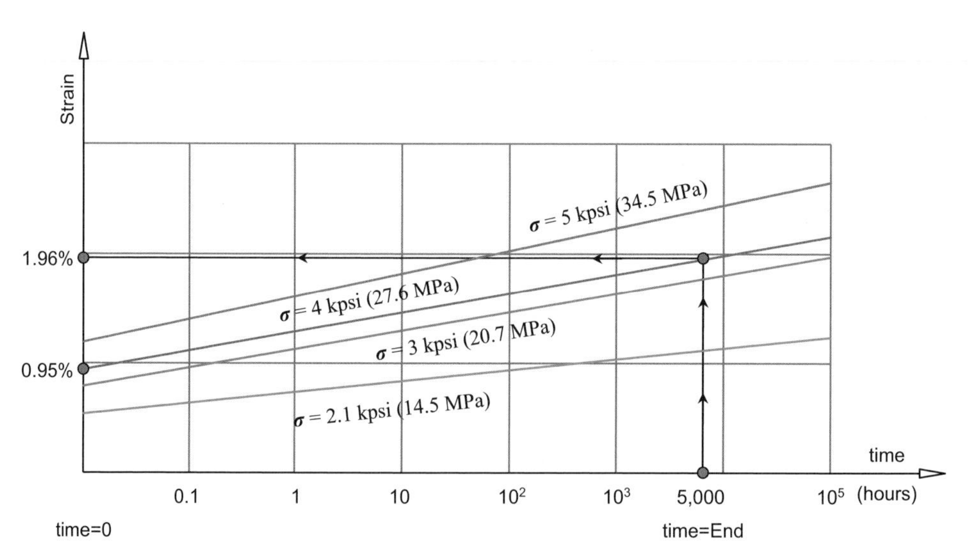

Figure 6.15 Creep curves at 23°C (73°F) for hub

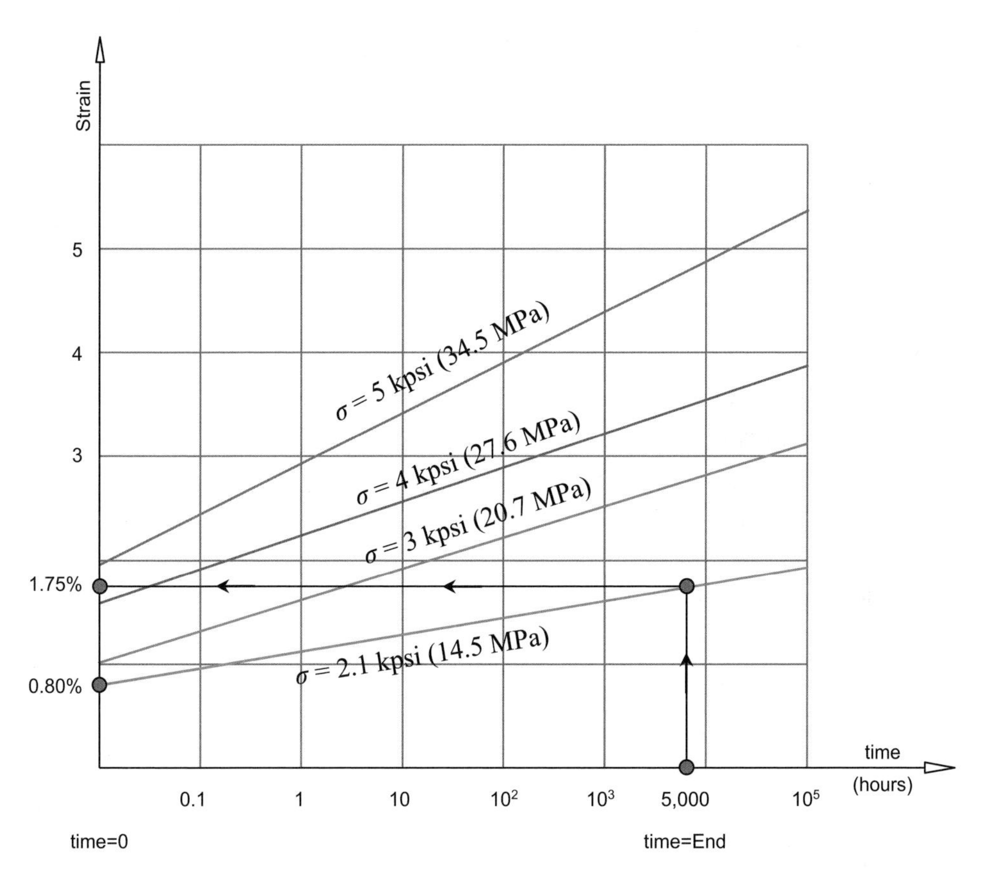

Figure 6.16 Creep curves at 93 °C (200 °F) for hub

6.9.4.7 Creep, Pulley at 93 °C

Exactly the same procedure is employed for the high-temperature (93 °C) conditions. Therefore, we find

$$\varepsilon_{\text{Final},t=\text{end},93^\circ\text{C}} = 1.75\% \tag{6.51}$$

and

$$E_{S_{\text{Final},t=\text{end},93^\circ\text{C}}} = 120 \text{ kpsi } (0.83 \text{ MPa}) \tag{6.52}$$

6.10 Solutions: Plastic Shaft, Plastic Hub

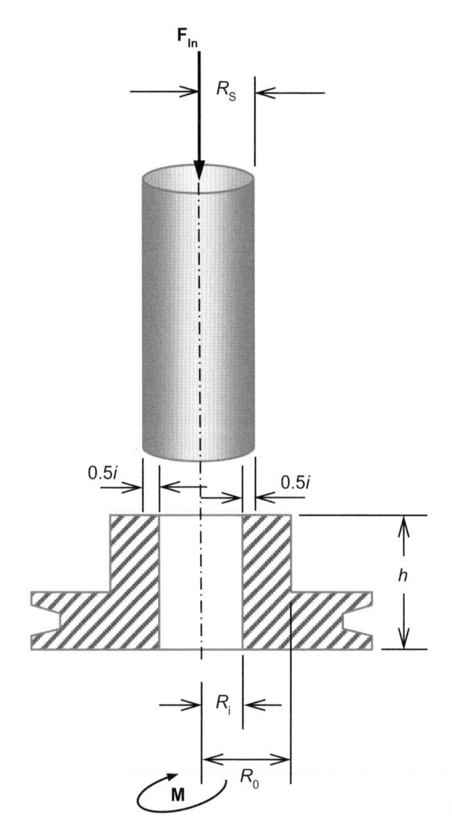

Figure 6.17
Hub/shaft assembly

6.10.1 Case A

For this case we use the first option mode described in detail in Table 6.1, which considers an assembly operating at a temperature of 23°C. The assumption for case A is that the product, a cassette player, is in use. An increase in temperature develops due to operation and friction. The heat is eliminated with the help of an air-conditioning unit or through ventilation, keeping the temperature of the part constant throughout the test. Also, because time = 0, one should also assume that the operating temperature will not be reached.

The unknown parameters are the amount of total interference required for the assembly, the insertion force needed to assemble the two parts, and the moment that the assembly will be able to carry.

The known parameters are

Shaft radius	$R_S = 0.25\text{ in.}\,(6.35\text{ mm})$
Hub outside radius	$R_H = 0.5\text{ in.}\,(12.7\text{ mm})$
Hub height	$h = 0.315\text{ in.}\,(8\text{ mm})$
Coefficient of friction	$\mu = 0.3$
Shaft Poisson's ratio	$\nu_S = 0.38$
Hub Poisson's ratio	$\nu_H = 0.40$
Hub secant modulus (time = 0; 23°C)	$E_{S_{\text{Design, Hub}}} = 400\text{ kpsi}$
Shaft secant modulus (time = 0; 23°C)	$E_{S_{\text{Design, Shaft}}} = 470\text{ kpsi}$
Final stress level, hub (time = 0; 23°C)	$\sigma_{\text{Final, Hub}} = 4\text{kpsi}$

The contact pressure between shaft and hub will be at the same level as the final stress level. Therefore:

$$p_c = \sigma_{\text{Final, Hub}} = 4\text{kpsi} \tag{6.53}$$

which represents the smallest value of the hub shaft combination of dimensions.

The geometric factor for the above combination is

$$\beta = \frac{R_{\text{Hub}}^2 + R_{\text{Shaft}}^2}{R_{\text{Hub}}^2 - R_{\text{Shaft}}^2} = 1.67 \tag{6.54}$$

Now we can calculate the amount of interference required:

$$i_{23^\circ\text{C}} = 2R_{\text{Shaft}} \frac{\sigma_{\text{Final}}}{\beta} \left(\frac{\beta + \nu_{\text{Hub}}}{E_{\text{Final, Hub}}} + \frac{1 - \nu_{\text{Shaft}}}{E_{\text{Final, Shaft}}} \right) \tag{6.55}$$

or

$$i_{23^\circ\text{C}} = 0.013\text{ in. }(0.33\text{ mm}) \tag{6.56}$$

The assembly force will be

$$\text{F}_{\text{In}} = 2\pi\mu h R_{\text{Shaft}} \sigma_{\text{Final}} = 594\text{ lbf }(2.6\text{ kN}) \tag{6.57}$$

and the moment that the assembly is able to carry will be

$$M = \text{F}_{\text{In}} R_{\text{Shaft}} = 148\text{ in.·lbf }(16.7\text{ N·m}) \tag{6.58}$$

Results evaluation

As a result of this option mode, the maximum interference fit within the safety-factor range selected was determined. Also, the maximum torque that can be transmitted and the press-in force necessary for assembly were established.

Alternating input data for material properties and geometry variables, the outcome presents a wide choice of tolerance and material combinations.

6.10.2 Case B

For this case we have the following option mode as described in detail in Table 6.1. The second option mode considers an assembly operating at a temperature of 93°C. This time we can also assume that the operating temperature will be reached within the time window, time = 0.

The unknown parameters are the amount of total interference required for the assembly, the insertion force needed to assemble the two parts, and the moment that the assembly will be able to carry.

The known parameters are

Shaft radius	$R_S = 0.25$ in. $(6.35$ mm$)$
Hub outside radius	$R_H = 0.5$ in. $(12.7$ mm$)$
Hub height	$h = 0.315$ in. $(8$ mm$)$
Coefficient of friction	$\mu = 0.3$
Shaft Poisson's ratio	$\nu_S = 0.38$
Hub Poisson's ratio	$\nu_H = 0.40$
Hub secant modulus (time = 0; 93°C)	$E_{S_{\text{Design, Hub}}} = 148$ kpsi
Shaft secant modulus (time = 0; 93°C)	$E_{S_{\text{Design, Shaft}}} = 293$ kpsi
Final stress level, hub (time = 0; 93°C)	$\sigma_{\text{Final, Hub}} = 2.1$ kpsi
Coefficient of thermal expansion, hub	$\alpha_{\text{Hub}} = 3.1\times10^{-5}$ in./in./°F
Coefficient of thermal expansion, shaft	$\alpha_{\text{Shaft}} = 7.1\times10^{-5}$ in./in./°F

The amount of interference for the high-temperature case is

$$i_{93℃} = i_{23℃} + 2R_{\text{Shaft}}\Delta T\left(\alpha_{\text{Shaft}} - \alpha_{\text{Hub}}\right) = 0.008 \text{ in. } (0.2 \text{ mm}) \tag{6.59}$$

If the assembly is to be dismantled at this temperature, the pull-out force is

$$\text{F}_{\text{Out}} = 2\pi\mu h R_{\text{Shaft}}\sigma_{\text{Final}} = 312 \text{ lbf } \left(1.4 \text{ kN}\right) \tag{6.60}$$

The torque the assembly is capable of transmitting at this temperature is

$$M = \text{F}_{\text{Out}} R_{\text{Shaft}} = 78 \text{ in.lbf } \left(8.8 \text{ N.m}\right) \tag{6.61}$$

Results evaluation

This case is of particular interest when the hub's coefficient of thermal expansion/contraction is higher than that of the shaft ($\alpha_{\text{Hub}} > \alpha_{\text{Shaft}}$). The hub's final stress level decreases at the high-temperature condition, and the press fit can become loose due to the thermal expansion.

Also of interest is the case of the hub's coefficient of thermal expansion being less than that of the shaft ($\alpha_{\text{Hub}} < \alpha_{\text{Shaft}}$). Due to the rise in temperature, we can say that the tangential stress, which is the contact pressure, multiplied by the geometric factor β, will equal the yield stress level.

6.10.3 Case C

For this case we have the following option mode as described in detail in Table 6.1. The third option mode considers an assembly operating at a temperature of 23°C. The operating temperature will not be reached within the time window time = end.

The unknown parameters are the amount of total interference required for the assembly, the insertion force needed to assemble the two parts, and the moment that the assembly will be able to carry.

The known parameters are

Shaft radius $R_S = 0.25$ in. (6.35 mm)

Hub outside radius $R_H = 0.5$ in. (12.7 mm)

Hub height $h = 0.315$ in. (8 mm)

Coefficient of friction $\mu = 0.3$

Shaft Poisson's ratio $\nu_S = 0.38$

Hub Poisson's ratio $\nu_H = 0.40$

Hub secant modulus (time = end; 23°C) $E_{S_{\text{Design, Hub}}} = 205$ kpsi

Shaft secant modulus (time = end; 23°C) $E_{S_{\text{Design, Shaft}}} = 373$ kpsi

Final stress level, hub (time = end; 23°C) $\sigma_{\text{Final, Hub}} = 4$ kpsi

Coefficient of thermal expansion, hub $\alpha_{\text{Hub}} = 3.1 \times 10^{-5}$ in./in./°F

Coefficient of thermal expansion, shaft $\alpha_{\text{Shaft}} = 7.1 \times 10^{-5}$ in./in./°F

The first unknown we solve for is the amount of contact pressure present in the assembly at this temperature:

$$p_c = \frac{i_{23°C}}{2R_{\text{Shaft}}\left(\dfrac{\beta + \nu_{\text{Hub}}}{E_{\text{Final, Hub}}}\right)} \tag{6.62}$$

or

$$p_c = 2.212 \text{ lbf (9.8 kN)} \tag{6.63}$$

Then the force required to disassemble the two parts is

$$F_{\text{Out}} = 2\pi\mu h R_{\text{Shaft}} \sigma_{\text{Final}} = 328 \text{ lbf (1.46 kN)} \tag{6.64}$$

and the transmitted torque at temperature is

$$M = F_{\text{Out}} R_{\text{Shaft}} = 82 \text{ in.·lbf (9.3 N·m)} \tag{6.65}$$

Results evaluation

Case C can be seen as a shelf-time depreciation that might apply to a spare assembly not in use. Material relaxation occurs and the resulting torque diminishes. It is of special interest for applications where the shelf life of a product needs to be estimated.

The procedure is not limited to press fits. It can also be effective in determining how components and products will respond to long periods of storage.

6.10.4 Case D

For this case we have the following option mode as described in detail in Table 6.1. The last option mode we will examine here involves an assembly operating at a temperature of 93 °C. The operating temperature will not be reached within the time window time = end.

The unknown parameters are the amount of contact pressure present at the hub/shaft interface in the assembly, the insertion force needed to assemble the two parts, and the moment that the assembly will be able to carry.

The know parameters are

Shaft radius $R_S = 0.25\text{ in.}\,(6.35\text{ mm})$

Hub outside radius $R_H = 0.5\text{ in.}\,(12.7\text{ mm})$

Hub height $h = 0.315\text{ in.}\,(8\text{ mm})$

Coefficient of friction $\mu = 0.3$

Shaft Poisson's ratio $\nu_S = 0.38$

Hub Poisson's ratio $\nu_H = 0.40$

Hub secant modulus (time = end; 93 °C) $E_{S_{\text{Design, Hub}}} = 120\text{ kpsi}$

Shaft secant modulus (time = end; 93 °C) $E_{S_{\text{Design, Shaft}}} = 251\text{ kpsi}$

Final stress level hub (time = end; 93 °C) $\sigma_{\text{Final, Hub}} = 2.1\text{ kpsi}$

Coefficient of thermal expansion, hub $\alpha_{\text{Hub}} = 3.1\times10^{-5}\text{ in./in./°F}$

Coefficient of thermal expansion, shaft $\alpha_{\text{Shaft}} = 7.1\times10^{-5}\text{ in./in./°F}$

The contact pressure present at the hub/shaft interface after 5,000 hours of use at elevated temperature is

$$p_c = \frac{i_{23^\circ\text{C}} + 2R_{\text{Shaft}}\Delta T\left(\alpha_{\text{Shaft}} - \alpha_{\text{Hub}}\right)}{2R_{\text{Shaft}}\left(\dfrac{\beta + \nu_{\text{Hub}}}{E_{S_{\text{Final, Hub}}}}\right)} \tag{6.66}$$

or

$$p_c = 1{,}051\text{ lbf (4.7 kN)} \tag{6.67}$$

The pull-out force required to dismantle the assembly after 5,000 hours of use at high temperature is

$$F_{Out} = 2\pi\mu h R_{Shaft}\sigma_{Final} = 151 \text{ lbf } (0.7 \text{ kN}) \tag{6.68}$$

and the transmitted torque the assembly is capable of handling at this condition is

$$M = F_{Out} R_{Shaft} = 32 \text{ in.·lbf } (4.3 \text{ N·m}) \tag{6.69}$$

Results evaluation

This is considered the "worst case" of thermal stress relaxation. It provides the remaining torque that can be transmitted at the end of the product cycle and the pull-out force. If the value of the transmitted moment falls out of the predetermined torque range, then the material and/or geometry can be changed.

When this happens it is necessary to perform iterative calculations in order to select the best possible material and geometry combination.

6.11 Case History: Metal Ball Bearing and Plastic Hub

This second case history will look at the interrelations between a metal part and a plastic component. The plastic component is an upper intake manifold (UIM) made of polyamide (PA) 6,6, 35% glass-reinforced polymer. The metal part is a ball bearing that initially was selected to be assembled by means of interference (press) fit to the UIM.

Before detailing the design algorithm, we will present over the next few pages the novel process by which the UIM was produced.

6.11.1 Fusible Core Injection Molding

Fusible core injection molding, also known as lost core injection molding, allows the production of three-dimensional hollow shapes of plastic by injection molding. The process requires one or more cores cast of a rather soft metal alloy (with low melting point temperature).

Besides the injection-molding machine, a core casting station is also used. Complex plastic components require multiple cores. Some metal cores are cast from the bottom up to eliminate flash. For these cases the core three-dimensional geometry needs to be oriented into the tool so the gate is located in the lowest possible place.

Vents are positioned in areas that will fill last (top) so that no air entrapment is possible during filling. When more than one gate is necessary to produce each core, precautions must be taken to ensure no splashing occurs during the liquid metal pouring. Any hot spots in the core tool caused by rib locations or sharp corners will adversely affect the casting cycle time. Also, flashing of the core must be avoided. Loose core flakes present in the plastic components will produce reject parts.

Once the cores are manufactured, an assembly operation is necessary for complex shapes. Cores are assembled with machine screws or bolts. During the assembly operation, metal shaving may take place when steel tools touch the cores. It should be avoided.

The cores pass through an oven to raise their temperature from ambient (23°C) to somewhere from 40°C to 80°C. A pick-and-place robot arm is used for all operations. Cores have considerable weight: from 5 kg (10 lb.) to more than 100 kg (200 lb.), depending on the design and size of the plastic parts.

Cores are placed into the vertical injection-molding tool by the robot arm. The polymer is then molded over the metal core. The melting-point temperature of the metal core is 138°C. Processing temperature of the polymer during the injection-molding process is between 295°C and 305°C. Poor gate location creates a thermal energy concentration that could melt locally the metal core. Another important aspect is that the cores must be well supported at specific locations to prevent them from deforming (shifting) during injection of the polymer over them.

Core shifting can be accounted for through analysis. Initially a flow analysis is conducted for the plastic component. The result will provide the pressure gradient to which the cores are exposed during overmolding of the polymer. Then a structural model of the core is created and subjected to the pressure gradient obtained earlier in the flow analysis. Subsequently the structural model, with its applied boundary conditions and loads, is passed through an FEA code. By applying a pressure gradient on the metal core through a finite element analysis, the amount and direction of deflection induced by the polymer injection is predicted. The component gate location, gate size, and filling patterns are crucial in predicting and minimizing the core shifting. A proper flow analysis could suggest a good gate location, its size, the last area to fill (predicting where the venting needs to be placed), the fiber orientation, and weld line positions.

After ejection from the tool, once the injection-molding cycle time is accomplished, the plastic component with the metal core inside is placed in a melt-out station. There are several ways of melting the metal cores. The first uses a hot oil van. For a low melting temperature alloy, the representative melt-out temperature is 160°C. The complete melt-out of the core is usually accomplished within 30 to 60 minutes. Another type of melt-out station is based on an induction-heated oven. This process is much faster. Within two to three minutes the melt-out is achieved. When a hot oil van is employed, the melted metal alloy is separated through gravitation from the hot oil. Then it is drained, captured, and returned to the core casting station to be reutilized. A loss of 0.01 to 0.1% of the melt-out is typical. The metal alloy used for cores is fairly expensive, about $20/lb.

The hollow plastic part obtained is then passed through a washing station that contains a glycol solution. A rinsing operation and a drying operation then follow. All of these steps are necessary so all of the melt-out still left in the part is completely removed.

The fusible core injection molding has been in use since the mid-seventies. It was developed for automotive applications such as air intake manifolds. The process is used today for other components such as aerodynamic bicycle wheels.

6.11.2 Upper Intake Manifold Background

Fiat Chrysler Automobile Group developed a new high-output (HO), 3.5-liter V6 engine, which was designed to compete with heavier, less efficient V8 engines. The HO unit produces 179 kW of power at 6,600 rpm and 329 N·m of torque at 4,000 rpm while providing a compression ratio of 9.1:1. This engine is used for the Chrysler 300M, LHS, and Eagle Vision vehicles.

A high output per liter resulted from maximizing all air-flow systems through extensive use of computer simulations. The engine was designed with a large throttle body, large intake valves, and a large throat area to support the high-flow intake port and chamber. Exhaust valve size and exhaust port, manifold, and catalytic converter entrance shapes were optimized for maximum flow exiting the engine. The diameter of the exhaust system was increased to decrease back pressure and allow more outward flow.

The HO 3.5-liter engine has a three-plenum intake manifold design with short runner valves and a manifold-tuning valve to regulate air flow and a higher lift camshaft for increased air flow.

By managing stiffness, using increased engine isolation, and by properly balancing the rotating parts, the engine noise, vibration, and harshness (also known as NVH) were significantly minimized. Two weak points in powertrain design creating NVH are the joint between the engine and the transmission and the upper intake manifold (UIM). Both this joint and the inside runner surface roughness of the UIM can originate unwanted noise. A significant development in the 3.5-liter engine, in terms of noise isolation, was a uniquely designed UIM. The noise is minimized because runners have a very smooth surface finish. This was made possible by using polyamide 6,6, 35% glass reinforced, as the UIM material instead of a typical aluminum alloy.

Figure 6.18 (a) High-output, 3.5-liter 60° V-6 engine: displacement: 3518 cm^3; bore × stroke: 96 × 81 mm; fuel injection: sequential multiport (SMPI), electronic; engine construction: aluminum block, semipermanent mold aluminum heads, PA 6,6 35% glass-reinforced upper intake manifold; compression ratio: 9.1:1; power: 179 kW at 6,600 rpm; torque: 329 N·m at 4,000 rpm; specific output (kW/L): 53; rpm, red line: 6,800 rpm; fuel type: unleaded midgrade, 89 octane (R+M)/2; emission controls: three-way catalyst, heated oxygen sensors, and electric EGR (exhaust gas recirculation); (b) computer solid model of the same engine *(Courtesy of Fiat Chrysler Automobile Group)*

There are several reasons why the thermoplastic polymer is used. The smooth surface of the inside walls of the plastic UIM, compared to those of an aluminum-cast UIM, increase the maximum engine power by 1.5 to 2%. Better engine performance, by 0.5 to 1%, especially noticeable after hot starts and at maximum power condition, is possible because the polymer's low thermal conductivity does not require high air temperature passing through the intake. A UIM made of polymers reduces the weight of the component up to 50% versus aluminum. Substantial cost reductions varying from 20 to 50% are possible when compared with aluminum UIM manufactured through casting and then machining. Polymer UIMs allow more design flexibility through integration of various features such as throttle body, quick connects, fuel rails, and so on.

From a manufacturing perspective, injection-molding tools have a longer life expectancy when compared with casting tools. Also, the fusible core injection molding allows more design freedom for overall component packaging and for optimum air flow geometry, such as runner location and its cross section geometry. Single-piece design of the manifold provides a higher structural integrity, necessary to overcome vibrations and burst pressures, when compared with manifolds also made of polymeric materials and assembled through welding techniques such as vibration welding.

One of the UIM assembly components is the short runner valve shaft (SRVS). During the engine operation, the SRVS activates six short runner valve plates (SRVP) mounted on it. Through their rotation, SRVPs moderate the amount of gasoline and air mixture that passes through the manifold. There are seven holes with a minimum diameter of 8.10 mm and a maximum diameter of 8.30 mm that are machined into the manifold to assemble the SRVS to it. The above tolerance will allow an overall leakage of 100 cc/min (cubic centimeters per minute) at an air pressure of 0.138 MPa, which is acceptable for this design.

Figure 6.19 Cross section of the V6 3.5-L engine assembly, including a cross section of the upper intake manifold *(Courtesy of Fiat Chrysler Automobile Group)*

The SRVS is oriented parallel to the longitudinal axis of the car. Towards the front end of the car (where the first of the seven holes is located) there were two possibilities to choose from in securing the free end of the shaft to the manifold. One option was to use a metal ball bearing (BB) with its outer diameter (OD) press fit against the manifold and its inner diameter (ID) against the SRVS. The other option was not to use a press fit BB and to allow the SRVS to be supported only by the plastic wall of the manifold, relying only on the resin's ability to act as a bearing surface during the SRVS rotation when the engine is running.

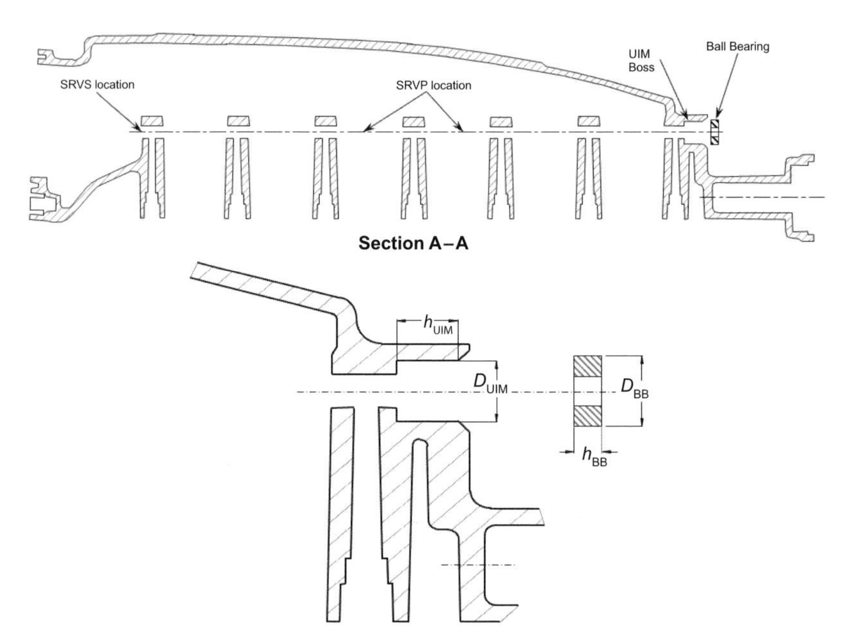

Figure 6.20 Cross section through the upper intake manifold at the short runner valve shaft *(Courtesy of Fiat Chrysler Automobile Group and Borg-Warner Automotive)*

The design algorithm detailed below will verify the compatibility of using a press fit ball bearing in this assembly to secure the SRVS at one end.

6.11.3 Design Algorithm

The plastic upper intake manifold (UIM) is made of BASF polyamide 6,6, 35% glass-reinforced polymer (Ultramid® A3HG7Q17). The ball bearing (BB) to be assembled by press fitting into the UIM has an OD of 16 mm and a height of 5 mm. The BB locates the SRVS in proper position, which in turn activates the SRVP.

To ensure proper assembly integrity the UIM needs to be machined. Over the next few pages we will determine what is required so the plastic UIM and the metal BB will perform over a temperature range from −40°C to 118°C for 5,000 hours.

Polyamide polymers, of which the UIM is made, are hygroscopic materials. There are two moisture conditions to which the UIM component is exposed. They are 50% relative humidity (or 2% water absorption by weight) and dry-as-molded.

The boss, in which the BB is to be press fitted, has a diameter after the molding process that varies between 14.4 mm and 14.5 mm.

Notations:

D_{BB}	BB diameter
D_{UIM}	UIM diameter
h_{BB}	BB height
μ	Coefficient of friction, metal to plastic
ν_{BB}	BB, Poisson's ratio
ν_{UIM}	UIM, Poisson's ratio
α_{BB}	BB, coefficient of linear thermal expansion
α_{UIM}	UIM, coefficient of linear thermal expansion
σ_{BB}	BB, tensile stress
$\sigma_{Y\,DAM}$	UIM, yield stress, dry-as-molded
$\sigma_{Y\,50\%RH}$	UIM, yield stress, 50% relative humidity
$\sigma_{D\,DAM}$	UIM, design stress, dry-as-molded
$\sigma_{D50\%RH}$	UIM, design stress, 50% relative humidity
E_{BB}	BB, Young's modulus
$E_{S_{DAM,UIM}}$	UIM, yield secant modulus, dry-as-molded
$E_{S50\%RH,UIM}$	UIM, yield secant modulus, 50% relative humidity
β	Geometric coefficient
n	Safety factor
T	Temperature

6.11.4 Material Properties

In the earlier case history (see Section 6.9), we used stress/strain curves and creep curves provided in a hardcopy form. This time we will use data provided in a computerized form.

Table 6.2 List of CAMPUS Licensees (as of July 2014)

Albis Plastic GmbH
A. Schulman GmbH
Arkema
BASF AG
BASF Polyurethanes

Table 6.2 List of CAMPUS Licensees (as of July 2014) *(continued)*

Bayer Material Science
DSM Engineering Plastics
DuPont Engineering Polymers
EMS Grivory
Evonik Industries
Lanxess AG
Mitsubishi Engineering Plastics Corporation
Momentive
Polyone
Radici Plastics
Rhodia, Member of Solvay Group
Solvay Engineering Plastics
Styrolution
Ticona
Versalis S.p.A.

6.11.4.1 CAMPUS

Computer-aided material preselection by uniform standards (or CAMPUS® for short) is a database that provides comparable data of different polymers from a variety of suppliers and includes a standard set of practical properties considered useful for the material preselection process.

Back in 1987 a cooperative agreement was reached between four German plastics companies–BASF AG, Bayer AG, Hoechst AG (now known as Ticona), and Hüls AG (now known as Evonik)–to develop CAMPUS. The introduction of the first version of CAMPUS took place in 1988 in Baden Baden, Germany. Five years later, two more plastics companies joined the founding members, Dow Chemical and DuPont Engineering Polymers, in the Managing Committee. Global launch of version 3 took place during NPE (National Plastics Exhibition) in 1994 in Chicago. Later the same year a working group started in Japan and successfully introduced it in Japan during IPF '96. In 1996 version 4.0 for Windows was introduced. Now over 19 international resin suppliers are providing technical data on their products with CAMPUS, and more than 130,000 copies have been distributed to end users worldwide (see Table 6.2).

The database is available free of charge to plastics end users directly from the resin producers. It is available in five international languages: English, German, French, Spanish, and Japanese.

CAMPUS provides both single-point data and multipoint data:

- Single-point data includes rheological properties, mechanical properties, thermal properties (including flammability characteristics), electrical properties, and material-specific properties as well as specimen molding conditions, additives, special characteristics, and processing options.
- Multipoint data includes viscosity vs. shear rate data for the recommended processing temperature range, secant modulus vs. strain curves, shear modulus versus temperature curve, tensile stress/strain curves for a wide temperature range, isochronous creep curves at different temperatures and stress levels, and creep modulus vs. time curves for different temperatures and stress levels.

There are a variety of national standards in use worldwide. For example, the American Society for Testing & Materials (ASTM) provides standards in the United States, Deutsches Institut für Normung (DIN) in Germany, British Standards Institution (BSI) in the United Kingdom, Association FranHaise de Normalisation (AFNOR) in France, Asociación Española de Normalización y Certificación (AENOR) in Spain, and Japanese Industrial Standards (JIS) in Japan. There are more than 20,000 grades of resins to choose from in the United States, over 20,000 grades in Europe, and nearly 10,000 grades in Asia/Pacific. The prevailing data for many products lacks sufficient information regarding test conditions and specimen details. Such lack of uniformity in the acquisition and data reporting, combined with lot-to-lot variability and interlaboratory variations, contributes to more frustration among material specifiers and design engineers. The standards used in CAMPUS eliminate the broad choice of specimen geometry, conditions for preparing test specimens, and test variables.

CAMPUS, by embracing five key international standards–ISO 10350, ISO 11403 parts 1 & 2, ISO 3167, and ISO 294–overcomes the prevailing lack of standardization in the acquisition and reporting format for material properties in the plastics industry. Its user base includes Audi, BMW, Bosch, Ford, GM, IBM, Mercedes, Renault spA, Volkswagen, and Volvo. These companies make the database–from CAMPUS participants–available to all their engineers through their computer servers. This was possible because of the program's unique concept that combines a uniform global protocol for acquisition and representation of data with user-friendly software. CAMPUS is one of the few database systems available in the marketplace that provides truly comparable data on polymers from different material suppliers.

6.11.5 Solution

To verify that it is possible to press fit a metal BB into the UIM, eight steps are necessary.

First, we need to calculate the amount of interference fit (IF) required when components are assembled at room temperature.

Next, we will look at how IF will diminish as the temperature gradient rises. When the engine is running, the upper temperature limit the UIM will be exposed to is 118°C. Plastics have a coefficient of linear thermal expansion greater than metals. In this case, for example, BB has a coefficient of linear thermal expansion (CLTE) of $6.5 \cdot 10^{-6}$ mm per degree Celsius, while UIM has a CLTE of $4.45 \cdot 10^{-5}$ mm per degree Celsius.

Table 6.3 Summary of Material Properties and Geometric Dimensions

						Units	Notation	Name
0	0	0	5,000	5,000	5,000	hr	t	time
-40	23	118	-40	23	118	°C	T	temperature
	16					mm	D_{BB}	BB diameter
	22					mm	D_{UIM}	UIM diameter
	5					mm	h_{UIM}	UIM height
	0.4					–	μ_{UIM}	UIM, coefficient of friction
	0.41					–	ν_{BB}	BB, Poisson's ratio
	0.3					–	ν_{UIM}	UIM, Poisson's ratio
	$6.5 \cdot 10^{-6}$					mm/°C	α_{BB}	BB, thermal expansion
	$4.5 \cdot 10^{-5}$					mm/°C	α_{UIM}	UIM, thermal expansion
	$2 \cdot 10^{5}$					MPa	σ_{BB}	BB, stress
264	209	100	60	36	25	MPa	$\sigma_{Y DAM}$	UIM, DAM yield stress

						Units	Notation	Name
236	144	79	38	23	16	MPa	$\sigma_{Y50\%RH}$	UIM, 50% RH yield stress
88	70	33	20	12	8	MPa	σ_{DDAM}	UIM, DAM design stress
79	48	27	13	8	5	MPa	$\sigma_{D50\%RH}$	UIM, 50% RH design stress
207,000	207,000	207,000	207,000	207,000	207,000	MPa	E_{BB}	BB, Young's modulus
8,515	7,491	2,512	5,000	3,600	2,500	MPa	E_{SDAM}	UIM, DAM yield secant modulus
9,516	4,948	2,160	4,800	3,200	1,600	MPa	$E_{S50\%RH}$	UIM, 50% RH yield secant modulus
3.2456	3.2456	3.2456	3.2456	3.2456	3.2456	–	β	geometric coefficient
3	3	3	3	3	3	–	n	safety factor

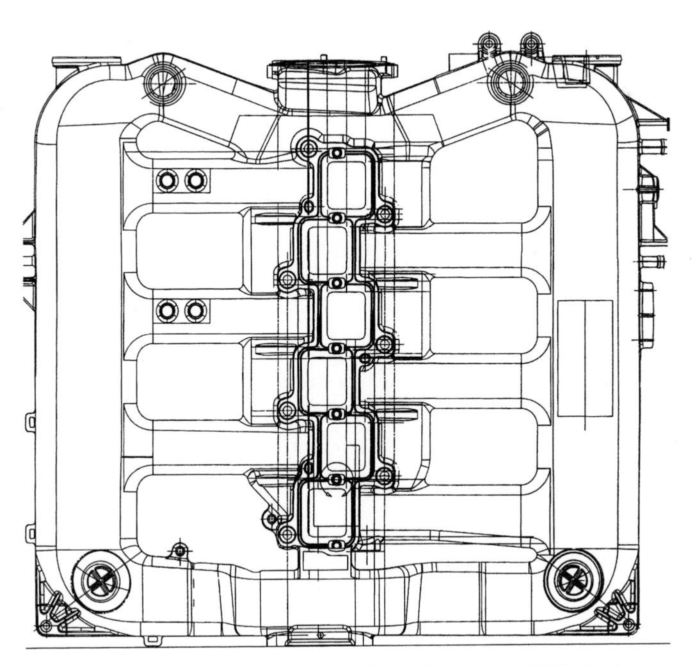

Figure 6.21 2-D drawing of the upper intake manifold *(Courtesy of ETS, Inc.)*

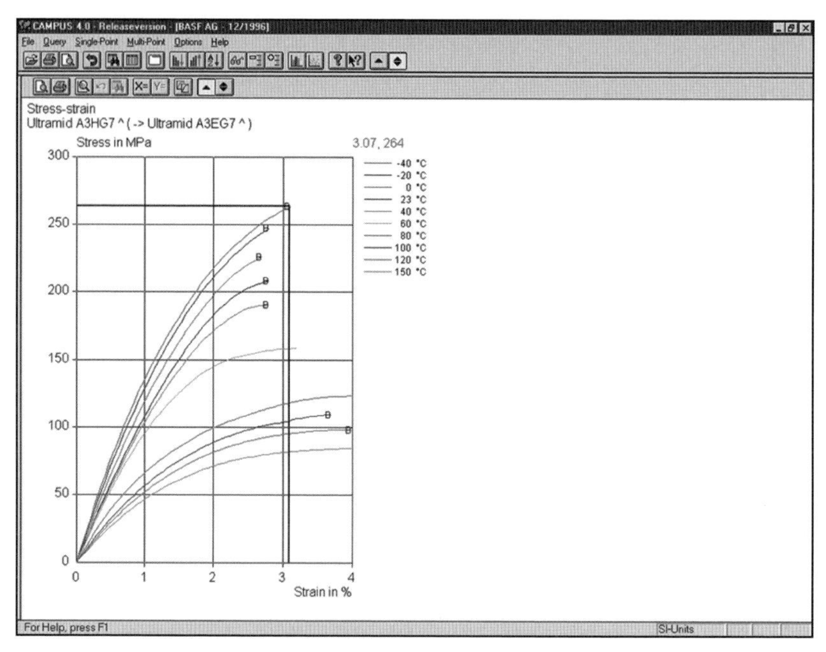

Figure 6.22 Polyamide 6,6, 35% glass reinforced *(BASF Ultramid A3HG7)*, yield strength and secant modulus at –40 °C dry-as-molded, as per CAMPUS version 4.0 for Windows *(Courtesy of BASF Corporation and CAMPUS® Consortium)*

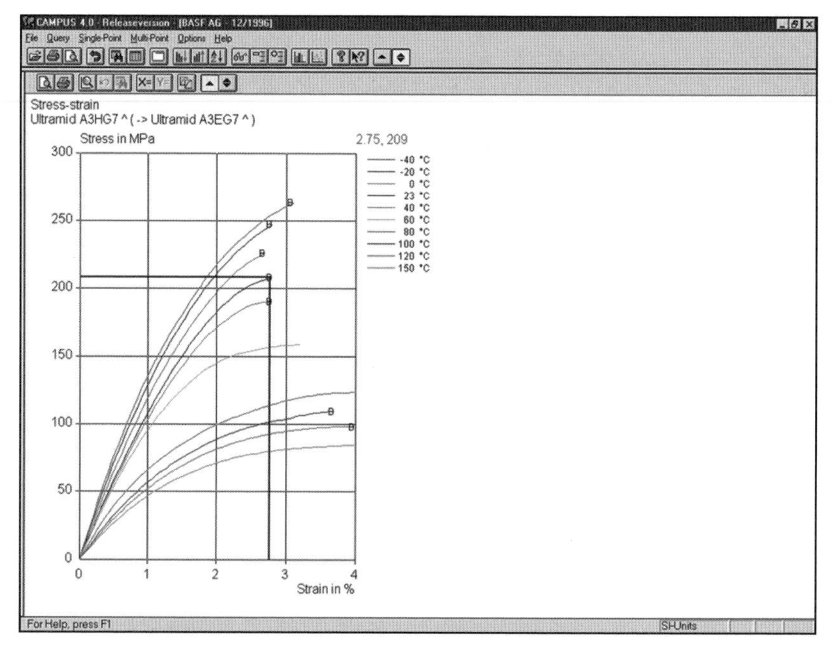

Figure 6.23 Polyamide 6,6, 35% glass reinforced *(BASF Ultramid A3HG7)*, yield strength and secant modulus at 23 °C dry-as-molded, as per CAMPUS version 4.0 for Windows *(Courtesy of BASF Corporation and CAMPUS® Consortium)*

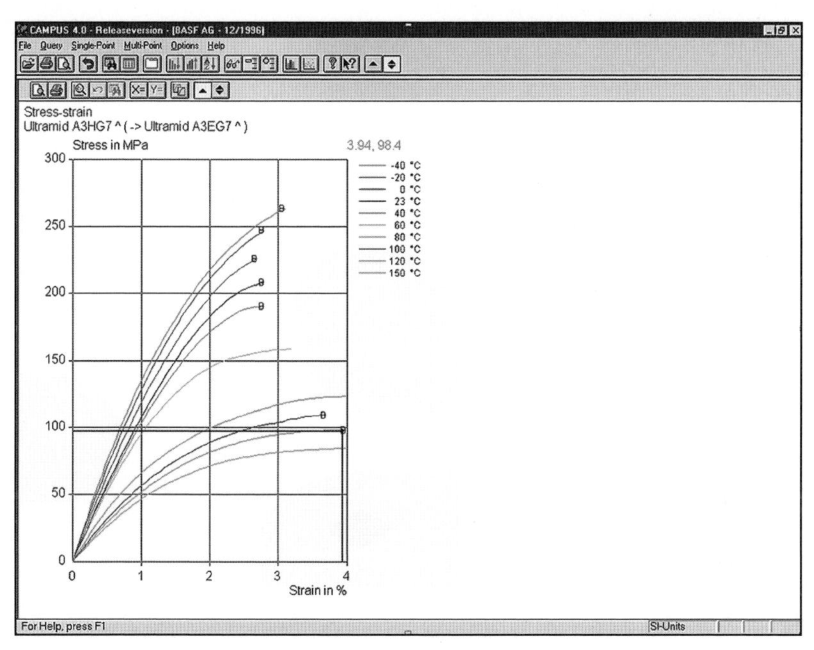

Figure 6.24 Polyamide 6,6, 35% glass reinforced *(BASF Ultramid A3HG7)*, yield strength and secant modulus at 120 dry-as-molded, as per CAMPUS version 4.0 for Windows *(Courtesy of BASF Corporation and CAMPUS® Consortium)*

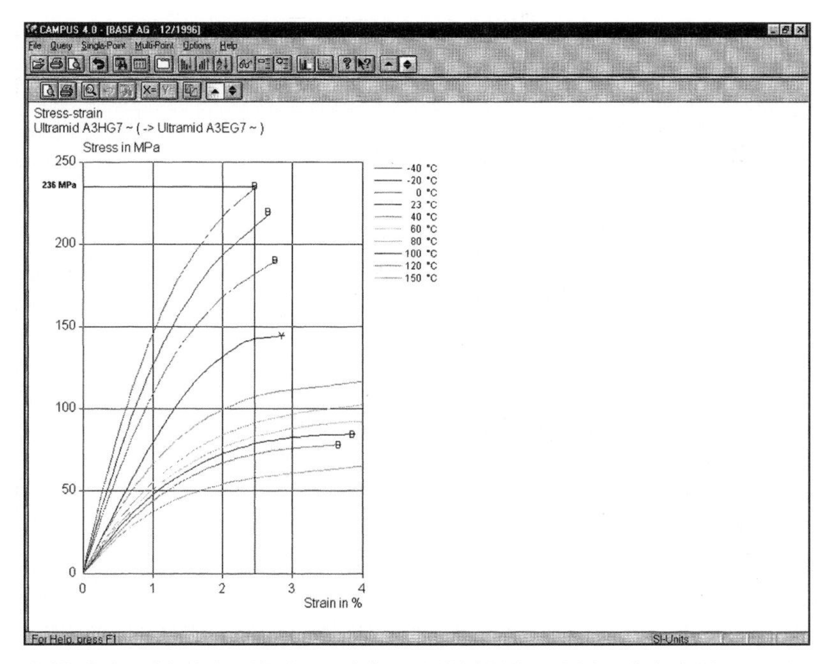

Figure 6.25 Polyamide 6,6, 35% glass reinforced *(BASF Ultramid A3HG7)*, yield strength and secant modulus at –40 °C 50% relative humidity, as per CAMPUS version 4.0 for Windows *(Courtesy of BASF Corporation and CAMPUS® Consortium)*

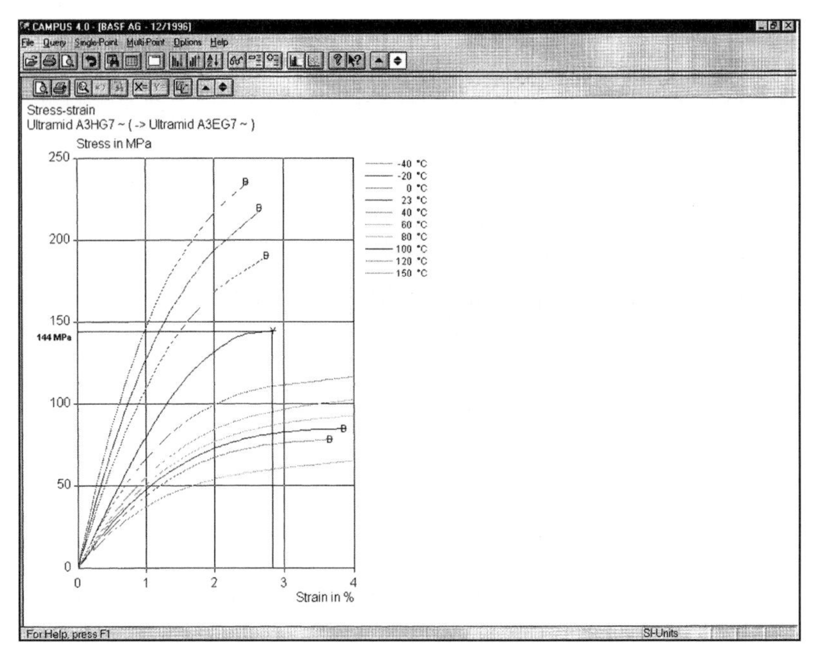

Figure 6.26 Polyamide 6,6, 35% glass reinforced *(BASF Ultramid A3HG7),* yield strength and secant modulus at 23°C relative humidity, as per CAMPUS version 4.0 for Windows *(Courtesy of BASF Corporation and CAMPUS® Consortium)*

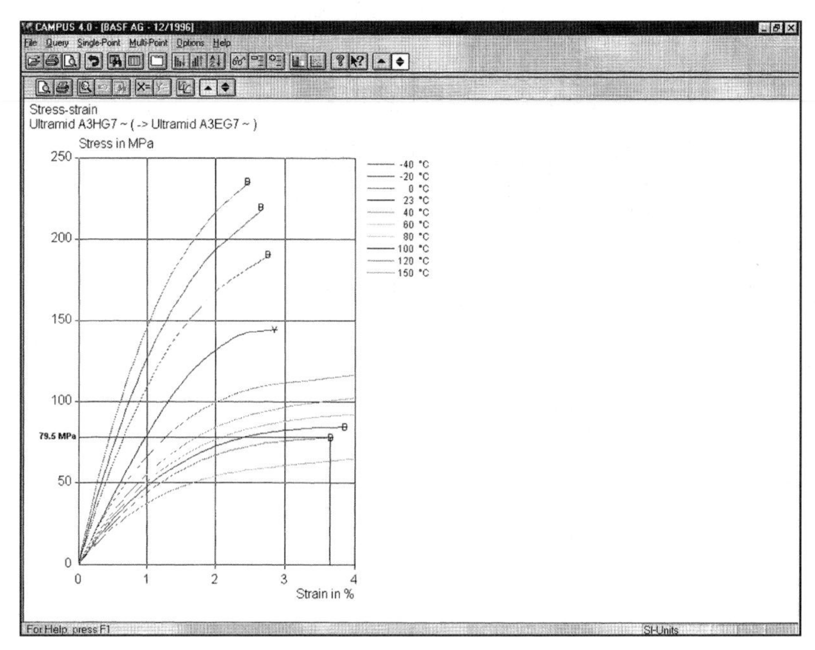

Figure 6.27 Polyamide 6,6, 35% glass reinforced *(BASF Ultramid A3HG7),* yield strength and secant modulus at 120°C relative humidity, as per CAMPUS version 4.0 for Windows *(Courtesy of BASF Corporation and CAMPUS® Consortium)*

The influence of thermal expansion or contraction of this assembly will be accounted for during the third step. As a guarantee that the assembly will perform over the stated temperature gradient range, the thermal expansion influence is incorporated into the IF solution.

Then for the same low-temperature considerations (-40°C) we will verify that the stress present at the interface between BB and UIM will not exceed the yield value of the polymer.

The last three steps will verify that in time (after 5,000 hours) the UIM will still perform as required. Because no stress relaxation curves are available for the polymer, we will use instead its creep curves. The amount of error associated with the lack of proper data is, in general, between plus and minus 15%.

6.11.5.1 Necessary IF at Ambient Temperature

The assembly operation takes place in a plant environment where the typical temperature is 23°C. Polyamides are hygroscopic (moisture-sensitive) resins, both before and after the injection-molding process. Therefore, to account for the necessary IF, two moisture conditions are considered.

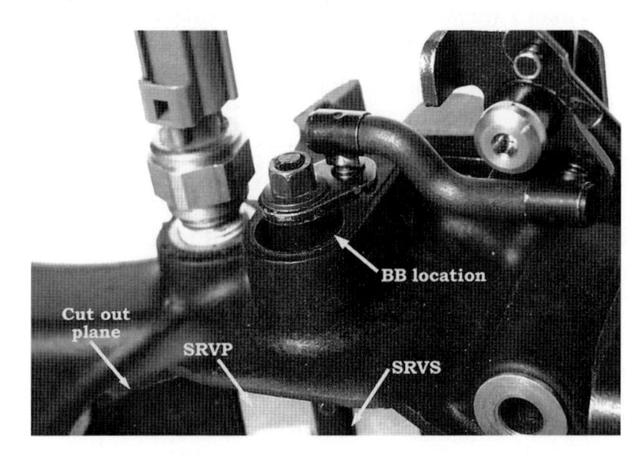

Figure 6.28
Detail of the ball bearing location showing the SRVS shaft and one of the six SRVP *(Courtesy of ETS Inc.)*

The first moisture condition is known as DAM, or dry-as-molded. The assumption is that the polymer is completely dry. In reality this condition is possible only immediately after the part is molded or when a component is annealed in an oven. The formula for interference is

$$i_{23℃,DAM} = D_{UIM} \frac{\sigma_{Y\,DAM,23℃,t=0}}{n\beta} \left(\frac{\beta + \nu_{BB}}{E_{S_{DAM,\,23℃,\,t=0}}} + \frac{1-\nu_{BB}}{E_{BB}} \right) \tag{6.70}$$

Replacing known values in the above equation, IF is

$$i_{23℃,DAM} = 0.1110 \text{ mm} \tag{6.71}$$

The second moisture condition represents a plastic part that has been in storage in open air for 6 months or so before the BB is press fitted into it. During this time the humidity present in the air has been absorbed by the hygroscopic polymer. This condition is known as 50% RH. For this condition the IF formula becomes

$$i_{23℃,50\%RH} = D_{UIM} \frac{\sigma_{Y50\%RH,23℃,t=0}}{n\beta} \left(\frac{\beta + \nu_{BB}}{E_{S_{50\%RH,23℃,t=0}}} + \frac{1-\nu_{BB}}{E_{BB}} \right) \tag{6.72}$$

The IF for a manifold that absorbed moisture from the air (50% RH) is lower:

$$i_{23℃,50\%RH} = 0.0473 \text{ mm} \tag{6.73}$$

Note: A safety factor of three is used to determine polymer's design point (UIM design strength) based on the resin's yield stress. The design strength, being a lower number than yield strength, accounts for imperfections created by the machining operation of the glass-reinforced, injection-molded UIM and the knit (weld) line present in the component.

6.11.5.2 IF Available at 118°C

From the two IF results calculated above we will select only the DAM condition because it has the highest value. When a temperature gradient is present, polymers expand several times more than steel, as was stated earlier. Therefore it is necessary to find out the amount of IF left in the assembly at 118°C using the formula

$$i_{118℃,DAM} = i_{23℃,DAM} + D_{UIM} \Delta T \left(\alpha_{UIM} - \alpha_{BB} \right) \tag{6.74}$$

Again, it is assumed that at this temperature no moisture is present in the air to be absorbed by the resin. Consequently, only the DAM condition is accounted for.

$$i_{118℃,DAM} = 0.0051 \text{ mm} \tag{6.75}$$

The assembly lost, with the rise in temperature, 0.106 mm of its original amount of interference fit present when press fitted at room temperature. Still, a small amount of IF is left in the assembly, but it is very unlikely that under road conditions (proving grounds) or on a shaker-table test that the component will last 5000 hours without becoming loose.

6.11.5.3 IF Verification at -40°C

By adding the necessary IF of 0.111 mm to the value of thermal contraction of the assembly from 23° to -40°C with the formula below,

$$i_{-40°C,DAM} = 0.1110 \text{mm} + D_{UIM} \left[23°\text{C} - (-40°\text{C}) \right] \left(\alpha_{UIM} - \alpha_{BB} \right) \tag{6.76}$$

the total IF at -40°C becomes 0.1806 mm.

Figure 6.29
Detail of SRVS and SRVP through a cut-off plane

6.11.5.4 Verification of Stress Level at –40°C, Time = 0

The amount of IF present in the assembly at –40°C (0.1806 mm) will create a certain stress level that will be present at the interface between the two components, BB and UIM. To calculate its amount we use the formula below:

$$p_{t-40℃,t=0} = \beta p_{UIM,-40℃,t=0} = \beta \frac{i_{-40℃}}{D_{UIM}\left(\dfrac{\beta+\nu_{BB}}{E_{S_{UIM,-40℃,t=0}}} + \dfrac{1-\nu_{BB}}{E_{BB}}\right)} \tag{6.77}$$

By replacing known values in the above equation it is found that UIM will be exposed to a stress level of 84.7 MPa (12,244 psi). When this value is compared with the polymer yield stress of 264 MPa (38,167 psi) for DAM and respectively 236 MPa (34,119 psi) for 50% RH, it is much lower. Therefore, the assembly will do well at this temperature.

6.11.5.5 Stress Level at –40°C, Time = 5,000 h

The assembly has to perform for at least 5,000 hours. During this time the polymer's mechanical properties will change. It is assumed that the metal BB will not change its mechanical properties. Stress relaxation curves for PA 6,6 not being available, creep curves are used instead.

The stress level present at the interface of two components is given by

$$p_{t-40℃,t=5,000} = \beta p_{UIM,-40℃,t=5,000} = \beta \frac{i_{-40℃}}{D_{UIM}\left(\dfrac{\beta+\nu_{BB}}{E_{S_{UIM,-40℃,t=5,000}}} + \dfrac{1-\nu_{BB}}{E_{BB}}\right)} \tag{6.78}$$

We obtain a stress level at 5,000 hours of 49.9 MPa (7,213 psi). Comparing it with the graphically extrapolated values extracted from the CAMPUS database of BASF, the above value is lower than the yield stress of 60 MPa (8,674 psi) for DAM condition. This means that the UIM assembly will be sustained for 5,000 hours if the component is exposed to a dry environment and a cold temperature. For the 50% RH condition the assembly will fail because the stress level present at the interface of the two components is much higher than the yield point of the resin of 37.5 MPa (5,421 psi) for this condition. This means there is a very high probability that UIM will exhibit stress cracks around BB or will take a permanent set (deformation) sometime before the 5,000 hours when the car is parked in a cold and humid climate for long periods of time.

6.11.5.6 Stress Level at 23°C, Time = 5,000 h

The press fit assembly of UIM will perform rather well when the car is parked at ambient temperature for long periods.

$$p_{23℃,t=5{,}000} = \beta p_{UIM,23℃,\ t=5{,}000} = \beta \frac{i_{23℃}}{D_{UIM,23℃,t=5{,}000}\left(\frac{\beta+\nu_{BB}}{E_{S_{UIM,\ 23℃,\ t=5{,}000}}}+\frac{1-\nu_{BB}}{E_{BB}}\right)} \tag{6.79}$$

A stress level of 22.1 MPa (3,196 psi) after 5,000 h will be present at the press fit interface. When compared with the graphically extrapolated values extracted from BASF's CAMPUS material database, the above number is lower than the yield stress of 36 MPa (5,205 psi) for DAM condition. For the 50% RH condition it is almost a wash (22.1 MPa vs. 23 MPa). Therefore, it is likely that in a humid environment the press fit may become loose.

6.11.5.7 Stress Level at 118°C, Time = 5,000 h

When the engine is running, the UIM operating temperature is 118°C. In these conditions the polymer UIM is made of will lose some of its mechanical properties and most likely will take a permanent set (deformation).

$$p_{118℃,t=5{,}000} = \beta p_{\mathrm{UIM},118℃,t=5{,}000} = \beta \frac{i_{118℃}}{D_{\mathrm{UIM},118℃,t=5{,}000}\left(\frac{\beta+\nu_{\mathrm{BB}}}{E_{S_{\mathrm{UIM},\ 118℃,\ t=5{,}000}}}+\frac{1-\nu_{\mathrm{BB}}}{E_{\mathrm{BB}}}\right)} \tag{6.80}$$

The stress level present in the assembly at this temperature is 15.9 MPa (2,324 psi), which is significantly higher than both yield stress levels at DAM of 8.33 MPa (1,205 psi) and at 50% RH of 5.2 MPa (753 psi).

Removing half of the 2% carbon black pigment needed to color the resin necessary for the manifold and replacing it with nigrosin black dye eliminated the need to use the ball bearing. When the glass-reinforced polyamide (PA) is machined to create the proper boss dimension to allow SRVS shaft assembly, the glass fibers are exposed to the surface. While in use, the shaft will wear against the exposed glass fibers in 90° back-and-forth rotations–similar to sanding paper–present in the boss during the engine's life. The 1% nigrosin, when motion is created by the shaft rotation in the machined boss, permeates to the polymer-machined surface, thus forming a lubricated bushing. By adopting this approach, Fiat Chrysler Automobile Group was able, by not using a ball bearing at all and by not employing additional labor to install it, to save around $1.6 million each year of production. The engine using the fusible core, injection-molded upper manifold was manufactured for a total of 9 years.

6.12 Successful Press Fits

Many companies now employ press fits successfully in their products. One of them is a company from Billund, Denmark, founded in 1932 by Kirk Kristiansen, a local carpenter. In 1947, this company bought its first injection-molding machine, and after two years of development, the first Lego plastic toys were released. In 2013 with sales in excess of US $4.6 billion and more than 11,700 employees, Lego Group had become the world's third-largest plastic toy manufacturer.

Over the years, in excess of 500 billion thermoplastic Lego bricks have been produced. All obey the classic design (see Fig. 6.30(c)), in which a very small amount of undercut in the draw direction of the mold core creates enough springiness that a small child can lock and unlock the bricks hundreds of times. Furthermore, a piece from a Lego set from the 1950s will interface with a brand-new set just as well as a new piece. The key of such precision is based on quality systems employed during the injection-molding cycle. Acrylonitrile butadiene styrene (ABS) thermoplastic polymer is heated to a temperature gradient of 232°C until it is fully melted. Then, it is injected into the tool. Depending on which type of brick element is being manufactured, the tool can be mounted in injection-molding presses ranging from 25 to 150 tons. The hold time to cool the ABS inside the tool is an average of seven seconds, after which the part is ejected. The production tools used by Lego Group have tolerances of less than 0.01 mm, representing a reject rate of just 0.0018%. The undercut shown in Fig. 6.30(c) is designed below the tensile yield strength of ABS using a robust safety factor, as discussed earlier in this chapter.

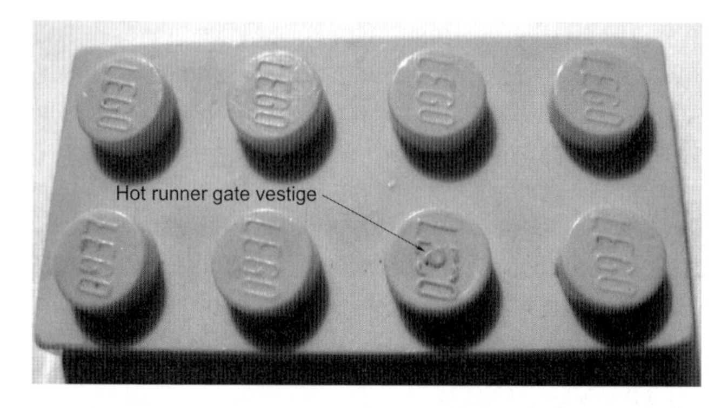

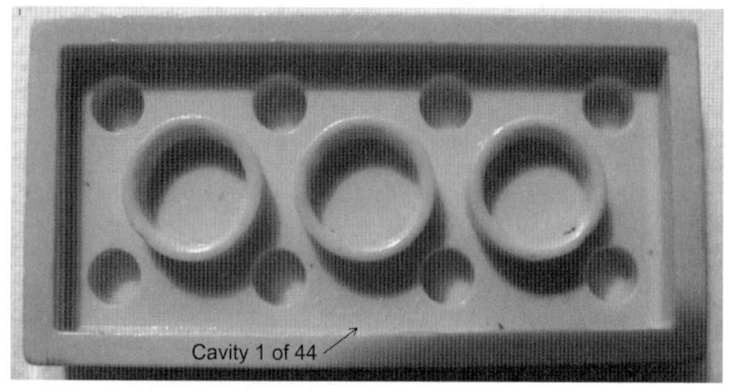

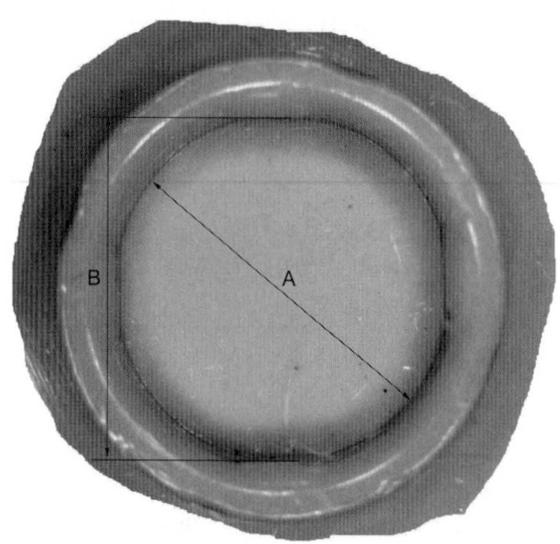

Figure 6.30
Lego brick: (a) top view with the hot runner gate vestige visible, (b) bottom view with imprint showing cavity 1 of 44, and (c) undercut detail, which has provided the press fit springiness over the years

Another company employing press fits successfully is in the medical field: Advanced Cardiovascular Systems (ACS) of Santa Clara, California. Angioplasty is the technique of mechanically widening narrowed or obstructed arteries in patients who have coronary disease as a result of atherosclerosis. An empty and collapsed balloon on a guide wire, known as a balloon catheter, is passed into the narrowed arteries near the patient's heart and then inflated to a fixed size using water pressures some 75 to 500 times normal blood pressure (6 to 20 atmospheres or 6 to 20 bar). The balloon forces expansion of the inner white blood-cell and clot-plaque deposits and the surrounding muscular wall, opening up the blood vessel for improved flow, and the balloon is then deflated and withdrawn.

ACS manufactures an Indeflator (see Fig. 6.31) assembly featuring a pressure gauge necessary to inflate and then deflate the angioplasty balloon.

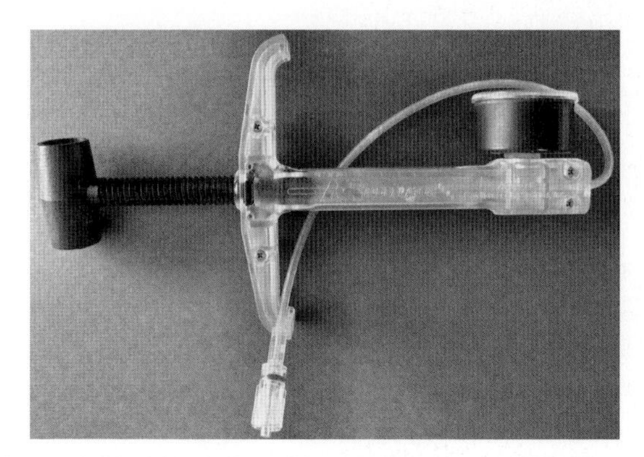

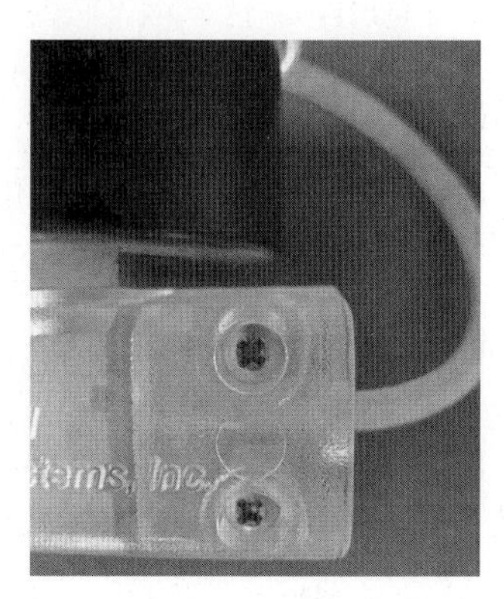

Figure 6.31 (a) Indeflator Plus, made from a medical grade of polycarbonate assembled using four thread-forming screws, (b) phillips screw head detail *(Courtesy ETS, Inc.)*

To improve the cost structure of the product, one of the goals was to replace the four thread-forming screws with press fit design features (see Figs. 6.32(a) and 6.32(b)). This objective was achieved by replacing the screws with four press fit features. Key characteristics of the press fit design were the small, 0.07 mm radial interference between the double-tapered pin on one component, which replaced the screw, and on the mating component, the hexagon, which replaced the round boss in which the thread-forming screw was fastened before. The initial alignment between the two halves of the Indeflator handle parts is achieved with the help of the generous 3° draft angle on the pin (see Fig. 6.32(a)). The handle parts pushed against each other until 80% of the pin length (L = 7.94 mm) is reached; then one point of the pin circumference and one of the sides of the hexagon make contact. The round pin and the hexagon boss are tangent.

The pin touches the hexagon in six tangential points. Then, with the final push, the two parts are press fitted. The initial six tangential contact points between round pin and hexagon-shaped boss become six lines of contact in a 2-D plane or a rectangular tangential contact shape in a 3-D space (see Fig. 6.32(c)). This ensures good grip. If the hole, instead of being hexagonal, was a round boss, the contact would occur all around the circumference, and the assembly would be stress-relaxing and losing its grip shortly. The hexagonal boss provides an elastic grip, which requires separating loads exceeding 100 lbf (approximately 440 N).

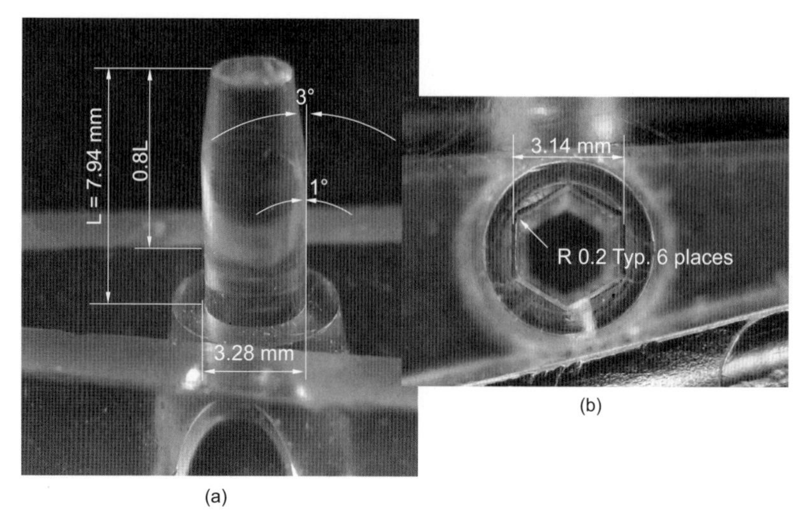

Figure 6.32 (a) Pin detail design, (b) hexagon detail design, and (c) six contact tangential points at 80% of *L* engagement become six lines of contact at 100% of *L* engagement (in 2-D plane); in 3-D, the lines of contact become rectangular contact areas *(Courtesy of ETS, Inc.)*

Finally, another company from Fremont, Michigan, Dura Automotive Systems, also employs press fit design features successfully in its automotive components. Since thermoplastic materials creep or stress-relieve under continued loading, loosening of the press fit–at least to some extent–should be expected. To prevent creep in a thermoplastic component due to the constant load exercised by the bolt head, a bronze bushing was used as a torque limiter (see Fig. 6.33(b)). Once the bolt is fully torqued, the load under the bolt head is carried out by the bushing, thus preventing the thermoplastic polymer from creeping over time and at various temperature gradients while in use. Testing under expected temperature cycles is recommended. The bronze limiter press fitted into a 33% glass-reinforced thermoplastic polyamide part is used in the transmission shifter mechanism for many Ford vehicles.

The key design feature in this case is the use of a curved beam design (see Fig. 6.33(a)). If the compression limiter bushing was press fitted into the plastic component without any radial space (curved beam design), the thermoplastic polyamide, even reinforced 33% with glass fibers, will stress-relax in time, even faster when exposed to temperature gradients, as in the change of summer to winter seasons. The curved beam (Fig. 6.33(c)) should be designed so that when the torque limiter is press fitted, the deformation in the polymer remains elastic, in the preyielding region, as discussed earlier in the chapter.

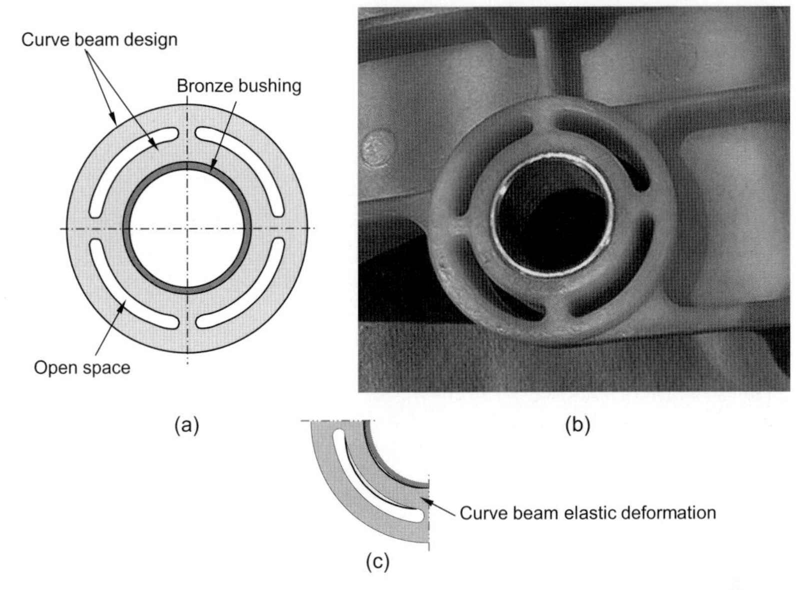

Figure 6.33 (a) Detail design of the compression limiter used in the transmission shifter; (b) actual image of the assembled component; (c) curved beam design detail *(Courtesy ETS, Inc.)*

6.13 Conclusion

There were two options to be considered: 1) using a metal ball bearing press fitted into the thermoplastic air manifold, and 2) relying on the polymer's ability to act as a bearing surface. The detailed design algorithm we reviewed above shows that the best option was to rely on the polymer's ability to act as a bearing surface. For this particular UIM the assembly procedure using a press fit to assemble a ball bearing into UIM is not recommended.

Even with tight machining tolerance of DIA = 15.85 mm to 15.88 mm, the BASF polyamide 6,6, 35% glass-reinforced polymer (Ultramid® A3HG7Q17) used for the 3.5-liter V6, 24-valve upper intake manifold, will be overstressed at low temperature. A high operating temperature could allow the ball bearing to become loose.

7 Living Hinges

7.1 Introduction

In many thermoplastic part designs, it is advantageous to create integral connecting members between parts that undergo relative movement or for parts to be made in one tool and then assembled. Such designs, using flexural soft materials, are commonly referred to as *living hinges.*

A living hinge or molded-in hinge is a very thin portion of polymeric material that bridges two heavy walls. It provides the ability for the part to open and close or flex without the use of a mechanical hinge.

The most common materials used for living hinge designs are polypropylene (PP) and polyethylene (PE). These materials are used because of their ability to flex many times without breaking, as well as their low cost and ease of processing. Most other plastic materials show superior mechanical, thermal, chemical, and electrical properties than do PP and PE, but they are seldom used in living hinge applications due to their low flexing capability. Polypropylene and polyethylene can flex a million cycles before failure; all other plastics only flex up to a few thousand times.

This chapter will cover a series of methods available to design and product engineers for use in living hinge design. These methods fall into two categories: designs for use with PP and PE, which are covered in Section 7.2; and designs for other materials, which are found in Section 7.3. These two categories comprise all living hinge design requirements in the marketplace.

The significance of the second category is that it gives three choices: a fully elastic hinge, which could flex for several thousand cycles; a fully plastic hinge capable of just a few flexing cycles; or a combination of elastic and plastic behavior, which is able to sustain several hundred flexing cycles. This gives the best mathematical model for approximation of the real material behavior in various applications.

7.2 Classic Design for PP and PE

The most common design structure for polypropylene and polyethylene living hinges can best be described as a bridge with a recess in the upper portion (Fig. 7.1(a)) and an arc in the lower portion. The use of these materials is only restricted by their physical limitations.

When the hinge is in the closed position (Fig. 7.1(b)), the recess helps create the necessary radius to prevent a possible crack that would eventually break the hinge. The arc in the lower portion of the hinge is designed to provide the orientation of the polymer molecules and enable flexing up to a million cycles.

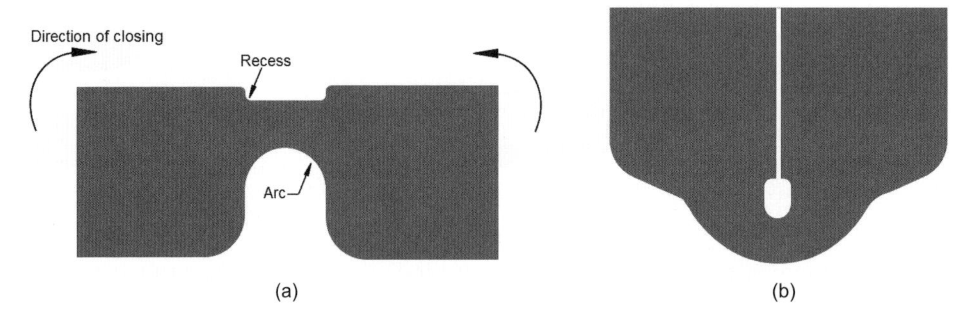

Figure 7.1 Living hinge design for PP and PE: (a) before, and (b) after bending 180°

To achieve the capacity for such a large number of cycles, a couple of quick flexes of the hinge should take place immediately after the ejection of the molded part from the tool. The still-high temperature of the part, combined with the few quick flexes applied by the molding operator, orients the material in the hinge area. This is true of polypropylene in all cases and polyethylene in certain cases.

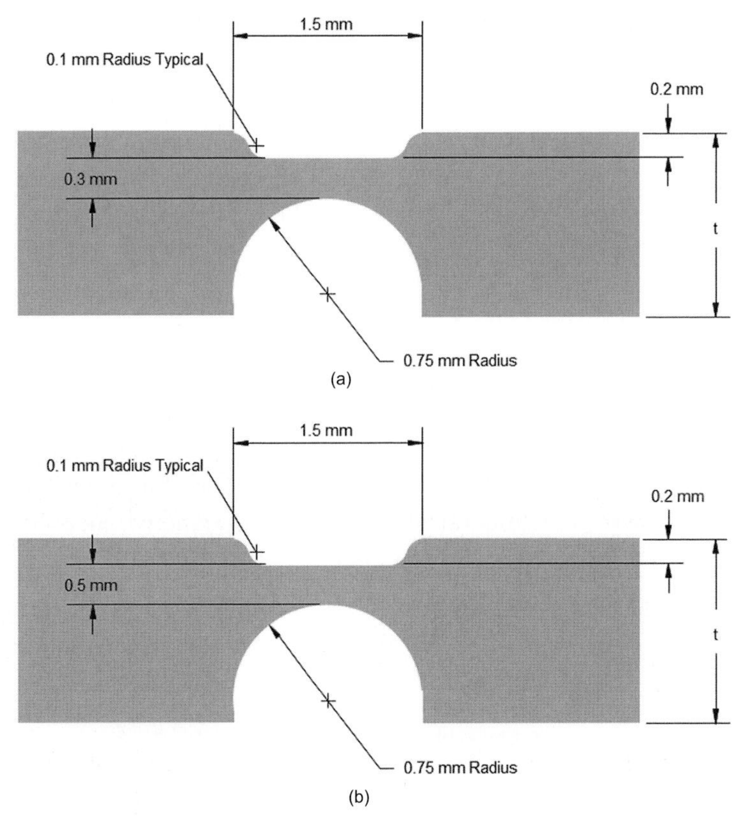

Figure 7.2 Typical values of living hinge design for PP and PE: (a) most applications; (b) maximum thickness

7.3 Common Living Hinge Design

For materials other than polypropylene and polyethylene, designs very similar to the one shown in Fig. 7.3 are still used in today's engineering community.

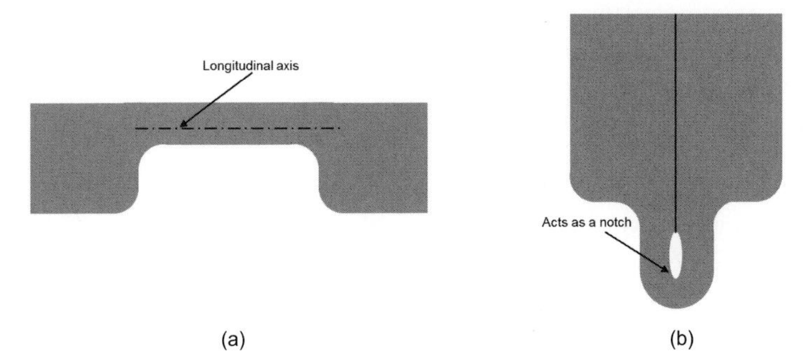

Figure 7.3 Most common living hinge design: (a) before, and (b) after bending 180°

This is unfortunate because, as Fig. 7.3(b) shows, this kind of design develops a stress concentration area in the closed position.

Figure 7.3(b) shows a cross section of the hinge and the neutral axis, which is described later. Once the hinge is closed, the material will be in compression above the neutral axis and in tension below it. This is an example of a 180° closing angle. Because there is no recess such as in Fig. 7.1, the outer upper portion of the hinge will act as a notch and force the hinge to break in tension.

7.4 Basic Design for Engineering Plastics

A more effective design for materials other than polypropylene and polyethylene is shown in Fig. 7.4. The only difference between this design and the previous one is the inclusion of the recess, which helps lower the stress concentration caused by the compression of the fibers above the neutral axis.

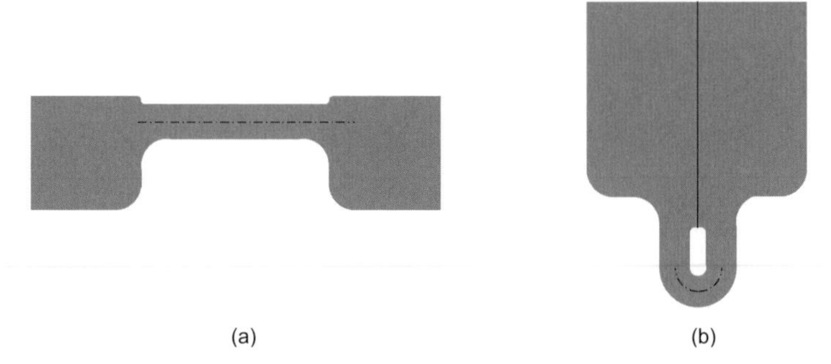

Figure 7.4 Recommended living hinge design with the neutral axis in the center: (a) before, and (b) at bending angle of 180°

Figure 7.4(b) shows the smooth radius the fibers in compression will follow. For the remainder of this chapter, we will develop and examine a mathematical model of this design and its application.

7.5 Living Hinge Design Analysis

The analysis begins with the consideration of a portion of the living hinge in closed position (see Figs. 7.5 and 7.6) with having a 180° closing angle.

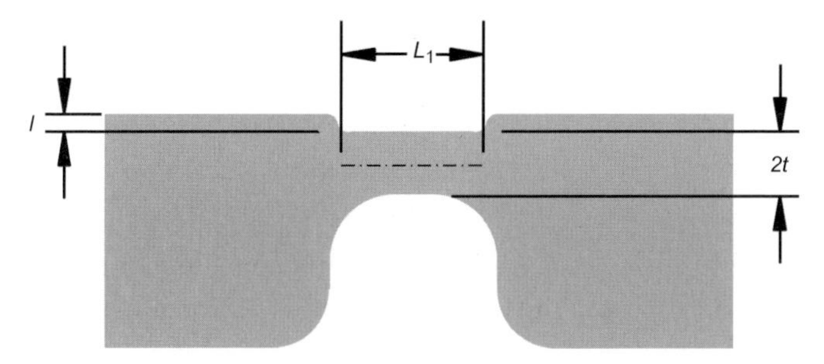

Figure 7.5 Cross section through the living hinge as molded, showing its principal dimensions

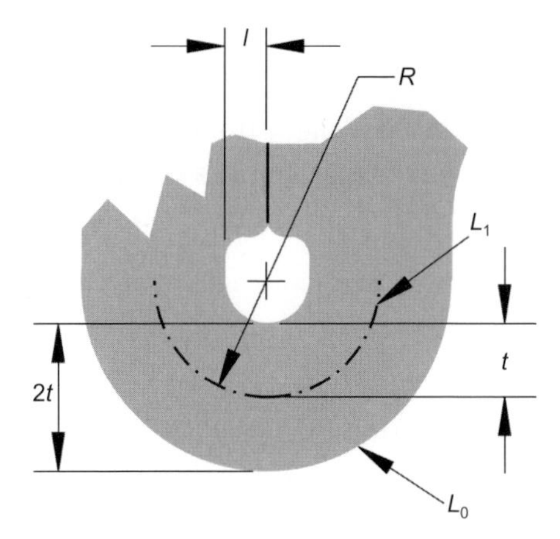

Figure 7.6
Cross section of a portion of a living hinge design and its neutral axis position

7.5.1 Elastic Strain Due to Bending

7.5.1.1 Assumptions

A. The hinge bends in a circle and the neutral axis coincides with the longitudinal hinge axis.

B. The outer fiber is under maximum tension.

C. The inner fiber is under maximum compression.

D. When the tension stress reaches the yield point, the hinge will fail by the design criteria employed.

We will use the following notations:

L_1 = length of the neutral axis of the living hinge

t = half of the hinge thickness

l = recess depth

R = hinge radius

L_0 = length of the outer lower fiber of the hinge

7.5.1.2 Geometric Conditions

The length of the neutral axis is the perimeter of the semicircle (hinge in closed position):

$$L_1 = \pi R \tag{7.1}$$

Based on the above assumptions, L_0 is calculated as being the length of a semicircle, which has a radius comprising the hinge radius and half the hinge thickness. Therefore:

$$L_0 = (R + t)\pi \tag{7.2}$$

7.5.1.3 Strain Due to Bending

Once the hinge is closed, the lower outer fiber changes its length from L_1 to L_0. The ratio between the variance of change of the lower outer fiber and the neutral axis (assuming the neutral axis does not change in length) provides the value of the strain due to bending (7.3).

$$\varepsilon_{\text{Bending}} = \frac{\Delta L_0}{L_1} = \frac{L_0 - L_1}{L_1} = \frac{\pi(t + R) - \pi R}{\pi R} = \frac{t}{R} \tag{7.3}$$

Because

$$R = \frac{L_1}{\pi} \tag{7.4}$$

the final form for the strain due to bending is

$$\varepsilon_{\text{Bending}} = \frac{\pi t}{L_1} \tag{7.5}$$

7.5.1.4 Stress Due to Bending

Stress due to bending is determined from the calculated strain. We have assumed pure bending where the neutral axis remains in the center. In practice, the neutral axis changes position above or below its initial position when tension or compression is applied. The amount of shift can be determined relative to the amount of tension stress/strain. Lab tests show this assumption to be acceptable in a majority of cases.

$$\sigma_{\text{Bending}} = \varepsilon_{\text{Bending}} E = \frac{\pi t}{L_1} E \tag{7.6}$$

Figure 7.7 presents the diagram of stress/strain through the thickness of the living hinge for the pure bending case.

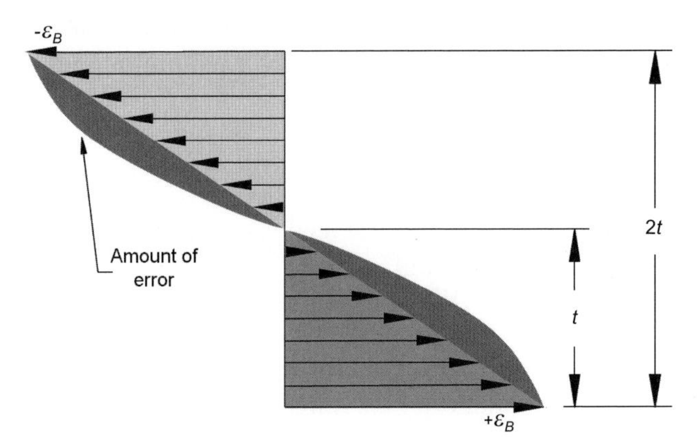

Figure 7.7 Strain diagram of pure bending

The secant modulus of the material is E. In order to avoid failure, the necessary condition is

$$\sigma_{\text{Bending}} < \sigma_{\text{Yield}} \tag{7.7}$$

and by replacing the value of the stress due to bending in the above inequation

$$\frac{\pi t}{L_1} E < \sigma_{\text{Yield}} \tag{7.8}$$

The condition to have elastic strain only for a minimum length of the fiber along the neutral axis is

$$L_1 > \frac{\pi t E}{\sigma_{\text{Yield}}} \tag{7.9}$$

7.5.1.5 Closing Angle of the Hinge

Rearranging the inequation, we can write the limit of the closing angle as:

$$\pi < \frac{L_1 \sigma_{\text{Yield}}}{tE} \tag{7.10}$$

If we define Φ as the closing angle

$$\phi = \frac{L_1 \sigma_{\text{Yield}}}{tE} \tag{7.11}$$

we can write the limiting condition as $\Phi < \pi$, so as to have only elastic strain.

7.5.1.6 Bending Radius of the Hinge

Replacing L_1 in 7.4 with the value from 7.9, the constraint for the bending radius of the hinge is

$$R > \frac{tE}{\sigma_{\text{Yield}}} \tag{7.12}$$

This is true for a hinge that experiences elastic strain only. To have an elastic hinge at the lower limit of the inequation domain, the minimum value of the needed radius will be

$$R = \frac{tE}{\sigma_{\text{Yield}}} \tag{7.13}$$

7.5.2 Plastic Strain Due to Pure Bending

The above considerations were for a fully *elastic* material. Now we will evaluate the fully *plastic* behavior of the walls and the hinge.

7.5.2.1 Assumptions

A. The material is fully plastic.

B. The hinge bends in a circle.

C. The neutral axis is in the hinge center.

D. Failure occurs when $\varepsilon_{\text{Bending}} = \varepsilon_{\text{Ultimate}}$.

7.5.2.2 Strain Due to Bending

In 7.5 the strain value due to bending was determined for an elastic case. To find the length of the neutral axis, the equation is rewritten:

$$L_1 = \frac{\pi t}{\varepsilon_{\text{Yield}}} \tag{7.14}$$

To find the limits of the hinge length, the *equal* sign has to be replaced with a *greater than* sign, and the *bending strain* is replaced with the *ultimate strain* in the above equation.

The ultimate strain is the value reached by the material strain at failure. As both the outer and lower fibers have plastic deformations, the condition necessary for no failure is

$$L_1 > \frac{\pi t}{\varepsilon_{\text{Ultimate}}} \tag{7.15}$$

As long as there is pure bending, a portion of the center of the living hinge never reaches plastic deformation. To determine the elastic portion of the living hinge thickness

$$h = \frac{L_1 \sigma_{\text{Yield}}}{2\pi E} \tag{7.16}$$

The length of the neutral axis L_1 for the pure bending region will vary between the limits

$$\pi(l+t) < L_1 < \frac{\pi t E}{\sigma_{\text{Yield}}} \tag{7.17}$$

At the lower limit and below, the living hinge behaves like a visco-elastic material. To explain this property mathematically, one must consider a tension strain that will actually model the necking effect that occurs. Therefore:

$$L_1 = \frac{\pi t}{\varepsilon_{\text{Bending}}} \tag{7.18}$$

Note that above the upper limit in 7.7, we have a completely elastic hinge.

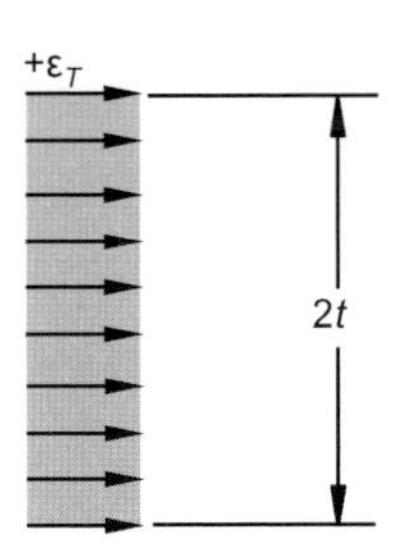

Figure 7.8
Strain diagram for pure tension

7.5.3 Plastic Strain Due to a Mixture of Bending and Tension

Most living hinges experience the necking behavior explained earlier. Considering a portion of the living hinge (Fig. 7.9), the following notations are used:

L_1 = length of the neutral axis

L_0 = length of the lower outer fiber

$2t$ = hinge thickness in open position

$2t'$ = hinge thickness in closed position

l = recess radius (open position)

l' = recess radius (closed position due to necking effect)

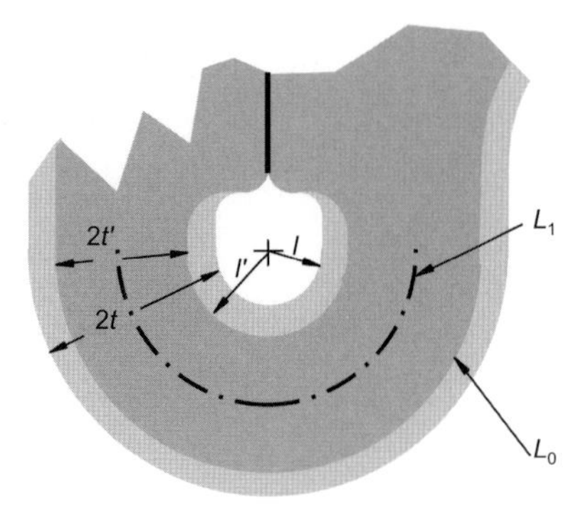

Figure 7.9
Detail of the living hinge design

Where L_0 is the length of the lower outer fiber with the hinge in the closed position, it can be considered the perimeter of a semicircle with a radius (l' + $2t'$):

$$L_0 = \pi\left(l' + 2t'\right) \quad (7.19)$$

or, using Equation 7.1:

$$L_0 = \frac{L_1}{R}\left(l' + 2t'\right) \quad (7.20)$$

7.5.3.1 Tension Strain

Figure 7.10 shows a cross-section diagram through the living hinge and its tension strain. Combining Figs. 7.7 and 7.8 creates a new diagram, which represents the sum of the two strains throughout the hinge thickness (Fig. 7.10). The two triangles ABC and CDE are proportional (using Euclidian geometry):

$$\frac{\varepsilon_{\text{Bending}} + \varepsilon_{\text{Tension}}}{\varepsilon_{\text{Bending}} - \varepsilon_{\text{Tension}}} = \frac{2t' - h}{h} \quad (7.21)$$

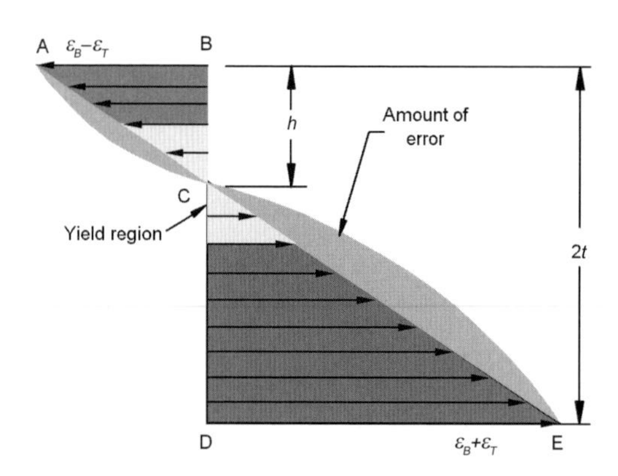

Figure 7.10
Strain diagram across the hinge thickness

Reworking 7.21 we can write

$$h\varepsilon_{\text{Bending}} + h\varepsilon_{\text{Tension}} = 2t'\varepsilon_{\text{Bending}} - 2t'\varepsilon_{\text{Tension}} - h\varepsilon_{\text{Bending}} + h\varepsilon_{\text{Tension}} \quad (7.22)$$

or

$$2h\varepsilon_{\text{Bending}} - 2t'\varepsilon_{\text{Bending}} + 2t'\varepsilon_{\text{Tension}} = 0 \quad (7.23)$$

$$\varepsilon_{\text{Bending}} = \frac{t'}{t' - h}\varepsilon_{\text{Tension}} \quad (7.24)$$

The sum of the two strains must equal the total strain:

$$\varepsilon_{\text{Bending}} + \varepsilon_{\text{Tension}} = \frac{L_0 - L_1}{L_1} = \frac{\pi\left(l' + 2t'\right) - L_1}{L_1} \quad (7.25)$$

Equations 7.24 and 7.25 form a system with two unknowns: $\varepsilon_{Tension}$ and $\varepsilon_{Bending}$. The elastic portion of the hinge thickness is

$$h = Rl' = \frac{L_1}{\pi} l' \tag{7.26}$$

Therefore, replacing the value of h in 7.24 we obtain

$$\varepsilon_{Bending} = \frac{t'}{t' - \frac{L_1}{\pi} + l'} \tag{7.27}$$

and rearranging 7.25

$$\varepsilon_{Bending} = \frac{\pi\left(l' + 2t'\right)L_1}{L_1} \varepsilon_{Tension} \tag{7.28}$$

To find the finite solutions of 7.27 and 7.28, there are two conditions for the denominators in 7.29 and 7.30:

$$\frac{t'\varepsilon_{Tension}}{t' - \frac{L_1}{\pi} + l'} = \frac{\pi\left(l' + 2t'\right) + L_1}{L_1} - \varepsilon_{Tension} \tag{7.29}$$

$$\frac{\pi t'\varepsilon_{Tension}}{\pi t' + \pi l' - L_1} = \frac{\pi\left(l' + 2t'\right) + L_1 - L_1\varepsilon_{Tension}}{L_1} \tag{7.30}$$

The first one

$$L_1 \neq 0 \tag{7.31}$$

and the second one

$$\pi t' + \pi l' - L_1 \neq 0 \tag{7.32}$$

Then

$$L_1\varepsilon_{Tension} = \pi t' + \pi l' - L_1 \tag{7.33}$$

and finally

$$\varepsilon_{Tension} = \frac{\pi}{L_1}\left(t' + l'\right) - l \tag{7.34}$$

The strain in the longitudinal direction is proportional to the strain in the transverse direction by a factor called Poisson's ratio, which defines the necking behavior:

$$\varepsilon_{Transversal} = \frac{\Delta t}{2t} = \nu\varepsilon_{Tension} \tag{7.35}$$

The variance in hinge thickness before and after closing will be:

$$\Delta t = 2t\nu\varepsilon_{\text{Tension}} \tag{7.36}$$

$$2t' = 2t\Delta t \tag{7.37}$$

$$t' = t - t\nu\varepsilon_{\text{Tension}} \tag{7.38}$$

$$l' = l - \frac{\Delta t}{2} \tag{7.39}$$

$$l' = l + t\nu\varepsilon_{\text{Tension}} \tag{7.40}$$

Replacing t' and l' from 7.38 and 7.40 in 7.34 we obtain

$$\varepsilon_{\text{Tension}} = \frac{\pi}{L_1}\left(t - t\nu\varepsilon_{\text{Tension}} + l + t\nu\varepsilon_{\text{Tension}}\right) - 1 \tag{7.41}$$

or, after rearranging the factors, $\varepsilon_{\text{Tension}}$ is obtained in a simpler form:

$$\varepsilon_{\text{Tension}} = \frac{\pi}{L_1}(t + l) - 1 \tag{7.42}$$

7.5.3.2 Bending Strain

Bending strain $\varepsilon_{\text{Bending}}$ is obtained by substituting $\varepsilon_{\text{Tension}}$ from 7.42 in 7.24:

$$\varepsilon_{\text{Bending}} = t' \frac{\left[\frac{\pi}{L_1}\left(l' + t'\right) - 1\right]}{t' + l' - \frac{L_1}{\pi}} \tag{7.43}$$

or

$$\varepsilon_{\text{Bending}} = \frac{t'\left[\frac{\pi}{L_1}\left(l' + t'\right) - 1\right]}{t' + l' - \frac{L_1}{\pi}} = \frac{\pi t'}{L_1} \tag{7.44}$$

The value of t' from 7.38 in the above equation can now be replaced:

$$\varepsilon_{\text{Bending}} = \frac{\pi}{L_1}\left(t - t\nu\varepsilon_{\text{Tension}}\right) \tag{7.45}$$

and by substituting the value of the tension strain from 7.42:

$$\varepsilon_{\text{Bending}} = \frac{\pi t}{L_1}\left[1 - \nu\frac{\pi t + \pi l - L_1}{L_1}\right] \tag{7.46}$$

$$\varepsilon_{\text{Bending}} = \frac{\pi t}{L_1} - \nu\frac{\pi t}{L_1}\frac{\pi t + \pi l - L_1}{L_1} \tag{7.47}$$

Using a common denominator and factorizing

$$\varepsilon_{\text{Bending}} = \frac{\pi t}{L_1} - \frac{\pi^2 \nu t l}{L_1^2} - \frac{\pi^2 \nu t^2}{L_1^2} + \frac{\pi \nu t}{L_1} \tag{7.48}$$

$$\varepsilon_{\text{Bending}} = \frac{\pi t}{L_1^2}\left(L_1 + \nu L_1 - \pi \nu t - \pi \nu l\right) \tag{7.49}$$

The resulting final value for the bending strain is

$$\varepsilon_{\text{Bending}} = \frac{\pi t}{L_1^2}\left[L_1(1+\nu) - \pi \nu (t + l)\right] \tag{7.50}$$

7.5.3.3 Neutral Axis Position

Given the calculated tension strain ($\varepsilon_{\text{Tension}}$ = the elongation in the outside fiber due to tensile strain) and the bending strain ($\varepsilon_{\text{Bending}}$ = the elongation in the inner and outer fiber due to bending strain) then h (Fig. 7.10), which represents the distance from the inside fiber to the neutral axis under the plastic bending and tension, can be calculated as

$$h = R - l' = \frac{L_1}{\pi} - l' = \frac{L_1}{\pi} - l - \nu t \varepsilon_{\text{Tension}} \tag{7.51}$$

$$h = \frac{L_1}{\pi} - l - \frac{\nu t}{L_1}\left(\pi t + \pi l - L_1\right) \tag{7.52}$$

Note: h will become negative if the tensile strain reaches high values, which means that the living hinge is under tension and there is no neutral axis within the living hinge thickness.

7.5.3.4 Hinge Length

From the condition that

$$\varepsilon_{\text{Tension}} + \varepsilon_{\text{Bending}} < \varepsilon_{\text{Ultimate}} \tag{7.53}$$

where ultimate strain represents the elongation at break, the hinge minimum length before failure can now be found. Thus the inequation can be rewritten as

$$\varepsilon_{\text{Tension}} + \varepsilon_{\text{Bending}} - \varepsilon_{\text{Ultimate}} < 0 \tag{7.54}$$

$$\varepsilon_{\text{Bending}} = \frac{\pi t}{L_1} - \frac{\pi^2 \nu t l}{L_1^2} - \frac{\pi^2 \nu t^2}{L_1^2} + \frac{\pi \nu t}{L_1} \tag{7.55}$$

By using the following notation

$$a = \frac{2t}{L_1} \tag{7.56}$$

and

$$\varepsilon_{\text{Tension}} = \frac{\pi}{L_1}(t - l) - 1 \tag{7.57}$$

and by rewriting the bending strain

$$\varepsilon_{\text{Bending}} = \frac{\pi}{2}a + \frac{\pi\nu}{2}a - \frac{\pi^2\nu}{4}a^2 - \frac{\pi^2\nu l}{2L_1}a \tag{7.58}$$

In a similar fashion the tension strain is obtained:

$$\varepsilon_{\text{Tension}} = \frac{\pi}{2}a + \frac{\pi l}{L_1} - 1 \tag{7.59}$$

Now the inequation will look like this:

$$\pi a + \frac{\pi\nu}{2}a + \frac{\pi l}{L_1} - \frac{\pi^2\nu}{4}a^2 - \frac{\pi^2\nu l}{2L_1}a - 1 - \varepsilon_{\text{Ultimate}} < 0 \tag{7.60}$$

Rearranging by the powers of a, by changing the sign and by dividing the inequality with the factor to the right

$$\frac{\pi^2\nu}{4}a^2 + \frac{\pi^2\nu l}{2L_1}a - \frac{\pi\nu}{2}a - \pi a - \frac{\pi l}{L_1} + 1 + \varepsilon_{\text{Ultimate}} < 0 : \frac{4}{\pi^2\nu} \tag{7.61}$$

we obtain

$$a^2 + \frac{2}{\pi}\left(\frac{\pi l}{L_1} - 1 - \frac{2}{\nu}\right)a - \frac{4}{\pi^2\nu}\left(\frac{\pi l}{L_1} - 1 - \varepsilon_{\text{Ultimate}}\right) < 0 \tag{7.62}$$

$$a^2 + \frac{2l}{L_1}a - \frac{2}{\pi}a - \frac{4}{\pi\nu}a - \frac{4l}{\pi\nu L_1} - \frac{4}{\pi^2\nu} + \frac{4}{\pi^2\nu}\varepsilon_{\text{Ultimate}} < 0 \tag{7.63}$$

$$a^2 + \frac{2}{\pi}\left(\frac{\pi l}{L_1} - 1 - \frac{2}{\nu}\right)a - \frac{4}{\pi^2\nu}\left(\frac{\pi l}{L_1} - l - \varepsilon_{\text{Ultimate}}\right) < 0 \tag{7.64}$$

The region where a has real values is

$$a \in (-\infty, a_1) \cup (a_2, +\infty) \tag{7.65}$$

The roots of the second-degree polynomial inequality are

$$a_{1,2} = -\frac{1}{\pi}\left(\frac{\pi l}{L_1} - 1 - \frac{2}{\nu}\right) \pm \frac{l}{\pi}\sqrt{\left(\frac{\pi l}{L_1} - 1 - \frac{2}{\nu}\right)^2 + \frac{4}{\nu}\left(\frac{\pi l}{L_1} - 1 - \varepsilon_{\text{Ultimate}}\right)} \tag{7.66}$$

And a_1 is

$$a_1 = -\frac{1}{L_1} + \frac{l}{\pi} + \frac{2}{\pi\nu} - \frac{1}{\pi}\sqrt{\left(1 - \frac{\pi l}{L_1}\right)^2 + \frac{4}{\nu^2}\left(1 - \nu\varepsilon_{\text{Ultimate}}\right)} \tag{7.67}$$

But since

$$1-\frac{\pi t}{L_1}=1-\frac{l}{R}=\frac{R-l}{R}=\frac{t}{R} \tag{7.68}$$

then the first value of a is

$$a_1=-\frac{1}{L_1}+\frac{l}{\pi}+\frac{2}{\pi\nu}-\frac{1}{\pi}\sqrt{\frac{\pi^2 l^2}{L_1^2}+1+\frac{4}{\nu^2}-\frac{2\pi l}{L_1}-\frac{4}{\nu}\varepsilon_{\text{Ultimate}}} \tag{7.69}$$

Using the following approximation, no excess error will take place:

$$\left(\frac{t}{R}\right)^2=0 \tag{7.70}$$

so, a_1 becomes

$$a_1=\frac{l}{L_1}+\frac{l}{\pi}+\frac{2}{\pi\nu}-\frac{2}{\pi\nu}\sqrt{1-\nu\varepsilon_{\text{Ultimate}}} \tag{7.71}$$

or

$$a_1=-\frac{l}{L_1}+\frac{l}{\pi}+\frac{2}{\pi\nu}\left[1-\sqrt{1-\nu\varepsilon_{\text{Ultimate}}}\right] \tag{7.72}$$

Because

$$\mathrm{a}<\mathrm{a}_1 \tag{7.73}$$

again, the domain for which a_1 has real values is

$$a\in(-\infty,a_1)\cup(a_2,+\infty) \tag{7.74}$$

and in order to have this condition fulfilled

$$\frac{2t}{L_1}<-\frac{l}{L_1}+\frac{l}{\pi}+\frac{2}{\pi\nu}\left[1-\sqrt{1-\nu\varepsilon_{\text{Ultimate}}}\right] \tag{7.75}$$

Finally, the smallest living hinge length should be

$$\frac{2t+l}{L_1}<\frac{l}{\pi}+\frac{2}{\pi\nu}\left[1-\sqrt{1-\nu\varepsilon_{\text{Ultimate}}}\right] \tag{7.76}$$

or

$$L_1>\frac{\pi\nu(2t+l)}{\nu+2\left[1-\sqrt{1-\nu\varepsilon_{\text{Ultimate}}}\right]} \tag{7.77}$$

7.5.3.5 Elastic Portion of the Hinge Thickness

The final calculation concern is the thickness (*W*) of the elastic portion of the total hinge thickness.

The strain diagram (Fig. 7.11) shows two proportional triangles: ABC and CDE. Applying Euclidian geometry theorems, one can write

$$\frac{\varepsilon_{\text{Yield}}}{\varepsilon_{\text{Tension}}+\varepsilon_{\text{Bending}}}<\frac{\frac{W}{2}}{2t-h} \tag{7.78}$$

Knowing that the yield strain is the ratio between yield stress and secant modulus, 7.78 becomes:

$$\frac{\sigma_{\text{Yield}}}{E\left(\varepsilon_{\text{Tension}}+\varepsilon_{\text{Bending}}\right)}<\frac{W}{2(2t-h)} \tag{7.79}$$

or

$$W=\frac{2(2t-h)\sigma_{\text{Yield}}}{E\left(\varepsilon_{\text{Tension}}+\varepsilon_{\text{Bending}}\right)} \tag{7.80}$$

$$W=\frac{2\left[2t-\frac{L_1}{\pi}+l+\frac{\nu t}{L_1}\left(\pi t+\pi l-L_1\right)\right]\sigma_{\text{Yield}}}{\left[\frac{\pi(t+l)-L_1}{L_1}+\frac{\pi t}{L_1}\left(L_1+\nu L_1-\pi\nu t-\pi\nu l\right)\right]E} \tag{7.81}$$

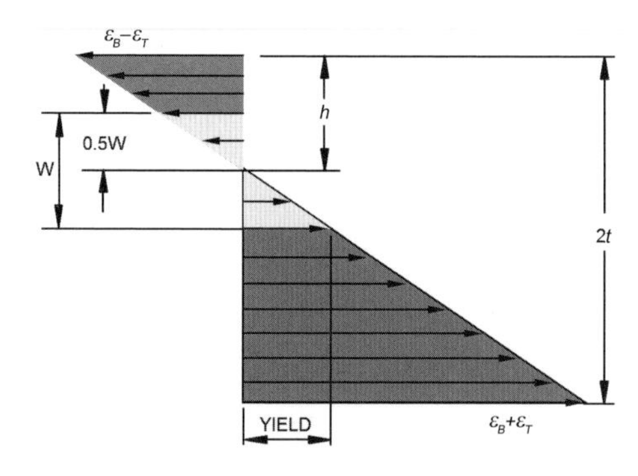

Figure 7.11 Strain diagram across the living hinge

7.6 Computer Flow Chart

Computer software that facilitates calculations of living hinge design is available in the marketplace [159, 160]. The following demonstrates the framework of such programs and how they can be developed further to suit specific applications.

7.6.1 Computer Notations

X = Hinge thickness

Y = Hinge length

Z = Hinge recess

U = Hinge new thickness

W = Hinge elastic strain portion

V = Distance to inner fiber

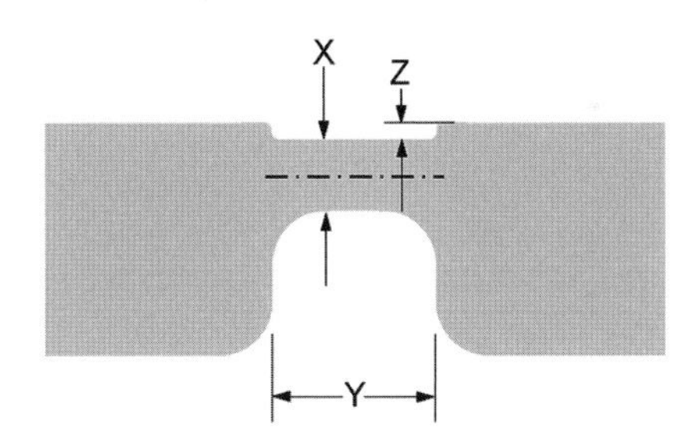

Figure 7.12
Computer variables X, Y, and Z

1. Select a material. The computer then selects the material properties from the built-in database.
2. Establish the minimum processing thickness X.
3. Select the hinge length Y based on space considerations.
4. Compute A.
5. IF
   a. IF $Y > A$, no failure: pure elastic hinge. STOP HERE.
      Computer lists:
      * Hinge thickness X
      * Hinge length Y
      * Hinge angle Φ
      * Hinge radius R
      * Maximum hinge strain $\varepsilon_{\text{Ultimate}}$

   b. If Y < A, go to next step.

6. Computer automatic selection of Z.
7. Compute B.
8. IF
   a. If $Y > B$, pure plastic bending. Go to next step.
      A. Compute C.
      B. IF
         a. If $Y > C$, no failure. STOP HERE.
            Computer lists:
            * Hinge thickness X
            * Hinge length Y
            * Hinge elastic strain portion W
            * Hinge maximum strain $\varepsilon_{\text{Ultimate}}$
         b. If $Y < C$, failure. STOP HERE.
   b. If $Y < B$, mixture of plastic bending and tension. Go to next step.
9. Compute D.
10. IF
    a. If $Y > D$, no failure. STOP HERE.
       Computer lists:
       * Bending strain
       * Tension strain
       * Distance from center axis to inner fiber V
       * Hinge thickness X
       * Hinge length Y
       * Hinge elastic strain portion W
       * Hinge new thickness U
    b. If $Y < D$, STOP HERE.

To illustrate the above, Fig. 7.13 shows the actual computer flow diagram that can be used.

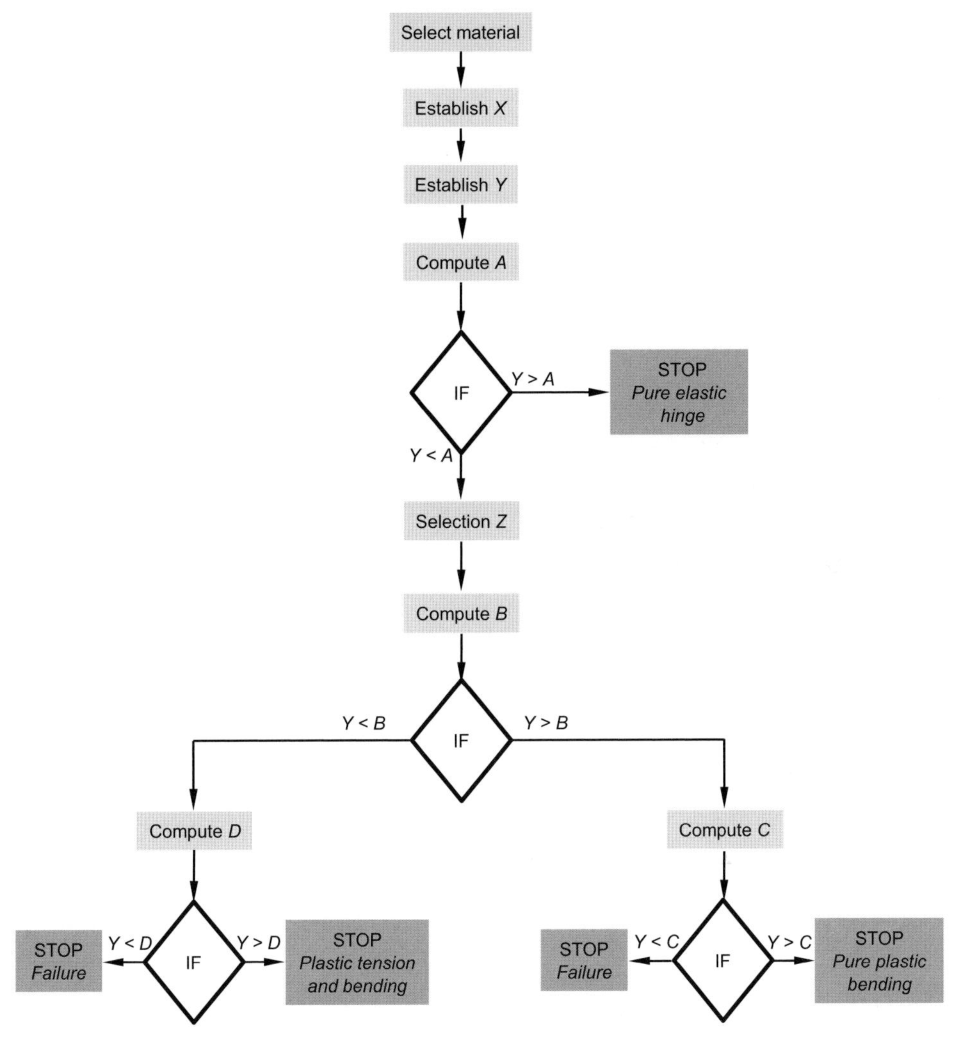

Figure 7.13 Computer flow chart

7.7 Computer Flow Chart Equations

The equations needed for the chart shown in Fig. 7.13 are as follows:

Thickness

$$X = 2t \tag{7.82}$$

Compute A

$$A = \frac{\pi t E}{\sigma_{\text{Yield}}} \tag{7.83}$$

Angle

$$\phi = \frac{Y\sigma_{\text{Yield}}}{tE} \tag{7.84}$$

Radius

$$R = \frac{tE}{\sigma_{\text{Yield}}} \tag{7.85}$$

Strain

$$\varepsilon = \frac{\pi t}{Y} \tag{7.86}$$

Compute B

$$B = \pi(t + l) \tag{7.87}$$

Recess

$$Z = l \tag{7.88}$$

Compute C

$$C = \frac{\pi t}{\varepsilon_{\text{Ultimate}}} \tag{7.89}$$

Elastic strain portion

$$W_p = \frac{Y\sigma_{\text{Yield}}}{2\pi E} \tag{7.90}$$

Compute D

$$D = \frac{\pi\nu(X + Z)}{\nu + 2\left[1 - \sqrt{1 - \nu\varepsilon_{\text{Ultimate}}}\right]} \tag{7.91}$$

New thickness

$$U = t(1 - \nu\varepsilon_{\text{Tension}}) \tag{7.92}$$

Bending strain

$$\varepsilon_{\text{Bending}} = \frac{\pi t}{Y^2}\left[Y(1 + \nu) - \pi\nu(t + Z)\right] \tag{7.93}$$

Neutral axis shift

$$V = \frac{Y}{\pi} + \nu t - l - \frac{\pi\nu t^2}{Y} - \frac{\pi\nu t l}{Y} \tag{7.94}$$

Elastic strain

$$W = \frac{2\left[2t\left(1-\nu\varepsilon_{\text{Tension}}\right)-h\right]\sigma_{\text{Yield}}}{\left[\pi t\left(2-\nu\varepsilon_{\text{Tension}}+l\right)-Y\right]E} \tag{7.95}$$

7.8 Example: Case History

This section will examine the actual cases of two plastic components that use living hinges. The first is a world-class connector, an automotive under-hood component that connects the onboard computer to the engine controls. The second case involves a bracket that organizes ignition cables between the spark plugs and the distributor. Two materials and designs will be analyzed for each case.

7.8.1 World-Class Connector

The original design consisted of three parts assembled. To reduce assembly labor costs, inventory, handling, and shipping costs, a new design was considered using living hinges, so that the three former components could be molded as one piece.

The material with the best properties for living hinges is PP. But this material is not able to survive the under-the-hood environment where temperatures of 250 to 300°F and possible spills of various chemicals such as gasoline and oil are commonplace. Therefore, the designer had to consider other plastic materials. The first to be considered was polyamide, which can withstand the temperatures and chemicals of the under-the-hood environment.

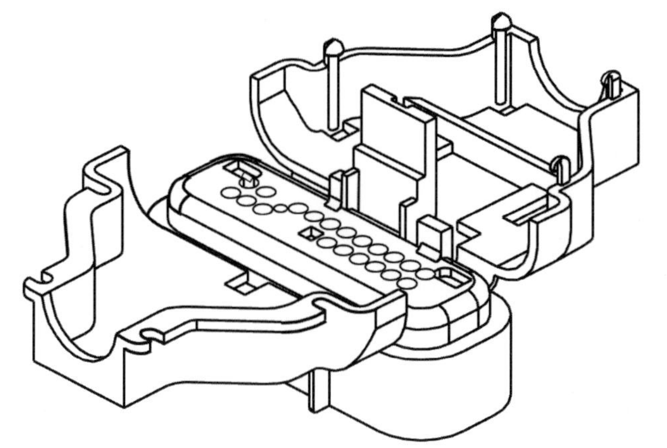

Figure 7.14
World-class connector (axonometric view)

Another requirement of the connector was that it withstand packing and transportation from the molding facility to the final electric component assembly. In transit, some connectors suffered breakage. Panels broke from the component when hinges cracked as a result of asymmetrical hinge design. The polyamide material was failing in this application because the panels, by moving through different positions during shipping, were causing the material to reach its elongation-at-break point.

A new material had to be considered. Polyester elastomer thermoplastic was chosen.

7.8.1.1 Calculations for the "Right Way" Assembly

To examine the details of the living hinge design and the materials used in this analysis, we will begin with what is called the "right way" assembly. The two panels in Fig. 7.15 are moved in the direction designated "right way" in Fig. 7.16. The panels rotate 90°, but point A on the hinge only needs to rotate 45°.

The analysis of the hinge in the example that follows begins by using the thermoplastic elastomer properties. Next, the polyamide case will be reviewed.

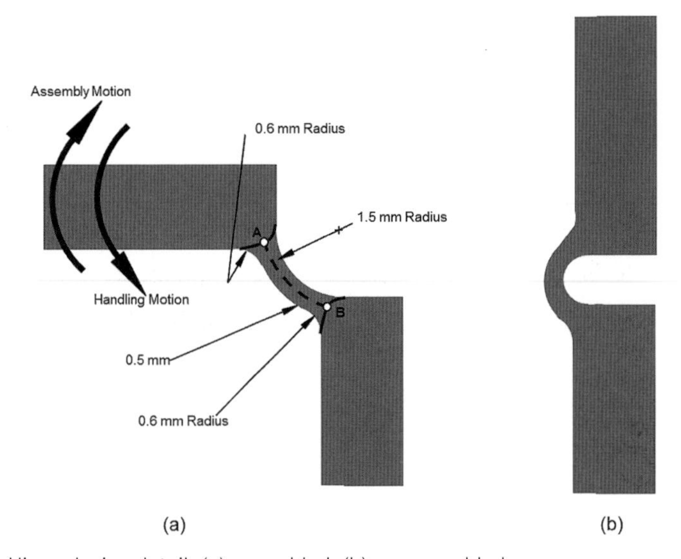

Figure 7.15 Hinge design detail: (a) as molded; (b) as assembled

Material properties for a PET elastomer are

σ_{Ultimate}	= 50 MPa	Tensile strength at break
σ_{Yield}	= 40 MPa	Tensile strength at yield
$E_{\text{Secant Yield}}$	= 236.5 MPa	Secant modulus for the yield point ($E_{S_{\text{Yield}}}$)
$\varepsilon_{\text{Ultimate}}$	= 39.7%	Ultimate strain (elongation at break)
$\varepsilon_{\text{Yield}}$	= 9.1%	Yield strain
v	= 0.45	Poisson's ratio

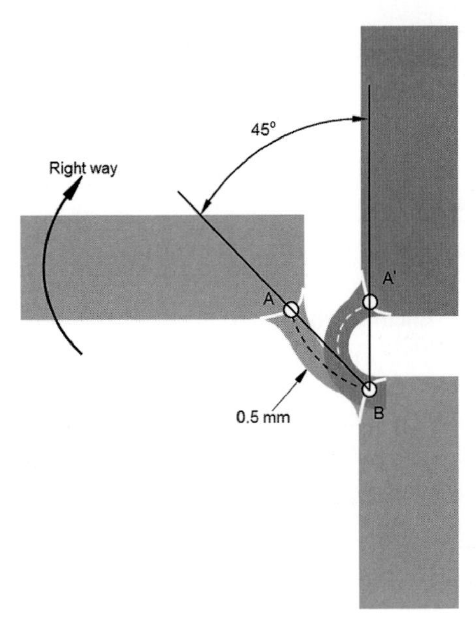

Figure 7.16
"Right way" assembly hinge detail

The values for hinge thickness, length, and recess for this design are

$X = 0.5$ mm Hinge thickness

$Y = 2.748$ mm Hinge length

$Z = 0.0$ mm Hinge offset or recess

The aim of the analysis that follows is to determine if the design is robust or not. To initiate the analysis, A is calculated, which represents the length of a fully elastic hinge. If $A < Y$, we will have a total elastic hinge. For a 45° bend angle ("right way" assembly), the value for A is

$$A = \frac{\pi t E_{S_{\text{Yield}}}}{4\sigma_{\text{Yield}}} = 1.16 \text{ mm} \tag{7.96}$$

$$A < Y \tag{7.97}$$

The 1.16 mm is the length necessary to have a completely elastic hinge.

Once a pure elastic hinge has been established, other hinge properties can be calculated:

Hinge radius

$$R = \frac{\pi t}{\sigma_{\text{Yield}}} = 2.74 \text{ mm} \tag{7.98}$$

Hinge maximum strain takes place in bending:

$$\varepsilon_{\text{Bending}} = \frac{\pi t}{L_1} = 0.071 = 7.1\% \tag{7.99}$$

Hinge angle, corresponding to the elastic portion of it, is

$$\phi = \frac{180 L_1 \sigma_{\text{Yield}}}{4\pi t E_{S_{\text{Yield}}}} = 27^{\circ} \tag{7.100}$$

7.8.1.2 Calculations for the "Wrong Way" Assembly

When the hinge bends in the opposite direction, only the angle changes. The angle of bending is 135° and not 45° as in the previous case. This time we can extend our elongation limits by using the ultimate strain. Because the hinge needs only to not break, permanent deformations can take place. The procedure will remain the same.

$$A = \frac{3\pi t E_{\text{SYield}}}{4\sigma_{\text{Yield}}} = 6.47 \text{ mm} \tag{7.101}$$

$$\mathrm{A} < \mathrm{Y} \tag{7.102}$$

Because A is greater than Y, B has to be calculated; B indicates the presence of the plastic bending within the hinge thickness.

$$B = \frac{3\pi t}{4} = 0.59 \text{ mm} \tag{7.103}$$

$$B < Y \tag{7.104}$$

Here, B is much smaller than Y. This means that the hinge undergoes plastic bending. Because C represents the hinge length that undergoes plastic bending, C has to be calculated:

$$C = \frac{3\pi t}{4\varepsilon_{\text{Ultimate}}} = 1.48 \text{ mm} \tag{7.105}$$

$$C < Y \tag{7.106}$$

Because C is smaller than Y, although the hinge undergoes plastic deformation, it will not fail through rupture. This is because the remaining portion of the hinge thickness is still elastic:

$$W = \frac{2\left[2t\left(l - \nu\varepsilon_{\text{Tension}}\right) - h\right] Y \sigma_{\text{Yield}}}{\left[\pi t\left(2 - \nu\varepsilon_{\text{Tension}}\right) - Y\right] E_{S_{\text{Yield}}}} = 0.26 \text{ mm} \tag{7.107}$$

One can obtain different results if the ultimate strain is substituted for the yield strain. For a 45° bend angle, A is

$$A = \frac{\pi t}{4\varepsilon_{\text{Ultimate}}} = 1.24 \text{ mm} \tag{7.108}$$

$$A < Y \tag{7.109}$$

Because A is smaller than Y, we have an elastic hinge. We can determine the other hinge parameters:

Hinge radius

$$R = \frac{t}{\varepsilon_{\text{Ultimate}}} = 0.63 \text{ mm} \tag{7.110}$$

Hinge maximum strain occurs in bending:

$$\varepsilon_{\text{Bending}} = \frac{\pi t}{4L_1} = 0.071 = 7.1\% \tag{7.111}$$

For a 135° bending angle, A is

$$A = \frac{3\pi t}{4\varepsilon_{\text{Ultimate}}} = 1.48 \text{ mm} \tag{7.112}$$

$$\text{A} < \text{Y} \tag{7.113}$$

Because Y exceeds A, we have an elastic hinge. Therefore, the hinge radius is

$$R = \frac{3t}{4\varepsilon_{\text{Ultimate}}} = 0.47 \text{ mm} \tag{7.114}$$

and the hinge maximum strain occurs in bending:

$$\varepsilon_{\text{Bending}} = \frac{3\pi t}{4L_1} = 0.213 = 21.3\% \tag{7.115}$$

7.8.2 Comparison Material

The initial material used for evaluation of the design was nylon (polyamide). The following shows the calculations performed to verify the design.

Polyamide material properties:

σ_{Ultimate}	= 93 MPa	Tensile strength at break
σ_{Yield}	= 65 MPa	Tensile strength at yield
$E_{\text{Secant}_{\text{Yield}}}$	= 1,599 MPa	Secant modulus for the yield point ($E_{\text{S}_{\text{Yield}}}$)
E_{Ultimate}	= 19%	Ultimate strain (elongation at break)
E_{Yield}	= 2.46%	Yield strain
v	= 0.42	Poisson's ratio

Given data (hinge principal dimensions):

X = 0.5 mm	Hinge thickness
Y = 2.748 mm	Hinge length
Z = 0.0 mm	Hinge offset

7.8.2.1 "Right Way" Assembly

Calculation for a 45° bending angle:

$$A=\frac{\pi t E_{S_{\text{Yield}}}}{4\sigma_{\text{Yield}}}=7.98\text{ mm} \tag{7.116}$$

$$A>Y \tag{7.117}$$

$$B=\frac{\pi(t+l)}{4}=0.2\text{ mm} \tag{7.118}$$

$$B<Y \tag{7.119}$$

$$C_{\text{Ultimate}}=\frac{\pi t}{4\varepsilon_{\text{Ultimate}}}=1.03\text{ mm} \tag{7.120}$$

$$C<Y \tag{7.121}$$

This hinge design will not fail. Now by using elongation at yield instead of the strain at break, C will be

$$C_{\text{Yield}}=\frac{\pi t}{4\varepsilon_{\text{Yield}}}=7.98\text{ mm} \tag{7.122}$$

$$C>Y \tag{7.123}$$

This hinge design will fail because C is greater than Y.

The safety factor can be defined as the ratio between the designed length of the hinge Y and the calculated value of C. Using the ultimate value for the strain, the safety factor is

$$n_{\text{Ultimate}_{45^o}}=\frac{Y}{C_{\text{Ultimate}}}=2.7\text{ times} \tag{7.124}$$

A safety factor of 2.8 times, based on ultimate material values, explains why the hinge can bend once having plastic deformation. The hinge will be able to return partially towards its initial position after it has been closed.

On the other hand, by using the yield strain instead of the elongation at break, the safety factor will be less than 1, meaning that this hinge will break (or initiate a crack) during the final assembly of the connector.

$$n_{\text{Yield}_{45^o}}=\frac{Y}{C_{\text{Yield}}}=0.34\text{ times} \tag{7.125}$$

7.8.2.2 "Wrong Way" Assembly

For the 135° bending angle, the calculations are similar:

$$A = \frac{3\pi t E_{S_{Yield}}}{4\sigma_{Yield}} = 14.5 \text{ mm} \tag{7.126}$$

$$A >>> Y \tag{7.127}$$

$$B = \frac{\pi t}{4} = 0.196 \text{ mm} \tag{7.128}$$

$$B < Y \tag{7.129}$$

$$C_{Ultimate} = \frac{3\pi t}{4\varepsilon_{Ultimate}} = 3.1 \text{ mm} \tag{7.130}$$

$$C > Y \tag{7.131}$$

This hinge design will fail. If, instead of the elongation at break, the yield strain is used

$$C_{Yield} = \frac{3\pi t}{4\varepsilon_{Yield}} = 23.94 \text{ mm} \tag{7.132}$$

$$C > Y \tag{7.133}$$

This hinge will fail. The safety factors based on the above assumptions are

$$n_{Ultimate_{135°}} = \frac{Y}{C_{Ultimate}} = 0.88 \text{ times} \tag{7.134}$$

Finally, for bends in the opposite direction (135°), using either ultimate strain (elongation at break) or yield strain will cause the hinge to break.

$$n_{Yield_{135°}} = \frac{Y}{C_{Yield}} = 0.11 \text{ times} \tag{7.135}$$

7.8.3 Ignition Cable Bracket

The second case history is also from the automotive field. The ignition cable bracket is used to organize the cables that connect the distributor to the spark plugs. There are two brackets used per engine. This specific component is used for V6 engines, each bracket holding three cables.

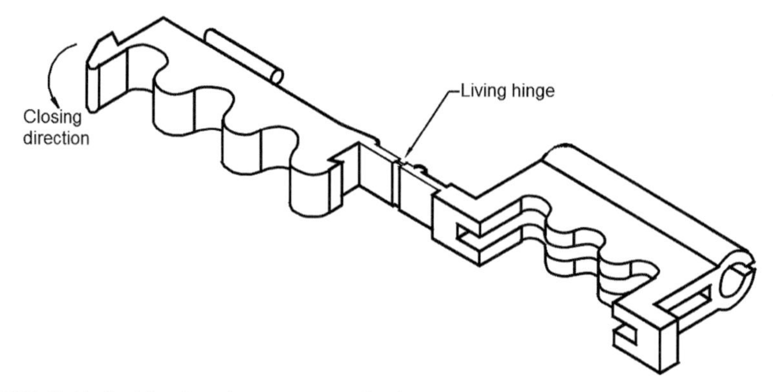

Figure 7.17 Cable ignition bracket axonometric view

Figure 7.17 shows the bracket in a 3-D drawing. The arrow indicates the direction of closure to bring the component to its final position. Cables are placed in the indentations shown on the right side of the drawing. Next, the bracket is closed by rotating the left side around until it snaps in place, holding the cables in position. The rotation is made possible by a living hinge.

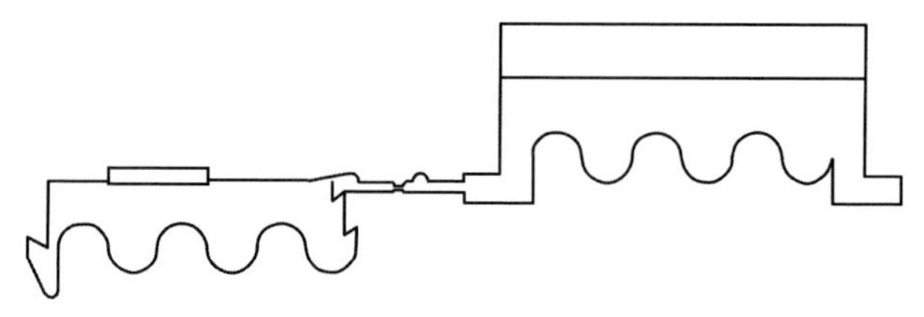

Figure 7.18 Cable ignition bracket side view

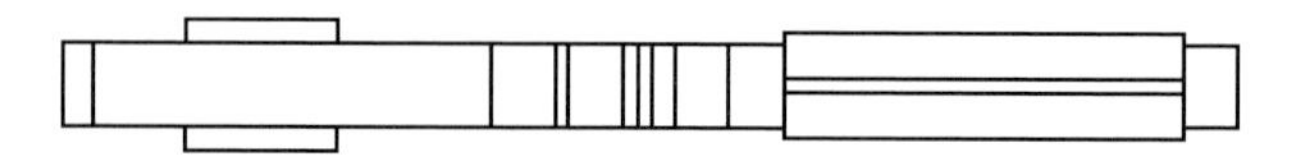

Figure 7.19 Cable ignition bracket top view

7.8.3.1 Initial Design

For the initial design of this bracket, a living hinge as shown in Fig. 7.20(a) was used. This is a common hinge design for polypropylene. It is customary in the PP molding process to orient the material in the hinge area by giving it a few quick flexes immediately after the part is ejected from the mold.

But this component is made from polyamide. Using the same design as for PP, the hinge failed in testing through fracture because polyamide does not orient in the hinge area. Moreover, nylon parts with such living hinge designs can break if they are flexed immediately after molding. The material is in a state known as DAM (dry-as-molded). During this state, when no moisture is present in the part, the material has very low strain levels. After molding, the material starts absorbing moisture from the air and reaches saturation within a few months after it is molded. (For details about the water absorption for different materials, see Section 1.7.10).

7.8.3.2 Improved Design

Because of the unpredictable failure of the initial design, the hinge was redesigned as shown in Fig. 7.20(b). The fibers in compression follow a smooth radius created by the recess. This design is explained in detail in Sections 7.4 and 7.5 and Fig. 7.4.

For this case history the calculations for the improved design follow.

Material properties for the polyamide polymer used are

$\sigma_{Ultimate}$ = 101 MPa Tensile strength at break

σ_{Yield} = 73 MPa Tensile strength at yield

$E_{Secant_{Yield}}$ = 2,800 MPa Secant modulus for the yield point ($E_{S_{Yield}}$)

$\varepsilon_{Ultimate}$ = 18% Ultimate strain (elongation at break)

ε_{Yield} = 2.6% Yield strain

v = 0.38 Poisson's ratio

Given data:

X = 0.5 mm Hinge thickness

Y = 4.5 mm Hinge length

Z = 0.25 mm Hinge offset

The analysis is initiated by calculating A, which represents the hinge length for a 180° closing angle:

$$A = \frac{\pi t E_{SYield}}{\sigma_{Yield}} = 30.12 \text{ mm} \tag{7.136}$$

$$A >>> Y \tag{7.137}$$

Because A is much greater than Y, B has to be calculated:

$$B = \pi(t + Z) = 1.6 \text{ mm} \tag{7.138}$$

$$B < Y \tag{7.139}$$

This hinge will undergo pure plastic bending; therefore, C has to be determined:

$$A = \frac{\pi t}{\varepsilon_{Ultimate}} = 4.36 \text{ mm} \tag{7.140}$$

With C being greater than Y, this hinge will not fail, but the hinge will have deformed plastically once it is in the closed position.

Figure 7.21 shows the hinge in a closed position. The offset Z creates the radius present on the inside, preventing the hinge from failing due to the notch effect.

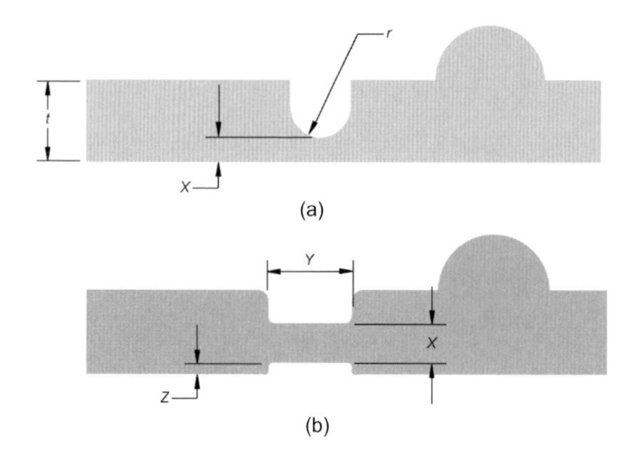

Figure 7.20
Hinge detail: (a) initial design; (b) improved design

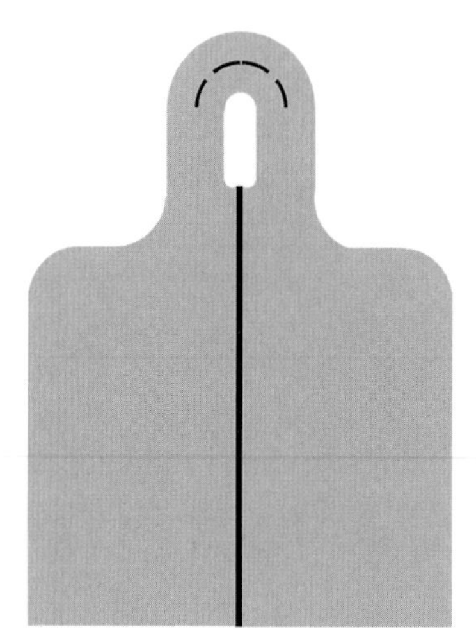

Figure 7.21
Hinge behavior after assembly

7.9 Processing Errors for Living Hinges

Many living hinge designs have met their doom as a result of improper mold processing. The key to successful living hinge design lies in a sound knowledge of the theory behind it. It is imperative that processing personnel understand the rheological activities that contribute to the production of successful hinges.

For living hinges to work, the molecular chains must be linearly aligned parallel to the direction of plastic flow across the hinge during the part fill. This does not mean that the chains are generally aligned.

For visualization purposes, imagine a bowl of cooked spaghetti being dumped from a bowl onto a table while the bowl is being dragged across the table surface. The spaghetti strands will be generally aligned in the direction the bowl is traveling, but the strands will remain tangled. Now imagine being able to secure the ends of all the spaghetti strands to the table before moving the bowl. A broad-toothed comb is placed at the rim of the bowl to add to the orderliness of the strands. When this bowl is dragged over the table surface, the strands of spaghetti will lie parallel to one another, aligned parallel to the direction of bowl movement. This second example represents the ideal orientation of plastic molecules in a living hinge.

The dynamic, thermal, and rheological processes for injection molding make it possible to align the molecules with a strong linear orientation. Sometimes a polymer melt is forced from a thick wall section through a thin wall section. It is possible in these cases to "snag" a section of molecules in the semiplastic region or frozen layer. If the flow channel between the frozen layers is thin enough, the sections of molecules of melt passing between the frozen layers will cause an alignment of the molecular chains, like the strands of spaghetti.

With proper material choice and processing, this creates an extremely durable hinge, which is highly resistant to fatigue.

Good living hinge designs can be seriously compromised by incorrect processing conditions or improperly designed fill patterns.

Polymer structures can be radically altered through the injection-molding process. The strong covalent bonds in the polymer chain can be ruptured if the shear rate for a given material is excessive. When this occurs, the polymer's molecular weight, as well as many physical properties, will be lessened.

Another processing problem that causes hinge failure is the application of excessive pressures to the melt stream after the filling phase has been completed. The filling phase is the period from the moment polymer is first introduced into the cavity, until the cavity has been filled. This is followed by a pressurization phase, during which an additional 15% of material can be added. The last phase is the compensating period or holding time, during which an additional 25% material can be added. When the end of the filling phase is reached and the pressurization phase begins, there is an abrupt pressure increase within the mold cavity and a sudden reduction of the velocity of plastic flow. If pressures are too high at this point, the velocity drops. This can bring about the risk of thermal conduction, causing the material to reach solidification before the part is optimally filled.

The problem often becomes compounded when processors attempt to fill the part by adding even more pressure. But the additional pressure forces the gelled material ahead of the hinge area into the living hinge itself, adding excessive internal stresses. In extreme cases, the material can fold itself into the hinge area.

Indirectly, the type of junction between part and hinge can lead molders to over-pressurize the mold cavity. The inlet to and exit from the hinge area should be blended into the part by generous radii. These radii will significantly affect the pressure drop across the hinge, thus allowing lower pressures to fill the part. They also aid in the process of linear molecular orientation in the hinge area.

The molding process of a part containing a living hinge is a system that consists of three distinct phases: filling, pressurization, and compensating. The process must be designed to be rheologically correct in each of these phases. While there must always be some compromise between hinge function and part aesthetics, these compromises should be addressed during product development and tool design.

7.10 Coined Hinges

Coining or *cold working* is a method of producing a living hinge in a part after the part has been molded. This process produces strong and durable hinges, with good material orientation.

Figure 7.22 shows the typical procedure for manufacturing coined hinges. The component, after molding (Fig. 7.22(a)), is set in a coining bed (Fig. 7.22(b)), also called a coining fixture. Then the coining head moves downwards under pressure exerted by a pneumatic cylinder. The pressure compresses the plastic to the desired thickness.

The polymer is forced out in the area contained between the coining head and bed. The part will be elongated in this area. The elongation also allows an orientation of the molecules of the polymer, producing a hinge of good strength and flexibility.

In order to produce a good coined hinge, the polymer should be exposed to pressures that will induce strains beyond the material's yield strain. This is necessary so that the hinge will be stable. If the strain rate that the polymer is exposed to is lower than its yield strength, the polymer will return partially to its original position once the pressure is removed. As was explained in Chapter 3, a resin will experience plastic deformation only beyond its yield point. If the component is subjected to deformations up to its yield strain level, a portion of its deformation will be elastic. This makes it possible for the polymer to return partially to its original position.

It also should be noted that the pressure applied in the coined area should not create strains in the part that reaches the ultimate strain. This can create a rupture at the molecular level between the chains. This may not be visible externally, but it can break the hinge.

There is some recovery present in the part once the pressure has been removed. This recovery takes place almost instantaneously.

Figure 7.23 shows how the hinge will look once it has been bent. The clearance between the coining head and bed should be much smaller than the final hinge thickness. This play can be controlled by using spacers, which will ensure a uniform thickness throughout the hinge.

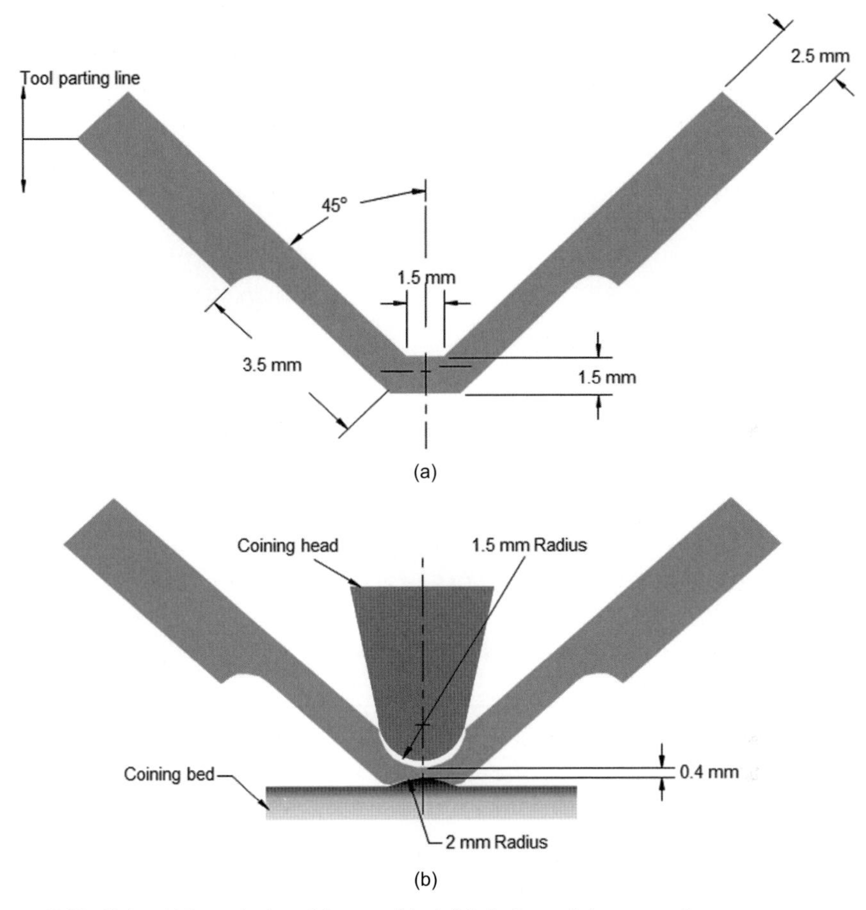

Figure 7.22 Coined hinge design: (a) as molded; (b) during coining operation

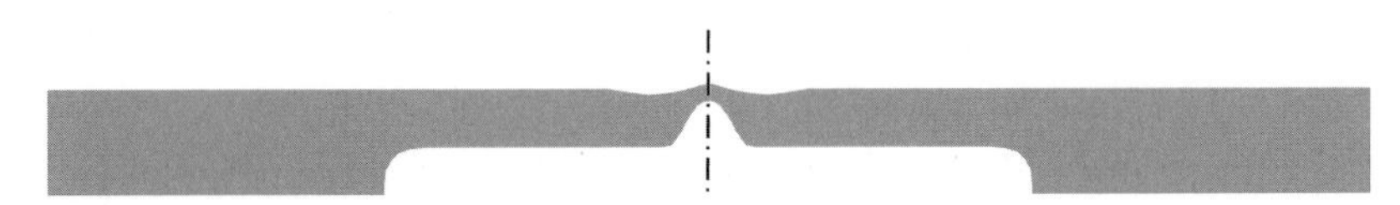

Figure 7.23 Coined hinge after being flexed

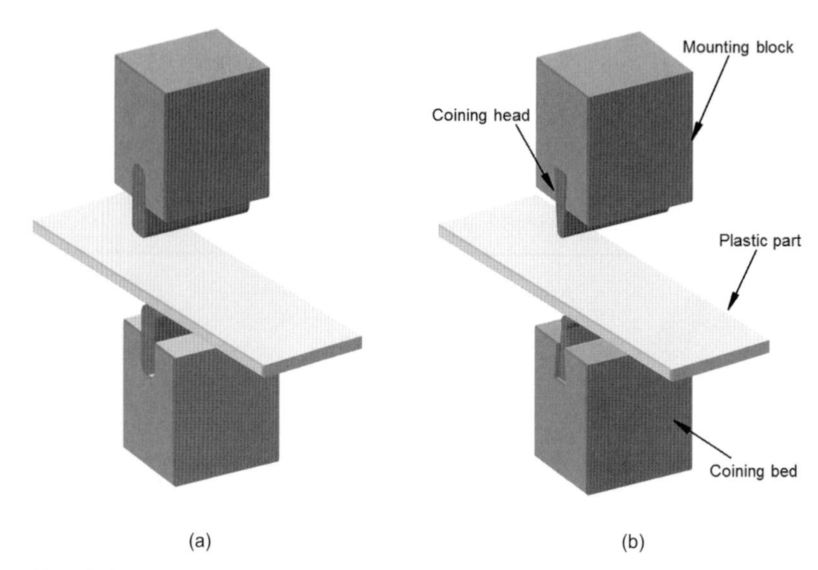

Figure 7.24 Coining process: (a) using parallel-design coining head; (b) using tapered-design coining head

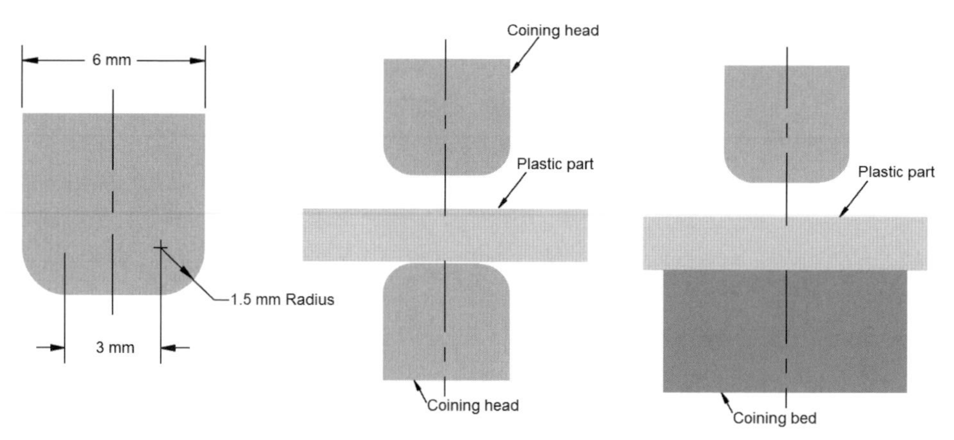

Figure 7.25 Coining head design: (a) parallel design, typical values; (b) coining process with two coining heads; (c) coining process using a coining bed

A hinge design for the coining process requires a thickness of 0.25 to 0.5 mm (0.01 to 0.02 in.). If this thickness range is exceeded, the compression of the fibers at the surface of the hinge area becomes too great. When the compression level reaches the ultimate strength of the polymer, the hinge will fail prematurely.

The coining process can take place at room temperature (known as cold forming), or it is possible to heat up the fixture or the plastic component itself before coining. This temperature should be well below the material glass transition temperature. The higher the temperature of the polymer or the fixture, the lower the pressure applied by the pneumatic cylinder should be.

In general, by heating the fixture, the polymer, or both results in final products with superior polymer properties.

Figures 7.24 and 7.25 show various combinations of the coining tools employed in this process. The flat portion of the coining head as shown in Fig. 7.25(b) should not exceed 3 mm (0.125 in.). Depending on the polymer used in the application, two coining heads usually produce more flexible hinges.

7.11 Oil-Can Designs

There are more than 800 patents worldwide regarding the design feature known as "oil-canning," which incorporates a number of living hinges. In Fig. 7.26 is a sketch of the design principle behind the oil-can feature. There are two beams of different lengths, L_1 and L_2, connected in the middle by a living hinge allowed to rotate on the Z axis and translate in the XY plane. The other end of each beam is also connected to a living hinge; however, they are allowed to rotate in the Z direction but with no translation in the XY plane.

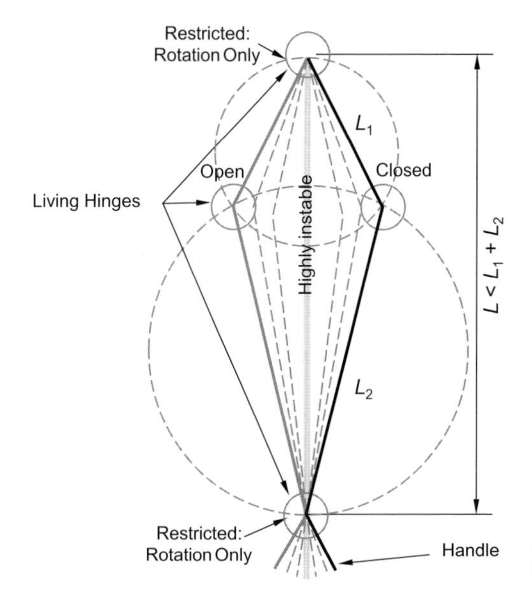

Figure 7.26
Oil-can design principle

Connected to the lower living hinge below the second beam is a small handle.

Applying a force F from right to left, the handle is rotated around the bottom living hinge, which is not allowed to translate in the XY plane, forcing the beam (L_2) to rotate from right to left as well. Connected at the other end of the L_2 beam is beam L_1. The connection of the two beams is achieved with the middle living hinge, which is allowed to rotate and translate. When L_2 is rotating from right to left, it is also forcing L_1 to rotate from right to left as well. The space between the lower and upper living hinges is L. This distance is less than the sum of L_1 and L_2. Once the rotation places the beams in the vertical, central position, the system is highly unstable because both beams are compression-loaded. The beams will be in a resting position, shown in black, on the right side (as closed), or in the other resting position shown in gray, on the left side (as open). They will never be anywhere in between. This represents the oil-can design feature used in many products.

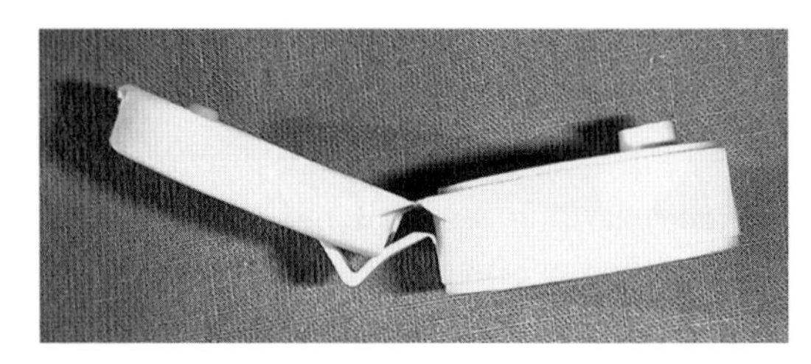

Figure 7.27
Lotion dispenser cap with beam oil-can feature

There are many design options available in creating the oil-can feature. Figure 7.27 shows an oil-can feature created with a single rigid beam and two additional living hinges on each side of the beam. The oil-can feature can be created using two offset planes, two triangles with curved surfaces (see Fig. 7.28), or even with two 3-D ellipsoidal surfaces intersecting (see Fig. 7.29).

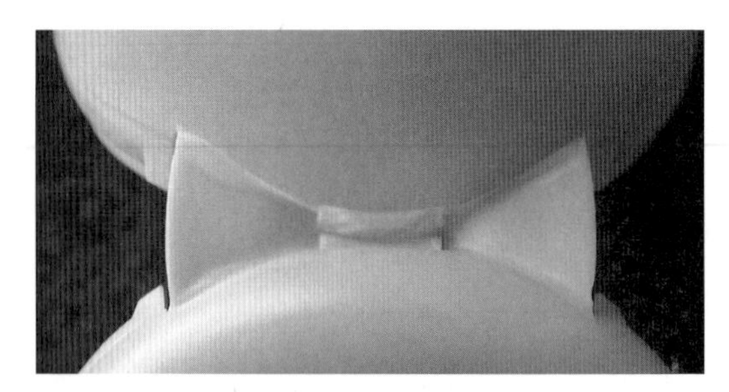

Figure 7.28
Another lotion dispenser having the oil-can feature made with two triangle-shaped planes

Figure 7.29
Fluid-dispensing closure with the oil-can feature created by the intersection of two ellipsoidal surfaces

In the packaging industry, this design is used frequently for toothpaste caps, lotion caps, and so on. Once you open the cap past the midpoint, the cap continues to open by itself due to the high instability of compression loads.

In the automotive industry, the oil-can design is used to create the mechanism inside rearview mirrors that flips the mirror from “day” position to “night” position.

7.12 Conclusion

There are two different design options when designing injection-molded living hinges: designs for commodity polymers, such as polypropylene (PP) and polyethylene (PE) (shown in Figs. 7.1 and 7.2), and designs for “engineering polymers” or other resins (shown in Figs. 7.4 and 7.5). The difference between the two design options relies on the fact that for PP and PE resins the longitudinal cross section through the living hinge appears as an arc, while for other polymers the living hinge walls are parallel.

The gate location has a major effect on living hinge performance. It is important to communicate with the toolmaker and the molder to ensure proper gate location and processing conditions. Whenever possible do not have cooling lines close to the living hinge area in an injection-molded part.

7.13 Exercise

A component contained within a consumer product needs to move at least one million times. To achieve such performance from a plastic part the designer selects injection-molded living hinges as integral features. The production drawing of the plastic part, called a flex element, is shown in Fig. 7.30. The flex element has four living hinges. The free body diagram of the flex element while in operation is shown in Fig. 7.31. A 20-pound force is exercised in a vertical plane during the one million cycles of desired operation. Figure 7.32 shows the operation of the flex element: it is moving left and right or back and forth, depending from which angle the flex element is observed, 11° per side.

The designer relied upon a toolmaker to build the tool to meet the performance criteria of the product. The gating scenario chosen by the toolmaker is shown in Fig. 7.33. Figure 7.34 shows the runner system and a cross section through the runner itself. The tool is a three-plate tool (also known as self-degating tool), which allows the molder to automatically separate the runner system from the part when parts are ejected from the tool cavity. Figure 7.35 shows the pinpoint gate design that allows the molten polymer to reach the tool cavity and fill the part.

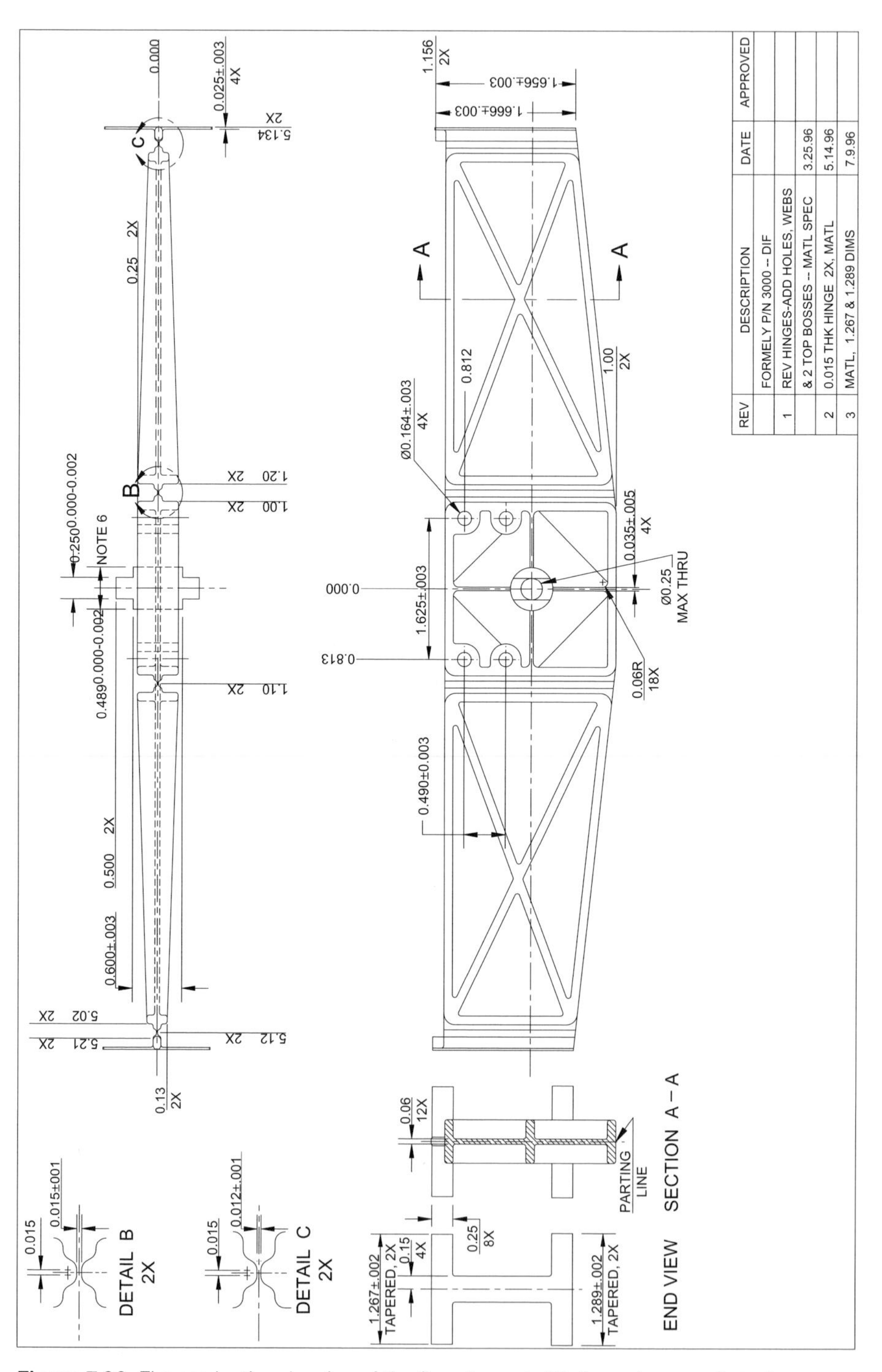

Figure 7.30 The production drawing of the flex element. All dimensions are in inches

Table 7.1 Polypropylene Material Properties

PP Grade	PD-626	PD-702	Units
Yield stress	4,500	4,600	psi
Yield strain	13	12	%
Flexural modulus @ 1% secant	160,000	170,000	psi
Rockwell hardness	86	89	R scale
HDT (Heat deflection temperature) @ 66 psi	181	203	°F
Notched Izod impact @ 73°F	0.7	0.6	ft·lbs/in.
Density	0.902	0.9	g/cm^3
Melt flow rate	12	35	dg/min

When the product is in use, the living hinges labeled "Hinge 2" and "Hinge 3" (Fig. 7.31) fail prematurely. Instead of having a life expectancy of one million flexing cycles or better, the two living hinges initiate microcracks around 500,000 cycles and break soon thereafter.

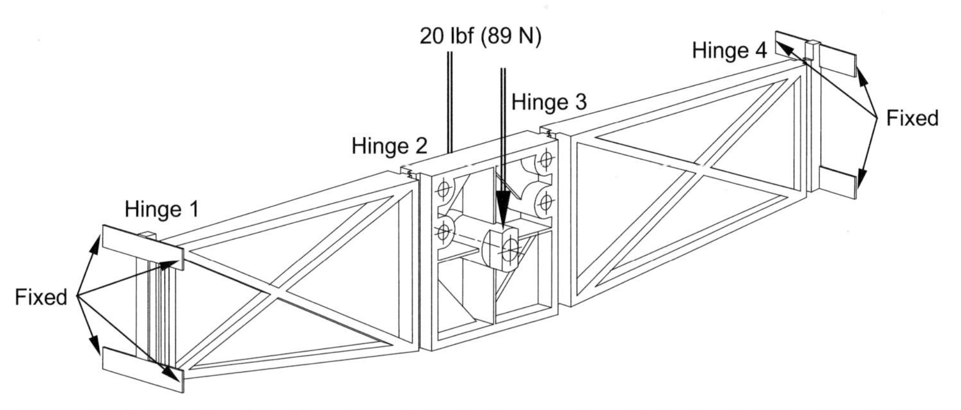

Figure 7.31 3-D view of the flex element. During its operation, the horizontal thin plates having fixed ends hold the flex element in its location. While in operation (when flexing in and out of the paper plane), the component is loaded with a 20-pound force vertical load as shown *(Courtesy of ETS, Inc.)*

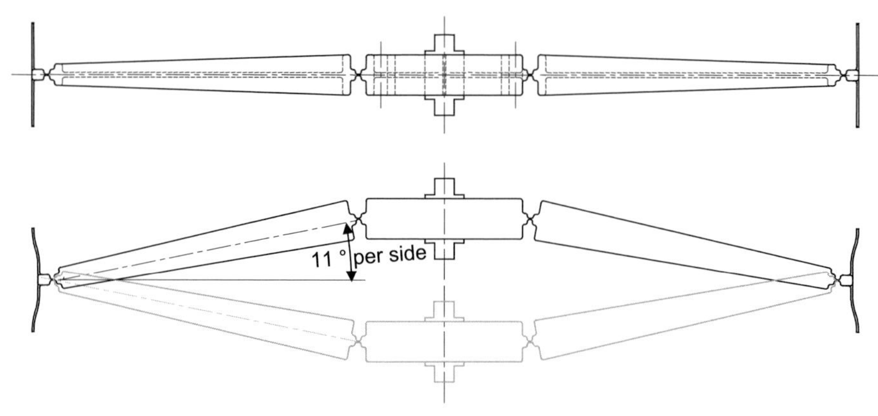

Figure 7.32 During operation, flex element moves left and right (or in and out as in Fig. 7.31), 11° per side *(Courtesy of ETS, Inc.)*

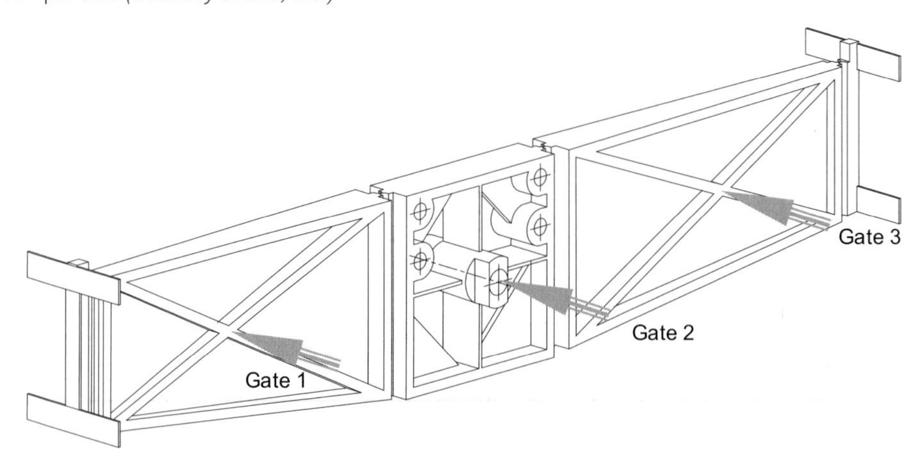

Figure 7.33 The location of the three gates *(Courtesy of ETS, Inc.)*

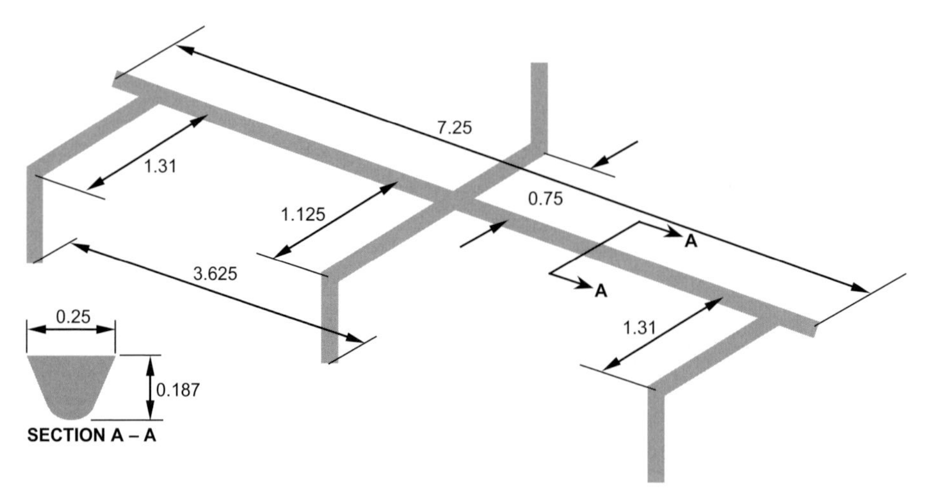

Figure 7.34 The runner system used in production. All dimensions are in inches *(Courtesy of ETS, Inc.)*

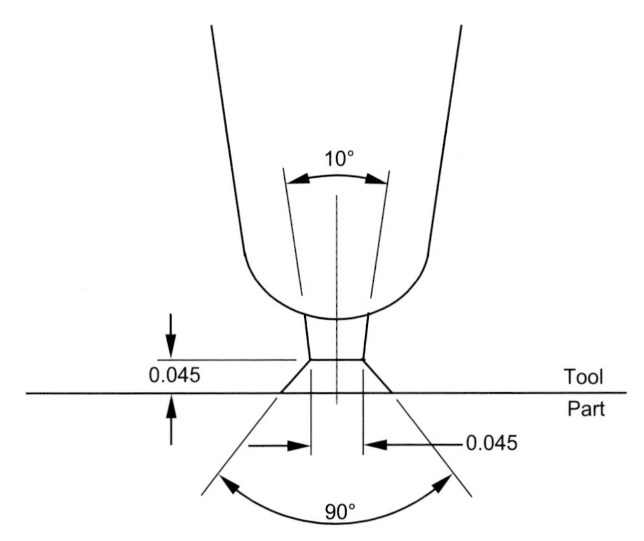

Figure 7.35 Pinpoint gate design utilized. All dimensions are in inches *(Courtesy of ETS, Inc.)*

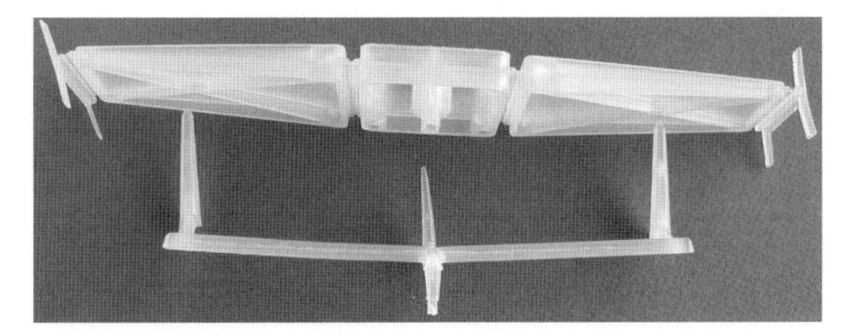

Figure 7.36 Actual production part and its runner system *(Courtesy of ETS, Inc.)*

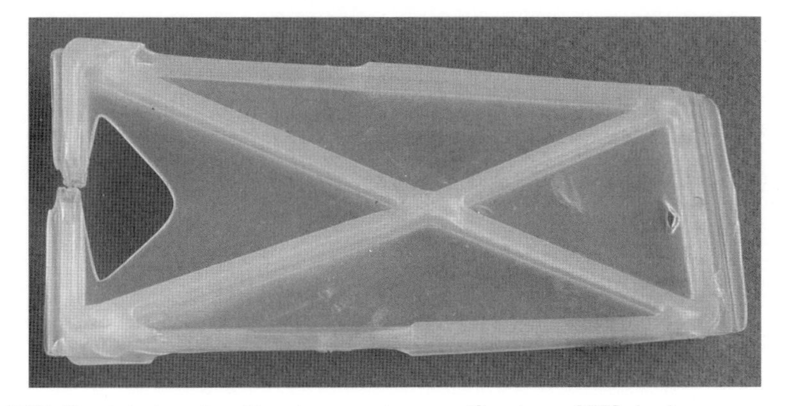

Figure 7.37 Short shot made with only one gate open *(Courtesy of ETS, Inc.)*

Using the information provided (Figs. 7.30 through 7.36 and all of the information described in the textbook until now) and selecting one of the two polymers shown in Table 7.1, solve for:

1. Achieve at least one million-cycle performance for the living hinges, which were initially failing around 500,000 cycles.
2. Using a very simple structural design criterion (also known as "design trick") further improves the performance of the living hinges from one million cycles to at least three million cycles.

To successfully solve this exercise the production drawing and the gating scheme should be reviewed and used.

(Hint: No calculations are required whatsoever.)

8 Snap Fitting

8.1 Introduction

Like the press fit, the *snap fit* is a simple assembly method that joins two parts without using any additional components or fasteners. A snap fit, also referred to as a lock arm, consists of a hook and a groove. During assembly, the hook is deflected or partially deflected by the mating part. Once inside the groove, the hook returns to its original position. The hook-and-groove interaction gives the snap fit its gripping force.

Snap fits can be used to join parts made of dissimilar polymers, or completely different materials, such as plastic and metal. They are used in a variety of industries to assemble power tools, computer cases, electronic components, packaging boxes, toys, automotive parts, medical devices, and thousands of other products. There is a trend to simplify manufacturing costs, and snap fits do this by acting as fasteners that are built into the parts themselves. They preclude additional tool investment and facilitate faster assembly. Depending on the design used, the snap-in features of the parts can be hidden from view in the final product. Successful snap fits rely on precise engineering and, although they have been in use for many years, only recently have the demands of manufacturing forced engineers to develop more reliable snap fit designs.

There are two major categories of snap fits. *Permanent* or one-time assembly snap fits, mostly used for disposable consumer products, are assembled in the manufacturing process, never to be removed. *Multiple* snap fits are used in applications, such as pen caps and bottle caps, that may be opened and closed many times, and products like automotive parts that have to be disassembled for servicing.

Both categories include several snap fit design families. A *cantilever beam* is a basic hook-and-groove joint with a beam that fits axially into a slot in the mating part. A *curved beam* is a variation of the cantilever type, which includes a bend in the beam. The *annular* snap fit is a round or oval joint found in products such as pen and bottle caps. The *spherical* snap fit features a dome-shaped protrusion that snaps into an indentation in the mating part. A *torsional* beam snap fit uses the shear stress of a second beam to hold it in position.

Snap fits can greatly benefit the manufacturing process. By reducing the number of parts, they can save on warehousing costs, reduce labor costs, cut inventory, reduce the number of suppliers required, and cut down on shipping and handling and all other costs associated with additional parts. They also save time in assembly.

But snap fits rely more on up-front engineering than other processes. Snap fits that are not designed properly can break in assembly or even before assembly.

This chapter will look closely at three snap fit families, present ways of selecting material, review geometry, and perform detailed design analysis.

■ 8.2 Material Considerations

Materials have a great influence upon the snap fit design. Polymers can be generally characterized as either stiff or flexible. Depending on the application requirements, both characteristics may be desirable for snap fits.

Because snap fits are often used to hold components that have show surfaces, the material must satisfy both functional and aesthetic criteria. Depending on the application, the engineer should consider a number of factors.

Ultraviolet color retention. Show surfaces may require UV stabilizers, otherwise the material's color and mechanical properties may deteriorate in time, resulting in eventual breakage.

Paintability. If the part will be painted, a material compatible with the paint should be selected, or the paint may attack the polymer.

Molded-in color. Some color additives, particularly cadmium red, reduce the polymer's material properties.

Thermal extremes. The parts can expand or contract at different rates due to temperature change. The variance in rates can be 1 to 2 times in cases of plastic-to-plastic assemblies, or 5 to 10 times in cases of plastic-to-metal assemblies.

Crack propagation resistance. It is important when designing snap fits to select materials with low crack-propagation properties.

Chemical resistance. Chemicals to which the part will be exposed must also be considered. For example, under-hood automotive components must be able to withstand spills of oils, antifreeze, and other chemicals.

Sometimes, in dynamic applications such as power tools and washing machines, noises such as rattles and squeaks are generated within the snap fit joint. This can be prevented by taking full advantage of the material's elastic properties. Designing a built-in preload, an interference that sustains pressure between the two parts, will use the material's elasticity to prevent the unwanted noise. In some cases, the preload may increase the material creep or stress relaxation over the life of the assembly. It should also be noted that the material's elastic properties are very helpful in absorbing tolerances, thereby creating a more robust design (a design by which a product or a process yields very little variance, also called noise, from one unit to another).

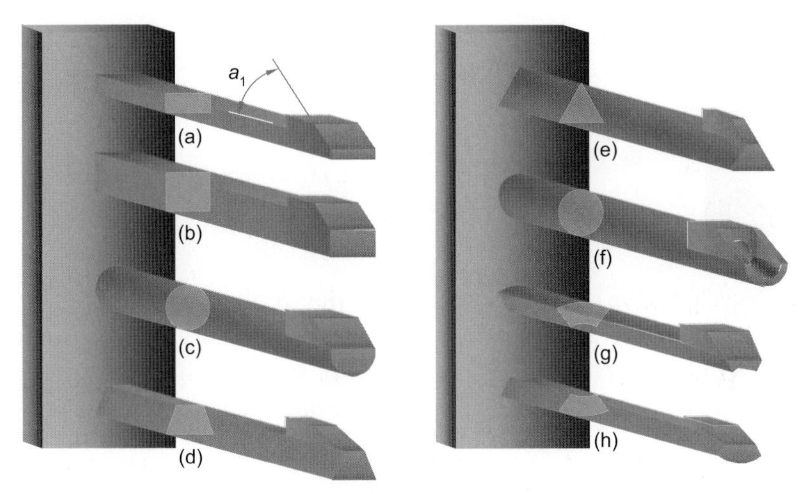

Figure 8.1 One-way snap fit, continuous beam: (a) rectangular cross section; (b) square cross section; (c) round cross section; (d) trapezoidal cross section; (e) triangular cross section; (f) round hollow cross section; (g) convex cross section; (h) concave cross section

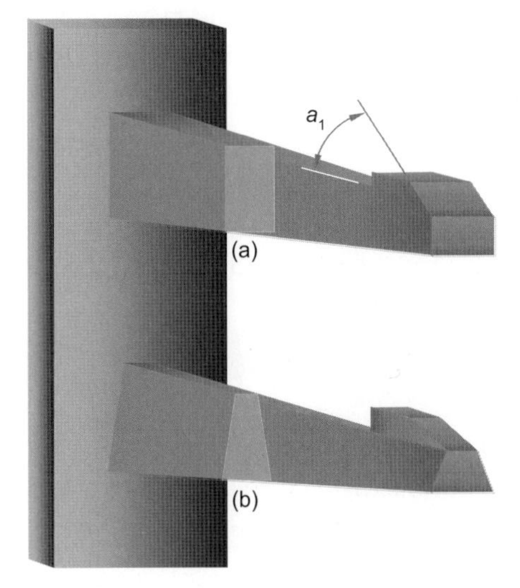

Figure 8.2
One-way snap fit, vertically tapered beam: (a) rectangular cross section; (b) trapezoidal cross section

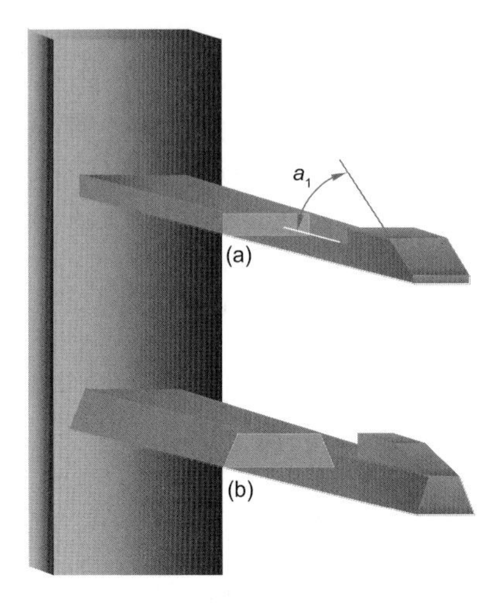

Figure 8.3
One-way snap fit, horizontally tapered beam: (a) rectangular cross section; (b) trapezoidal cross section

Regrind (recycled scrap or used parts) added to the virgin polymer reduces the material's mechanical properties. Therefore, materials containing smaller percentages of regrind should be selected for parts that feature snap fits. If the design has been tested using a given percentage regrind and the mechanical properties prove satisfactory, then this should not be a concern.

Other problems may arise when the material properties important to the specific design are not considered thoroughly. Examples include using compressive strain for a torsional beam instead of the torsion strain, designing a snap (hook) that is either too stiff or too short for the given application, having excessive deflection of the snap and possibly reaching the plastic deformation region of the material, not having a fillet at the beam root, using ribs that excessively stiffen the lock arm, and neglecting the consideration of impact loading in dynamic applications.

The environmental conditions to which the part will be exposed, such as gamma radiation, chemical spills, and operating temperature, must be considered.

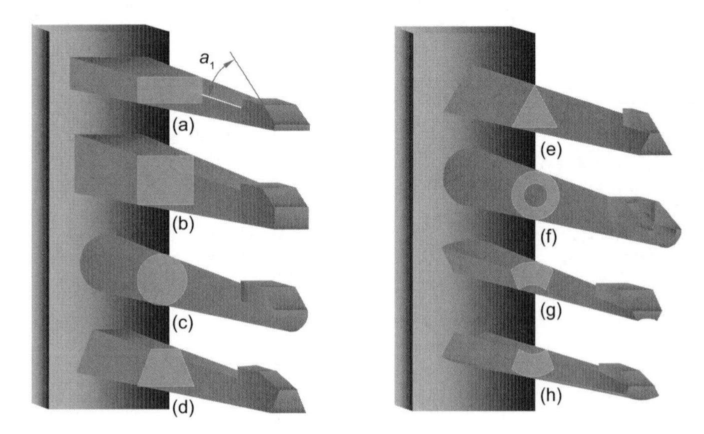

Figure 8.4 One-way snap fit, double-tapered beam: (a) rectangular cross section; (b) square cross section; (c) round cross section; (d) trapezoidal cross section; (e) triangular cross section; (f) round hollow cross section; (g) convex cross section; (h) concave cross section

■ 8.3 Design Considerations

There are a number of issues to be considered when designing a snap fit.

The packaging requirement refers to the amount of space surrounding the snap fit joint. There should be enough room for the function and motions of the parts themselves, as well as for hands and tools to reach the parts when assembling and disassembling the system.

Components may also require the inclusion of symbols that direct servicing or removal of the parts from the assembly.

Another early design consideration is the workloads the snap fit assembly will experience during usage and during shipping from the part producer to the assembly location. Some of the workloads include mass loads, operational loads, and impact loads.

Some applications require the snap fit to perform a function in addition to joining the parts. Snap fits can be designed to seal against water, dust, or even air. In some of these cases proper O-rings are required; other similar parts should be employed to obtain a suitable assembly that can be used as a seal.

When loads are carried by snap fits, the joint should be designed to ensure that the parts are nesting together by means of mechanical interference. The snap fit should only exist to maintain the nested relationship between parts.

In some cases, two parts made of rigid plastics or metal, which cannot deflect in a hook-and-groove type design, have to be joined. To accomplish this, a third piece can be designed to snap on or around the two parts, holding them together.

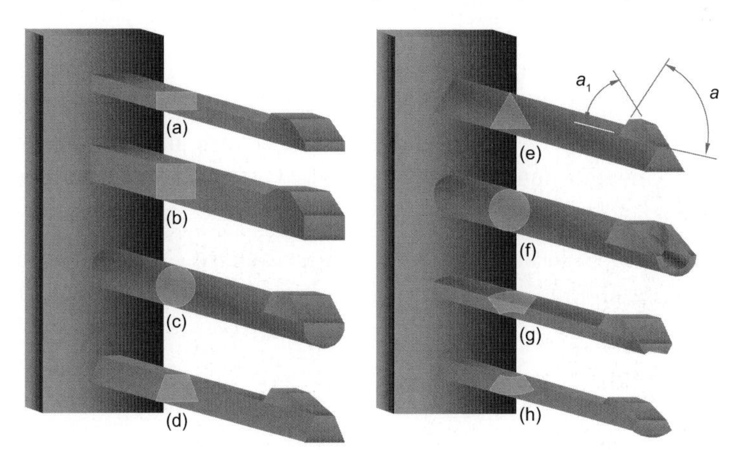

Figure 8.5 Two-way snap fit, continuous beam: (a) rectangular cross section; (b) square cross section; (c) round cross section; (d) trapezoidal cross section; (e) triangular cross section; (f) round hollow cross section; (g) convex cross section; (h) concave cross section

The determination of the load requirements for assembling parts is an important step in snap fit design. Whether the assembly part will be assembled manually or automatically, as well as the amount of load the joint will carry in its operation, must be considered. Positioning of the parts during assembly must be determined at the design stage. For manual and automatic assemblies, locators for positioning the parts should be incorporated in the components themselves. Locating pins can also be included in the assembly fixtures for automatic assemblies.

The friction coefficient is a factor that influences the engagement and disengagement forces needed to assemble and disassemble the part. Ergonomic studies show that injuries based on repetitive manual work take place when the maximum force of 27 N (6 lbf) for hands, 11 N (2.4 lbf) for thumbs, or 9 N (2 lbf) for fingers is exceeded. The motion associated with repetitive manual assembly work should be linear, preferably push rather than pull, and for vertical assemblies, from the top down. The position associated with this kind of work should be in a normal standing or sitting position.

Theoretically, the friction coefficient of two materials can vary between 0 and 1, without reaching the extremes. In practice, values range between 0.1 and 0.6. The static coefficient of friction is greater when tested statically than dynamically. Besides the material itself, the friction coefficient varies also based on the surface roughness of the mating parts. The rougher the surface, the higher the coefficient of friction. In certain applications, where the parts are immersed in various oils, the difference between the static and dynamic coefficients of friction is insignificant.

Before presenting a detailed procedure for snap fit analysis, a simplified guide can be used to give a preliminary profile of a proposed snap fit joint. The lock arm width can be approximated as less than half the lock arm length. For a tapered lock arm, the height at base can be approximated as 1.25 to 1.4 times the height of the tip. When the lock arm extends in the same plane as the part wall (the wall also deflects when the hook is engaged) a finite element analysis (FEA) needs to be performed. The fillet at the base of the lock arm (where the lock arm joins the rest of the part) should be approximately one-third the base height. These calculations are shown in detail in Section 8.4.

It is important to avoid locating the lock arm near sharp corners, sink marks, the part gate, or the weld line, as these can all cause internal stress concentrations in the part.

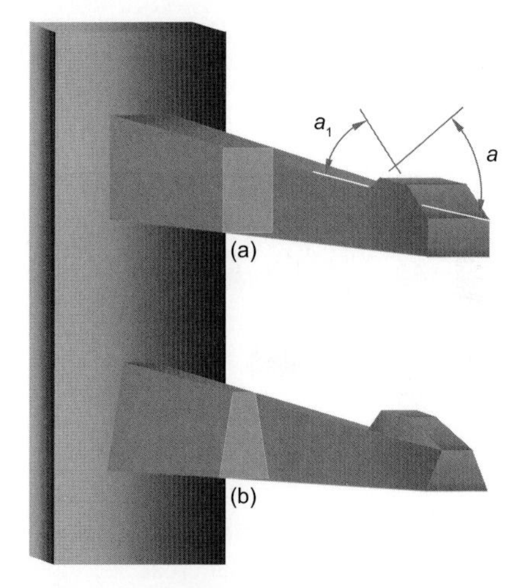

Figure 8.6
Two-way snap fit, vertically tapered beam: (a) rectangular cross section; (b) trapezoidal cross section

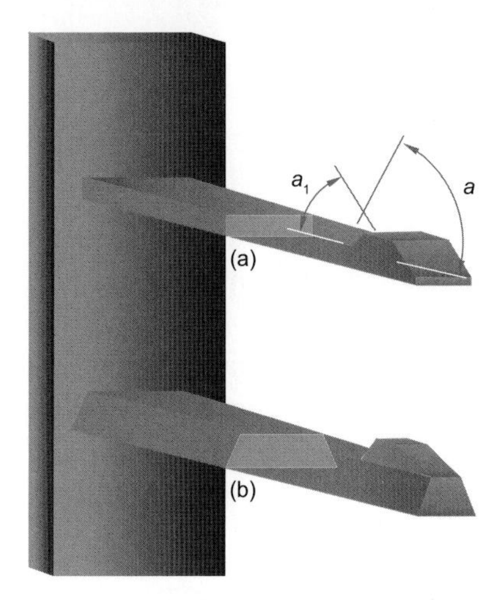

Figure 8.7
Two-way snap fit, horizontally tapered beam: (a) rectangular cross section; (b) trapezoidal cross section

8.3.1 Safety Factors

Recommended safety factors should be used to calculate the design stress limit. Materials that have a distinct yield point in parts that will be engaged only once should have a safety factor of 1.5 based on yield strength. For multiple assemblies and disassemblies with a distinct yield point the safety factor is 2.5.

Parts made from materials that do not exhibit a distinct yield point, such as fiber-reinforced polymers with only one assembly, will have a safety factor of 2. For similar materials in multiple assembly applications, the recommended safety factor is 3.25.

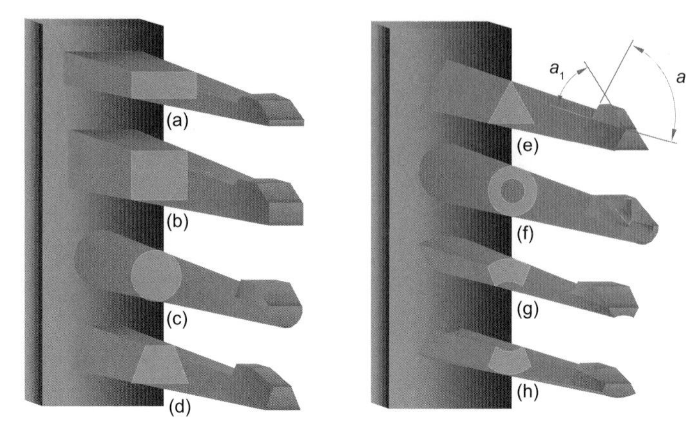

Figure 8.8 Two-way snap fit, double-tapered beam: (a) rectangular cross section; (b) square cross section; (c) round cross section; (d) trapezoidal cross section; (e) triangular cross section; (f) round hollow cross section; (g) convex cross section; (h) concave cross section

8.4 Snap Fit Theory

In this section, equations and formulas are developed for closed-loop solutions. The equations presented are rather complicated. References 46, 51, 52, 54, 133, 162, and 163 show computer software available in the marketplace to automate this process for certain snap families. There will always be designs for which hand calculations or FEA will be required. This section will detail how formulas should be developed in general. It will be applicable when procedures for other less-used snap fits are needed.

The math presented here is for a one-way, double-tapered cantilever beam.

8.4.1 Notations

The following notations will be used to develop formulas for cantilever beam snap fits.

M = Bending moment

L = Length of element

b_{Base} = Maximum width at root

b_{Tip} = Minimum width at tip

h_{Base} = Maximum height at root

h_{Tip} = Minimum height at tip

P = Snap-in force

F = Engagement force

ϕ = Angle of deflection

y = Deflection

y_{Max} = Maximum deflection at tip

I_Z = Moment of inertia relative to the z axis

I_0 = Moment of inertia for the root cross section

C_G = Geometric constant

$\varepsilon_{\text{Design}}$ = Design or permissible strain

$\varepsilon_{\text{Ultimate}}$ = Ultimate strain

$\sigma(x)$ = Instantaneous stress

σ_{Design} = Design or permissible stress

σ_{Ultimate} = Ultimate stress

n = Safety factor

μ = Coefficient of friction

E_S = Secant modulus of elasticity

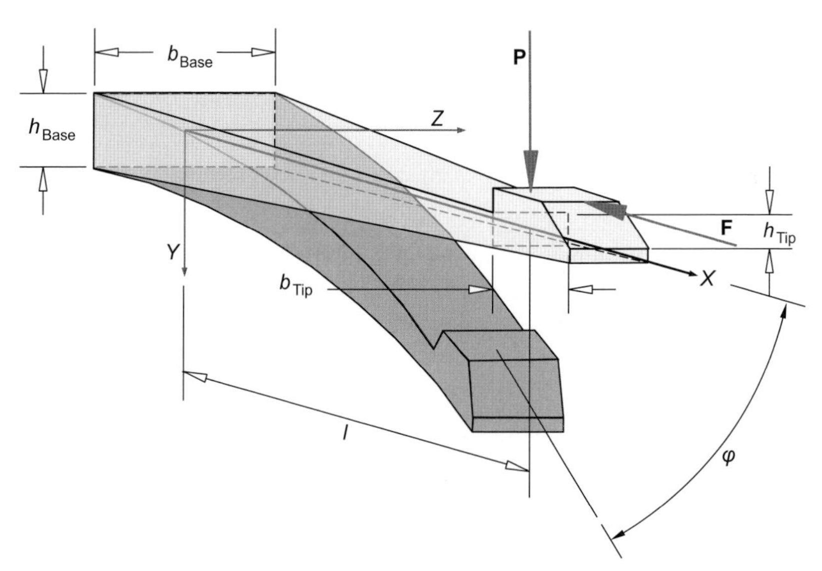

Figure 8.9 Boundary conditions for a one-way, double-tapered cantilever beam with rectangular cross section

8.4.2 Geometric Conditions

Considering a one- or two-way, double-tapered cantilever beam, two ratios, K_1 and K_2, can be determined; K_1 represents the ratio between the beam width at base (root) and the width at the tip of the beam:

$$K_1 = \frac{b_{\text{Tip}}}{b_{\text{Base}}} \tag{8.1}$$

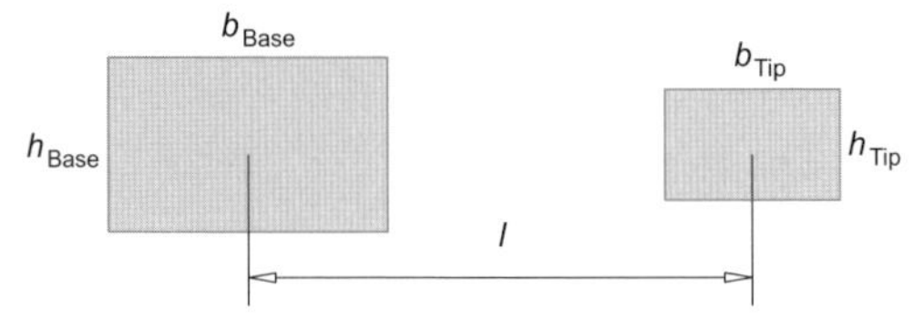

Figure 8.10 Cross section of the cantilever beam at the root and at the tip (undercut)

Similarly, a ratio between the beam height at the base (root) and the height at the tip of the beam is determined:

$$K_2 = \frac{h_{\text{Tip}}}{h_{\text{Base}}} \tag{8.2}$$

In order to further simplify calculations, a parametric transformation using the letters a_1 and a_2 as parameters is

$$a_1 = \frac{K_1 - 1}{l} \tag{8.3}$$

$$a_2 = \frac{K_2 - 1}{l} \tag{8.4}$$

Based on the above parameters we can determine the instantaneous beam width and height of the cantilever beam for any location of the section plane along the beam length l, which varies from 0 at the base (root) to 1 at the beam tip. Therefore, the instantaneous width is

$$b(x) = b_{\text{Base}}(1 + a_1 x) \tag{8.5}$$

The corresponding instantaneous height is

$$h(x) = h_{\text{Base}}(1 + a_2 x) \tag{8.6}$$

Based on 8.5 and 8.6, the moment of inertia at the root is

$$I_0 = \frac{b_{\text{Base}} h_{\text{Base}}^2}{12} \tag{8.7}$$

8.4.3 Stress/Strain Curve and Formulae

The graph in Fig. 8.11 shows a stress/strain curve for a typical plastic material. Two points are identified on the curve. The point on the right is designated as Yield. The other point to the left is known as the design point. The ratio between the stress or strain values between the two points is referred to in the technical literature as a safety factor (see Chapter 2). Therefore, the safety factor is

$$n = \frac{\varepsilon_{\text{Yield}}}{\varepsilon_{\text{Design}}} \tag{8.8}$$

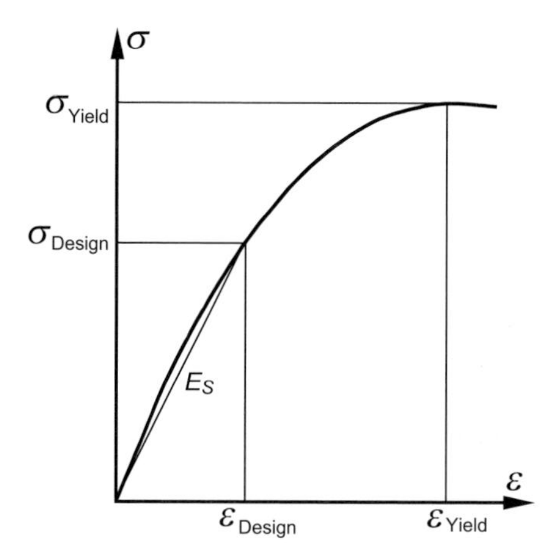

Figure 8.11
Stress/strain diagram

Now, the stress can be found by using the secant modulus of elasticity and the design strain limit for a given material:

$$\sigma_{\text{Design}} = E_S \varepsilon_{\text{Design}} \tag{8.9}$$

The instantaneous stress along the beam is a function of moment, radius of inertia, and moment of inertia:

$$\sigma(x) = \frac{M(x) z_{\text{Max}}}{I_Z(x)} \tag{8.10}$$

Using the condition at limit, the stress level of the cantilever beam at the base (root) is found. Therefore, for $x = 0$

$$\sigma(0) = \frac{M(0) z_{\text{Max}}}{I_Z(0)} = \frac{Pl}{I_0} z_{\text{Max}} \tag{8.11}$$

and further we have that

$$\frac{Pl}{E_S I_0} = \frac{\varepsilon_{\text{Design}}}{z_{\text{Max}}} \tag{8.12}$$

The instantaneous bending moment (the moment that is present in any given section of the beam) is

$$M(x) = -P(l-x) \tag{8.13}$$

Based on the above equations, the deflection of the beam is written as

$$y = \frac{d^2y}{dx^2} = -\frac{M(x)}{EI_Z(x)} \tag{8.14}$$

Therefore

$$\frac{d^2y}{dx^2} = \frac{P(l-x)}{EI_Z(x)} \tag{8.15}$$

where

$$I_Z(x) = I_0 f(x) \tag{8.16}$$

Using the notation shown in 8.16, the deflection y is rewritten as

$$y = \frac{P(l-x)dx}{EI_0(x)} = \frac{P}{EI_0}\frac{(l-x)dx}{f(x)} \tag{8.17}$$

We calculate the above double integral, integrating it by parts in the following form:

$$y = F_1(x) + F_2(x) \tag{8.18}$$

Functions $F_1(x)$ and $F_2(x)$ are defined as

$$F_1(x) = \int \frac{(l-x)dx}{f(x)} + C_1 \tag{8.19}$$

$$F_2(x) = \int F_1(x) + C_2 \tag{8.20}$$

8.4.4 Instantaneous Moment of Inertia

The instantaneous (at any given cross section) moment of inertia is

$$I_Z(x) = \frac{b(x)h^3(x)}{12} \tag{8.21}$$

By replacing b and h with their parametric form one obtains

$$I_Z(x) = \frac{b_0 h_0^3 (1 + a_1 x)(1 + a_2 x)^3}{12} \tag{8.22}$$

or

$$I_0 = \frac{b_0 h_0^3}{12} \tag{8.23}$$

Finally, the equation for the moment of inertia is

$$I_Z(x) = I_0(1+a_1x)(1+a_2x)^3 \tag{8.24}$$

8.4.5 Angle of Deflection

Using the instantaneous moment of inertia determined in 8.24 and replacing it in the angle of deflection equation, one obtains

$$\varphi(x) = \int \frac{P(l-x)dx}{EI_Z(x)} \tag{8.25}$$

or, with the constants placed outside the integral signs

$$\varphi(x) = \frac{P}{EI_0}\int \frac{(l-x)dx}{(1+a_1x)(1+a_2x)^3} \tag{8.26}$$

8.4.6 Integral Solution

Equation 8.26 can be separated into two simpler integrals as follows:

$$\varphi(x) = \frac{P}{EI_0}\int \frac{dx}{(1+a_1x)(1+a_2x)^3} - \frac{P}{EI_0}\int \frac{xdx}{(1+a_1x)(1+a_2x)^3} \tag{8.27}$$

Using the notation J_1 and J_2 for the two integrals we obtain a simplified equation for the angle of deflection:

$$\varphi(x) = \frac{P}{EI_0}J_1 - \frac{P}{EI_0}J_2 \tag{8.28}$$

where J_1 is

$$J_1 = \frac{a_1}{(a_2-a_1)(1+a_2)} - \frac{l}{(a_2-a_1)(1+a_2x)^2} - \frac{a_1}{(a_2-a_1)^3}\ln\left|\frac{1+a_1x}{1+a_2x}\right| \tag{8.29}$$

and J_2 is

$$J_2 = \int \frac{xdx}{(1+a_1x)(1+a_2x)^3} \tag{8.30}$$

or

$$J_2 = \int \frac{Adx}{1+a_1x} + \int \frac{Bdx}{1+a_2x} + \frac{Cdx}{(1+a_2x)^2} + \int \frac{Ddx}{(1+a_2x)^3} \tag{8.31}$$

By multiplying the numerators by the common denominator, and by setting the condition that the numerator equals x

$$A(1+a_2x)^3 + B(1+a_2x)^2 + C(1+a_1x)(1+a_2x) + D(1+a_1x) = x \tag{8.32}$$

To find out A, B, C, and D, a system of four equations with four unknowns can be formed as follows:

$$a_2A + a_1B = 0$$

$$3a_2A + (2a_1 + a_2)B + a_1C = 0$$

$$3a_2A + (a_1 + 2a_2)B + (a_1 + a_2)C + a_1D = 1$$

$$A + B + C + D = 0 \tag{8.33}$$

Once solved, the solutions are

$$A = \frac{a_1^2}{(a_1 - a_2)^3} \tag{8.34}$$

$$B = \frac{a_1a_2}{(a_1 - a_2)^3} \tag{8.35}$$

$$C = \frac{a_2}{a_1 - a_2} \tag{8.36}$$

$$D = \frac{1}{a_1 - a_2} \tag{8.37}$$

Therefore, the integral (8.31) becomes

$$J_2 = \frac{A}{a_1}\ln(1 + a_1x) + \frac{B}{a_2}\ln(1 + a_2x) + \frac{C}{a_2(1 + a_2x)} + \frac{D}{2a_2(1 + a_2x)} \tag{8.38}$$

or further

$$J_2 = \frac{a_1}{(a_2 - a_1)^3}\ln\left(\frac{1 + a_1x}{1 + a_2x}\right) - \frac{1}{(a_2 - a_1)^2(1 + a_2x)} + \frac{1}{2(a_2 - a_1)(1 + a_2x)^2} \tag{8.39}$$

Now we can replace the two coefficients J_1 and J_2 of the angle of deflection in the 8.28 as follows:

$$\varphi(x) = \frac{P}{EI_0}\left[\frac{a_1l}{(a_2 - a_1)(1 + a_2x)^2} - \frac{1}{2(a_2 - a_1)(1 + a_2x)^2} - \frac{a_1^2l}{(a_2 - a_1)^2}\ln\left|\frac{1 + a_1x}{1 + a_2x}\right|\right] -$$

$$-\frac{P}{EI_0}\left[\frac{1}{(a_2 - a_1)^2(1 + a_2x)} - \frac{1}{2a_2(a_2 - a_1)(1 + a_2x)^2} + \frac{a_1}{(a_2 - a_1)^2}\ln\left|\frac{1 + a_1x}{1 + a_2x}\right|\right] + C_1 \tag{8.40}$$

The constant of integration C_1 is found by applying the conditions at limit: when the length of the beam is zero or $x = 0$, the cantilever beam angle of deflection will also be zero, or $\phi(0) = 0$. So, by using the condition that the deflection angle at the root must be zero, the constant of integration is

$$C_1 = \frac{(a_2 - a_1)(1 - a_2 l) - 2a_2(1 - a_1 l)}{2a_2(a_2 - a_1)^2} \tag{8.41}$$

8.4.7 Equation of Deflection

By integrating the angle of deflection, 8.40, the amount of beam deflection (undercut) is found:

$$y(x) = \int \varphi(x)dx \tag{8.42}$$

Equation 8.42 represents the equation for deflection for a two-way, double-tapered cantilever beam.

8.4.8 Integral Solution

To calculate $y(x)$ it is necessary to integrate 8.42. By replacing $\phi(x)$ in 8.42 and integrating, we have

$$y(x) = \frac{\mathrm{P}}{EI_0}\left[\frac{1 + a_2 l}{(a_2 - a_1)^3}\frac{1}{a_2}\ln(1 + a_2 x) + \frac{1 + a_2 l}{2a_2^2(a_2 - a_1)^2}\right] -$$

$$-\frac{\mathrm{P}}{EI_0}\left[\frac{1 + a_1 l}{(a_2 - a_1)^3}a_2(1 + a_1 x)\ln(1 + a_1 x)\right] -$$

$$-\frac{P}{EI_0}\left[\frac{1 + a_1 l}{(a_2 - a_1)^3}\frac{a_1}{a_2}(1 + a_2 x)\ln(1 + a_2 x)\right] + C_1 x + C_2 \tag{8.43}$$

The second constant of integration, C_2, is defined by using a condition at limit that $z(0) = 0$ for $x = 0$. Therefore C_2 is

$$C_2 = \frac{1 + a_2 l}{2a_2(a_2 - a_1)} \tag{8.44}$$

Then by replacing C_1 and C_2 in 8.43, the deformation of the cantilever beam will have the following form:

$$y(x)=\frac{\mathrm{P}}{EI_0}\left[\frac{1+a_1 l}{(a_2-a_1)}\frac{1}{a_2}\ln(1+a_2 x)+\frac{1}{1+a_2 x}\frac{1+a_1 l}{2a_2^2(a_2-a_1)^2}\right]-$$

$$-\frac{\mathrm{P}}{EI_0}\frac{1+a_1 l}{(a_2-a_1)^3}\left[(1+a_1 x)\ln(1+a_1 x)+\frac{a_1(1+a_2 x)\ln(1+a_1 x)}{a_2}\right]+$$

$$+\frac{\mathrm{P}}{EI_0}\left[\frac{(a_2-a_1)(1+a_2 l)-2a_2(1+a_1 l)}{2a_2(a_2-a_1)}-\frac{1+a_2 l}{2a_2^2(a_2-a_1)}\right] \tag{8.45}$$

8.4.9 Maximum Deflection

Equation 8.45 represents the amount of deflection a cantilever beam is capable of sustaining. To determine the maximum deflection (undercut) possible, we need to replace $x = l$ in 8.45 and a_1 and a_2 with their parametric substitutions (8.46 through 8.48):

$$a_1=\frac{K_1-1}{l} \tag{8.46}$$

$$a_2=\frac{K_2-1}{l} \tag{8.47}$$

$$a_2-a_1=\frac{K_2-K_1}{l} \tag{8.48}$$

Then the maximum deflection that can be sustained by a cantilever beam with a rectangular cross section is

$$y_{\mathrm{Max}}=\frac{\mathrm{P}l^3}{EI_0}\left[\frac{K_1\ln K_2}{(K_2-1)(K_2-K_1)^2}+\frac{1}{2(K_2-1)^2(K_2-K_1)}\right]-$$

$$\frac{\mathrm{P}l^3}{EI_0}\frac{K_1(K_1-1)}{(K_2-K_1)^3}\left[\frac{K_1\ln K_1}{(K_1-1)}-\frac{K_2\ln K_2}{K_2-1}\right]+$$

$$-\frac{\mathrm{P}l^3}{EI_0}\left[\frac{(K_2-K_1)K_2-2K_1(K_2-l)}{2(K_2-l)(K_2-K_1)^2}-\frac{K_2}{2(K_2-l)(K_2-K_1)}\right] \tag{8.49}$$

After simplifications, 8.49 becomes

$$y_{\mathrm{Max}}=\frac{\mathrm{P}l^3}{EI_0}\frac{2}{(K_2-K_1)^3}\left[K_1^2\ln\frac{K_2}{K_1}+\frac{1}{2}(K_2-K_1)^2-K_1(K_2-K_1)\right] \tag{8.50}$$

By defining a geometric constant C_G as

$$C_G = \frac{2}{(K_2 - K_1)^3}\left[K_1^2 \ln\frac{K_2}{K_1} + \frac{1}{2}(K_2 - K_1)^2 - K_1(K_2 - K_1)\right] \tag{8.51}$$

the final form for the maximum deflection (undercut) of a one-way, double-tapered cantilever beam with rectangular cross section is

$$y_{\text{Max}} = \frac{Pl^3}{EI_0} C_G \tag{8.52}$$

The geometric constant C_G takes different forms for different cantilever beam cross sections, as shown in Table 8.1.

Table 8.1 Geometric Constant Values for Cantilever Beam Cross Section: Triangular, Square, Rectangular, Trapezoidal, Round, Elliptical, Round Hollow [164, 165, 166]

Cross Section	Beam Type	Geometric Constant C_G
(triangular)	Continuous	$C_G = \frac{2}{3}$
	Double Tapered $K_1 \neq K_2$	$C_G = \frac{2}{3(K_2 - K_1)^3}\left[K_1^2 \ln\frac{K_2}{K_1} + \frac{1}{2}(K_2 - K_1)^2 - K_1(K_2 - K_1)\right]$
(square)	Continuous	$C_G = \frac{2}{3}$
	Double Tapered $K_1 = K_2$	$C_G = \frac{2}{3K_1}$
(rectangular)	Continuous	$C_G = \frac{2}{3}$
	Horizontal Tapering	$C_G = \frac{2}{(K_1 - 1)}\left[\frac{K_1}{(K_1 - 1)^2}(K_1 \log K_1 - K_1 + 1) - \frac{1}{2}\right]$
	Vertical Tapering	$C_G = \frac{2}{(K_1 - 1)^3}\left[\log K_2 + \frac{1}{2}(K_2 - 1)^2 - (K_2 - 1)\right]$
	Double Tapered $K_1 \neq K_2$	$C_G = \frac{2}{(K_2 - K_1)^3}\left[K_1^2 \log\frac{K_2}{K_1} + \frac{1}{2}(K_2 - K_1)^2 - K_1(K_2 - K_1)\right]$

Table 8.1 Geometric Constant Values for Cantilever Beam Cross Section: Triangular, Square, Rectangular, Trapezoidal, Round, Elliptical, Round Hollow *(continued)*

Cross Section	Beam Type	Geometric Constant C_G
Trapezoidal	Continuous	$C_G = \frac{1}{3}$
	Horizontal Tapering	$C_G = \frac{1}{K_1 - 1}\left[\frac{K_1}{(K_1-1)^2}(K_1 \log K_1 - K_1 + 1) - \frac{1}{2}\right]$
	Vertical Tapering	$C_G = \frac{1}{(K_2-1)^3}\left[\log K_2 + \frac{1}{2}(K_2-1)^2 - (K_2-1)\right]$
	Double Tapered $K_1 \neq K_2$	$C_G = \frac{1}{(K_2-K_1)^3}\left[K_1^2 \log\frac{K_2}{K_1} + \frac{1}{2}(K_2-K_1)^2 - K_1(K_2-K_1)\right]$
Round	Continuous	$C_G = \frac{2}{3}$
	Double Tapered $K_1 = K_2$	$C_G = \frac{2}{3K_1}$
Elliptical	Continuous	$C_G = \frac{2}{3}$
	Horizontal Tapering	$C_G = \frac{2}{K_1 - 1}\left[\frac{K_1}{(K_1-1)^2}(K_1 \log K_1 - K_1 + 1) - \frac{1}{2}\right]$
	Vertical Tapering	$C_G = \frac{2}{(K_2-1)^3}\left[\log K_2 + \frac{1}{2}(K_2-1)^2 - (K_2-1)\right]$
	Double Tapered $K_1 \neq K_2$	$C_G = \frac{2}{(K_2-K_1)^3}\left[K_1^2 \log\frac{K_2}{K_1} + \frac{1}{2}(K_2-K_1)^2 - K_1(K_2-K_1)\right]$
Round Hollow	Continuous	$C_G = \frac{2}{3}$
	Double Tapered $K_1 = K_2$	$C_G = \frac{2}{3K_1}$

8.4.10 Self-Locking Angle

The self-locking angle (SLA) is a characteristic that must be avoided in two-way snap fits, joints that must be easily assembled and disassembled many times. The SLA is the angle at pull-out (removal or disassembly) that will not allow the assembly to open (see Fig. 8.7(a), angle a_1). It is the designer's job to determine the self-locking angle for a given joint and avoid that angle in the design.

The SLA's value ranges between 0° and 90° without reaching the extremes. It is a function of the coefficient of friction (the angle of friction from the angled plane theory). The smaller the coefficient of friction, the higher the number of degrees is in the self-locking angle. It must be noted that in cases of manual disassembly (for servicing and so on) there will be a variance in the dynamic friction, which is a function of velocity. This will vary the friction coefficient. In cases where the friction coefficient increases as the disassembly speed decreases, this may change the self-locking angle, making the joint difficult or impossible to open without breaking the part or using additional tools.

The self-locking angle has the following formula:

$$\xi = \arctan \frac{1}{\mu} \tag{8.53}$$

where μ represents the coefficient of friction.

8.5 Case History: One-Way Continuous Beam with Rectangular Cross Section

Figure 8.12
OmniBook 300 was the first notebook that employed snap fits for preassembly of the LCD screen before was ultrasonically welded to prevent moisture getting in *(Courtesy of Hewlett-Packard Company, Corvallis Division)*

In certain applications, such as the one shown in Fig. 8.14, snap fits are used as temporary holders of the parts being assembled until a different assembly method is applied. In this case it is ultrasonic welding. Snap fits can be used as temporary holders for other assembly methods such as adhesive bonding or welding.

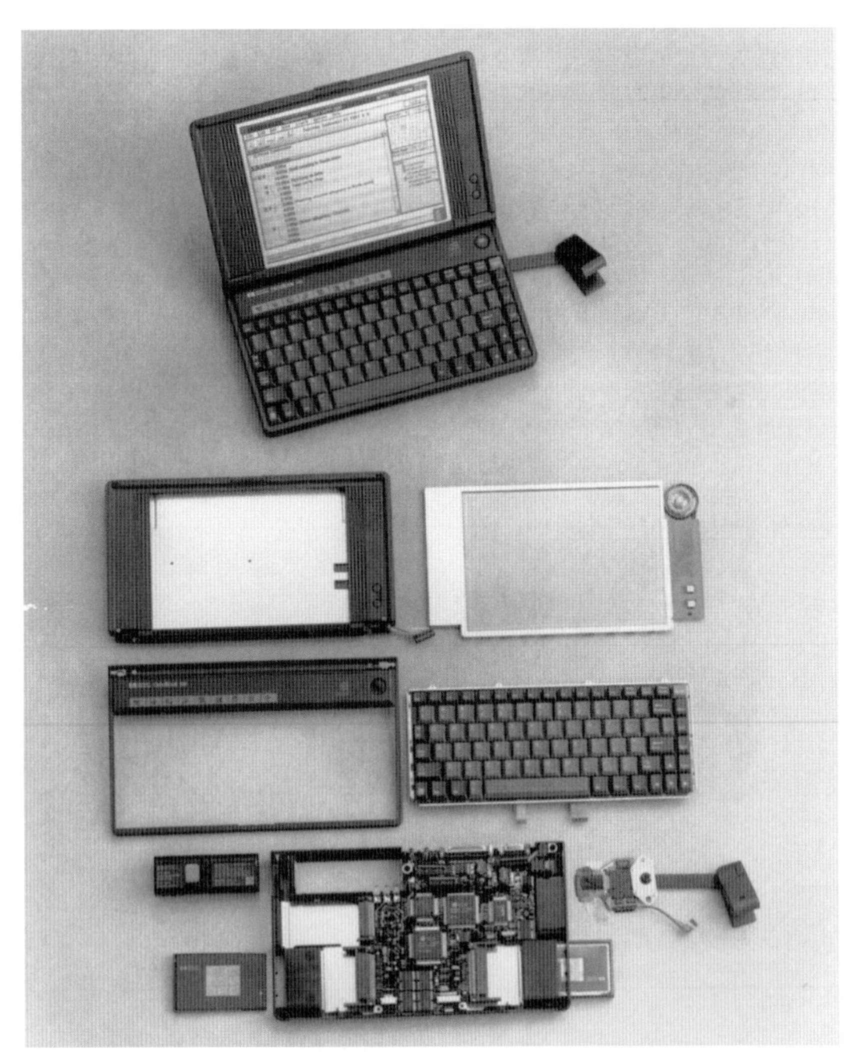

Figure 8.13 The notebook assembly was streamlined. The PC's display module snapped into the unit's case manually. Then the case was ultrasonically welded. The one-way, vertically tapered snap fit lugs ran equally along both sides of the case to hold it together in case the welds were fractured by dropping or by other mechanical stress *(Courtesy of Hewlett-Packard Company, Corvallis Division)*

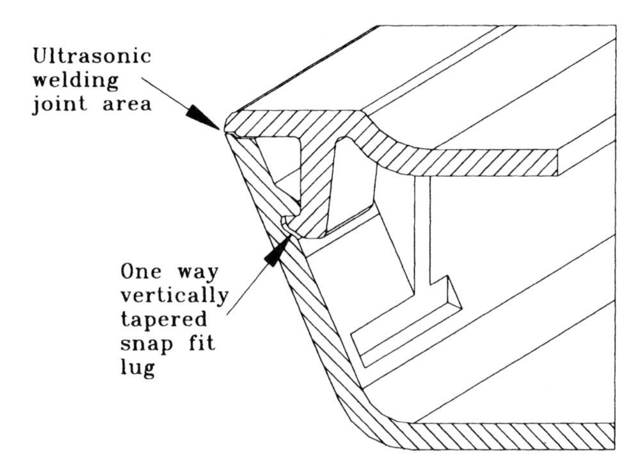

Figure 8.14
Detail of a one-way, vertically tapered cantilever beam snap fit that was used to join the display module to the case of Hewlett-Packard's OmniBook 300 notebook PC *(Courtesy of Hewlett-Packard Company, Corvallis Division)*

8.5.1 Geometrical Model

The following example will cover the calculations necessary to design a lock arm that has a rectangular cross section and is used to assemble an automotive interior door panel to the door frame. Because the return angle is 90°, this type of lock can be used only once.

Given data (see Fig. 8.15)

l = 0.5 in. (12.7 mm)

b = 0.25 in. (6.35 mm)

h = 0.08 in. (2 mm)

The material properties for the unreinforced polymer are

E_{Secant} = 312 kpsi (2,150 MPa)

$\varepsilon_{\text{Design}}$ = 2.23% (0.0223)

σ_{Design} = 6.957 kpsi (48 MPa)

Ideal Model

The cantilever beam theory, explained in Section 8.5, is applied to determine the maximum stress level and the maximum deflection force (see P in Fig. 8.15) that the cantilever beam in this example can sustain. Therefore, the stress level is

$$\sigma_{\text{Design}} = \frac{3hE_{\text{Secant}_{\text{Design}}}\, y_{\text{Tip}}}{l^2} = 6.95 \text{ kpsi } (47.92 \text{ MPa}) \tag{8.54}$$

In 8.54, y_{Tip} represents the maximum deformation the beam is capable of sustaining and has the following form:

$$y_{\text{Tip}} = 2l^2 \frac{\varepsilon_{\text{Design}}}{3h} = 0.04646 \text{ in. } (1.18 \text{ mm}) \tag{8.55}$$

The corresponding deflection force (see P in Fig. 8.15) is

$$\text{P} = \frac{bhE_{\text{Secant}_{\text{Design}}} y_{\text{Tip}}}{4l^3} = 3.711 \text{ lbf } (16.5 \text{ N}) \tag{8.56}$$

The results in 8.54, 8.55, and 8.56 were obtained using the closed-loop solution (theoretical approach) developed earlier in the chapter.

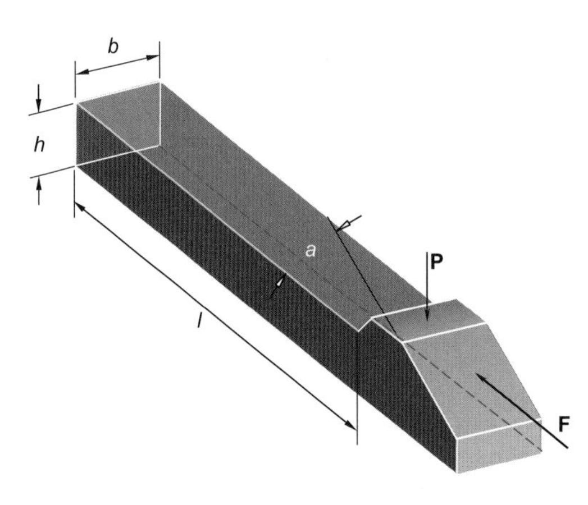

Figure 8.15
Geometric model for the door panel case history example

Another set of analyses were conducted, this time employing NISA II finite element analysis software (see Chapter 4) supplied by Cranes Software, Inc. of Troy, MI. Appendix A and Appendix B show the result files the software provides.

The first analysis conducted (Appendix A) was for an enforced displacement case. An enforced displacement of y = 0.046 in. (1.18 mm) was applied to the element nodes situated at the undercut for a solid element model (brick elements). A static linear analysis was conducted (see Fig. A.1 for the FEA model in Appendix A), and a von Mises stress plot was obtained (see Fig. A.2 in Appendix A). The maximum stress level of 7.112 kpsi (49 MPa) was reached. Material properties are exactly the same as for the closed-loop solution calculation. Therefore

$$\sigma_{\text{von Mises}} = 7.112 \text{ kpsi } (49 \text{ MPa}) \tag{8.57}$$

In the last set of analyses conducted for the same model there was one exception: instead of using an enforced displacement as a boundary condition of the model, a force of 3.711 lbf was equally divided between the nodes at the base of the undercut (see Fig. B.1, Appendix B) and applied to each node. The only exception was that the two nodes situated at the outside walls of the cantilever beam (top view) were each loaded with half the load or point force applied to the other nodes. This was done for reasons of symmetry.

This time the von Mises stress level present in the cantilever beam (see Appendix B, Nodal Principal Stresses) is

$$\sigma_{\text{von Mises}} = 6.514 \text{ kpsi } (47.67 \text{ MPa}) \tag{8.58}$$

Results review

It is known that real materials display nonlinear properties. It is also known that a closed-loop calculation, as an ideal model, uses small-displacement linear approximations, also called linear geometry. But even the stress/strain curve is nonlinear in shape, and Hooke's Law (see Chapter 3, linearity of the stress/strain variation) is no longer applicable in a large-displacement case. When the beam bending angle exceeds 8° the beam theory does not apply, and the plate theory should be applied.

Hand calculations proved to be time-consuming. The FEA modeling process also requires an investment in time directly proportional to the complexity of the model and its nonlinear factors. The solution is to compromise, taking advantage of both methods in a way that suits the particular application. Knowing that the solution can be at best an approximation, our goal is to establish parameters of fair error percentage relative to the amount of time to be invested.

The options under these circumstances are

1. to choose a magnitude of error (M) appropriate to the industry of the part application, and compare the results obtained with the M selected; or
2. if the result is above the selected M, then choosing a softer or more elastic polymer may give results within the selected error range.

Let's select M for an automotive interior plastic trim application, using the case history results:

$$M = 5\% \tag{8.59}$$

The stress values for the above example can now be compared to those obtained by two different methods, using the magnitude of error selected. The first method was the closed-loop solution for which we developed the theory. Comparing the results obtained by hand calculation with the magnitude of error selected, we have that

$$M = 2.3\% \tag{8.60}$$

The M value in 8.60 is well within the selected window. The calculations for other, more complex cross sections of tapered cantilever beams can be quite laborious.

Next, we can compare how the results obtained in the point force boundary condition fared:

$$M = 9.2\% \quad (8.61)$$

When a point force was applied at the tip of the cantilever beam, the least accurate results were obtained. The most precise analysis was the case for which an enforced displacement was applied at the tip on the beam. For this case the magnitude of error was

$$M = 0.5\% \quad (8.62)$$

In general, FEA is used where difficult geometries are to be analyzed. Finite element analysis can be a very effective method of predicting part behavior, but it is expensive and highly time-consuming.

8.6 Annular Snap Fits

Analysis types for annular snap fits are

1. unknown force/deflection (undercut) for a given geometry and material,
2. unknown dimension(s) for a given deflection (undercut)/forces and material, and
3. unknown material for a given force/deflection (undercut) and dimension (geometry).

The theory for this type of snap fit is very similar to the one developed in Section 8.4.

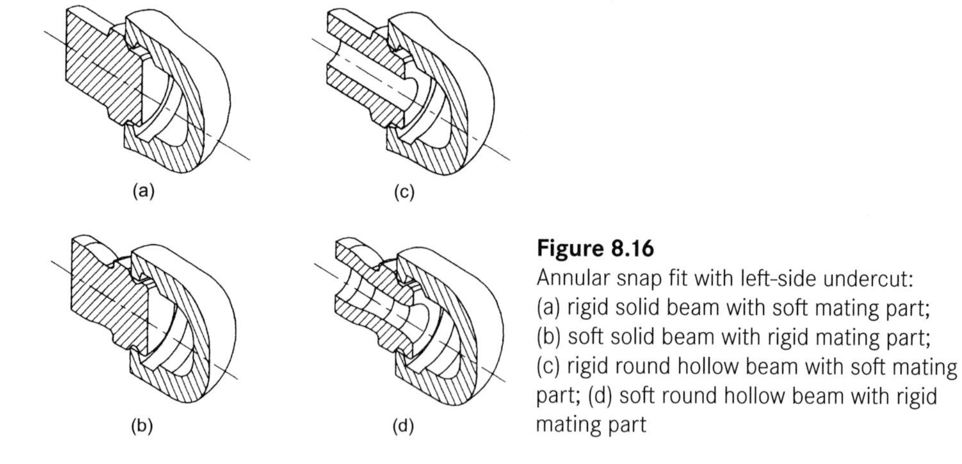

Figure 8.16
Annular snap fit with left-side undercut:
(a) rigid solid beam with soft mating part;
(b) soft solid beam with rigid mating part;
(c) rigid round hollow beam with soft mating part; (d) soft round hollow beam with rigid mating part

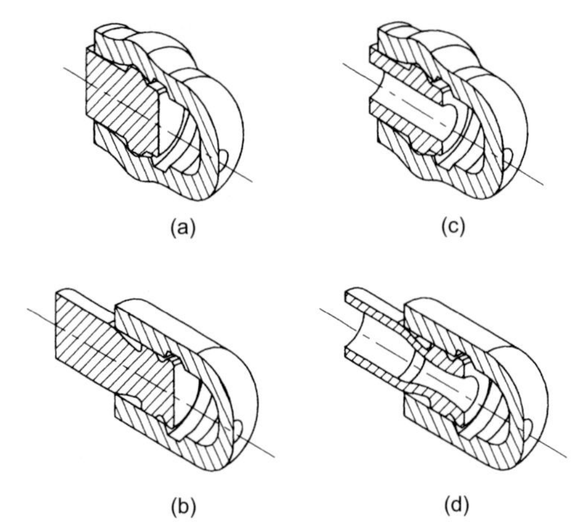

Figure 8.17
Annular snap fit with right-side undercut:
(a) rigid solid beam with soft mating part;
(b) soft solid beam with rigid mating part;
(c) rigid round hollow beam with soft mating part; (d) soft round hollow beam with rigid mating part

8.6.1 Case History: Annular Snap Fit, Rigid Beam with Soft Mating Part

This example deals with a pen and its cap. The formulas are shown without their derivations for reasons of space.

Figure 8.18 Example case history: pen assembly

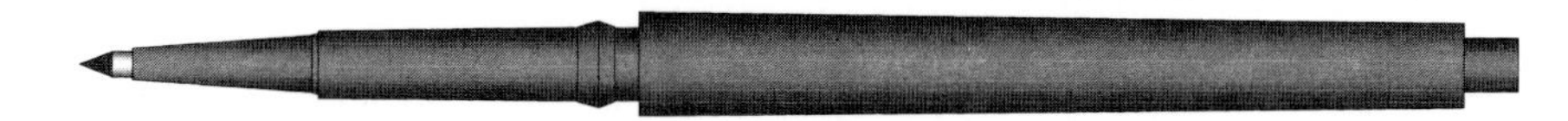

Figure 8.19 Pen

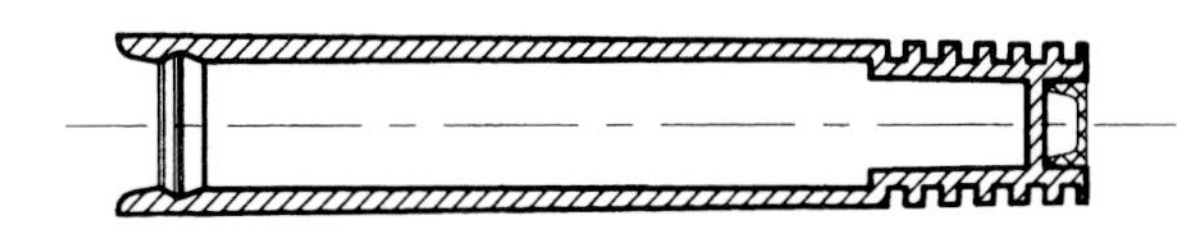

Figure 8.20 Pen cap cross section

8.6.2 Notations

D = Outside diameter cap

d = Inside diameter cap/outside diameter pen

f = Undercut

a_1 = Engagement angle

a = Pull-out angle

l = Undercut position, cap

P = Deflection force

k = Geometric coefficient

F_E = Engagement force

F_0 = Pull-out force

E_S = Secant modulus

ν = Poisson's ratio

ε_A = Allowable/permissible/design strain

α = Ratio of cap OD to cap ID

μ = Coefficient of friction

8.6.3 Geometric Definitions

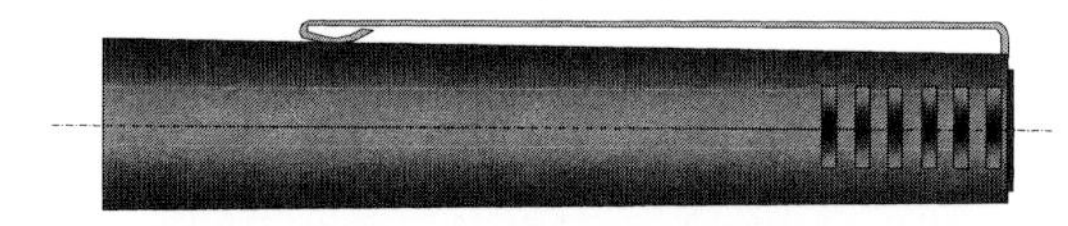

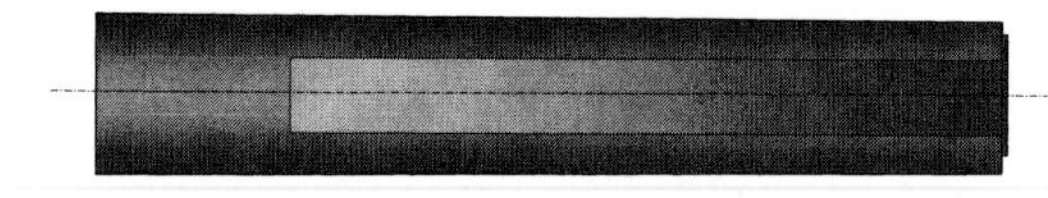

Figure 8.21
Cap views: side view (above), and top view (below)

1. The engagement angle has the same value as the pull-out angle. The force required to put the cap on is equal to the force needed to separate the assembly. Therefore

$$a_1 = a \tag{8.63}$$

which implies

$$F_E = F_0 \tag{8.64}$$

2. The engagement angle has a smaller value than the pull-out angle. This case requires that the engagement force is lower than the pull-out force:

$$a_1 < a \tag{8.65}$$

and the force has the following relation:

$$F_E < F_0 \tag{8.66}$$

3. In the last possible case, the engagement angle is greater than the pull-out angle. In this case the cap could be put on the pen, but it would fall off easily without any significant external force. The angle inequation is

$$a_1 > a \tag{8.67}$$

and the force inequality is

$$F_E > F_0 \tag{8.68}$$

8.6.4 Material Selections and Properties

Because throwaway or nonrefillable pens are considered commodities, the designer should start his material selection with families of resin that provide low cost per cubic inch as well as a very fast molding cycle.

In this case history, the material selected was polypropylene with carbon black pigment. The material mechanical properties required to conduct the calculations are

Secant modulus E_S = 497 MPa

Allowable strain ε_A = 6.24%

Poisson's ratio v = 0.43

Coefficient of friction μ = 0.2

Note: the coefficient of friction has a very low value for two reasons. Both the cap and the pen are made of the same material, and a smooth, polished surface of the tool cavity reduces surface roughness of the final part.

8.6.5 Basic Formulas

Deflection force:

$$P = kE_S d^2 \varepsilon_A \tag{8.69}$$

In 8.69, k represents a geometric factor, as shown by the following equation:

$$k = \frac{\pi(\alpha-1)\sqrt{\alpha^2-1}}{5\alpha^2(1-\nu)+5+5\nu} \tag{8.70}$$

The engagement force, or the force necessary to put the cap on, will have the following form:

$$F_E = P\frac{\mu+\tan a_1}{1-\mu\tan a_1} \tag{8.71}$$

The pull-out force, or the force necessary to take the cap off, is

$$F_0 = P\frac{\mu + \tan a}{1 - \mu \tan a} \tag{8.72}$$

After applying the geometric dimensions shown in Fig. 8.23 to Equations 8.69 through 8.72, the following results are obtained:

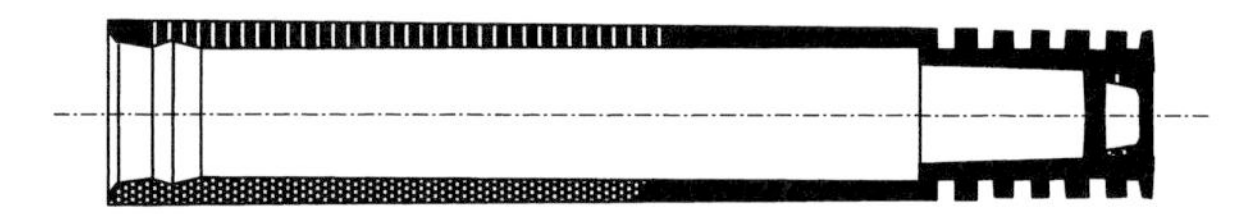

Figure 8.22 Pen assembly design detail

The geometric factor is

$$k = 0.052 \tag{8.73}$$

The deflection is

$$f = 0.5\ \text{mm} = 0.2\ \text{in.} \tag{8.74}$$

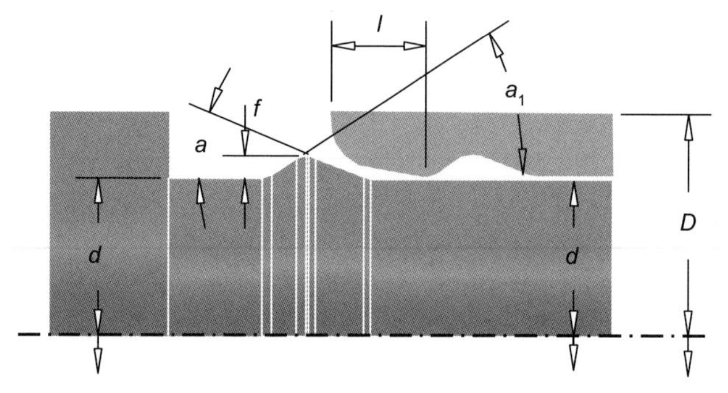

Figure 8.23 Dimensions for pen assembly detail

The deflection force, or the normal force to the joint axis, is

$$P = 103.21\ \text{N} = 23.17\ \text{lbf} \tag{8.75}$$

The engagement force is

$$F_E = 62.77\ \text{N} = 14.1\ \text{lbf} \tag{8.76}$$

The pull-out force is

$$F_E = 90.7\ \text{N} = 20.36\ \text{lbf} \tag{8.77}$$

8.6.6 Angle of Assembly

Figure 8.24 shows an angle of assembly of 0°. This is unachievable in real life, due to tolerance stack-up between fixtures and a pick-and-place robot arm. There will always be some misalignment, no matter how small. All of the calculations handled so far in the case history for annular snap fits assume that there is no misalignment. The results show large forces present at both assembly and disassembly. In reality, the assembly or disassembly forces are much lower than the calculations show.

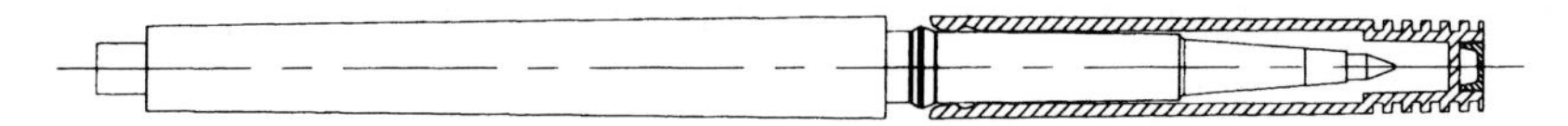

Figure 8.24 Angle of assembly: 0°

Figure 8.25 (top) shows an assembly angle of 1°. Results of tests performed for different experiment cases show that the theoretical loads obtained through calculations should be lowered by about 70 to 90%, depending on the size of the part being assembled. Typically, the larger the parts, the lower the assembly load will be.

Figure 8.25 (bottom) shows an angle for assembly of 2°. Tests show the calculated loads should be reduced by 50 to 70% in order to get comparison between theory and tests. Again, the larger the parts, the lower the actual insertion and pull-out forces.

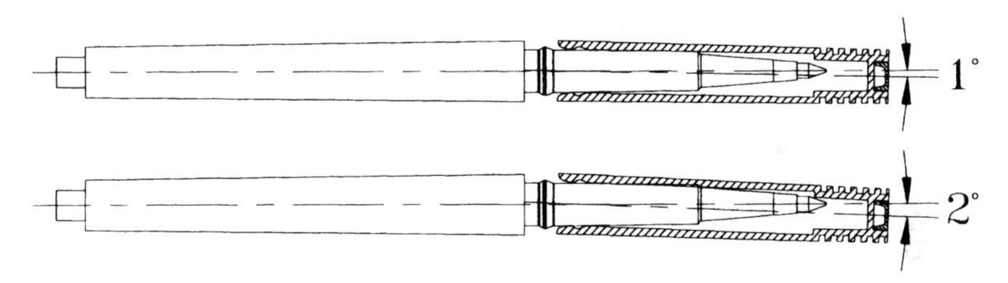

Figure 8.25 Angle of assembly: 1° (top), and 2° (bottom)

8.6.7 Case History: Digital Wristwatch

A product designer should be in charge of his or her design. When that is not the case, terrible things can happen. A U.S. company entrusted a molder in China to actually design its product, which ended up being recalled by the United States Consumer Product Safety Commission. Almost two million products have been recalled.

A three-year-old child, before he went to bed, washed his hands and face and brushed his teeth while wearing his Spider-Man digital wristwatch. Then he went to bed wearing on his wrist the toy watch he was in love with, received as a present from his mom. The child was covered with a comforter while in bed, which was located on the second floor of the house. While sleeping in the bed, the wristwatch became loose, and his arm pressed against the loose watch. When washing earlier, before he went to bed, a drop of water had reached the inside of the digital wristwatch. The contact with water initiated a chemical reaction within the watch, which caused the lithium button cell battery to leak and thus resulted in an acid burn injury on the child's arm (see Fig. 8.26).

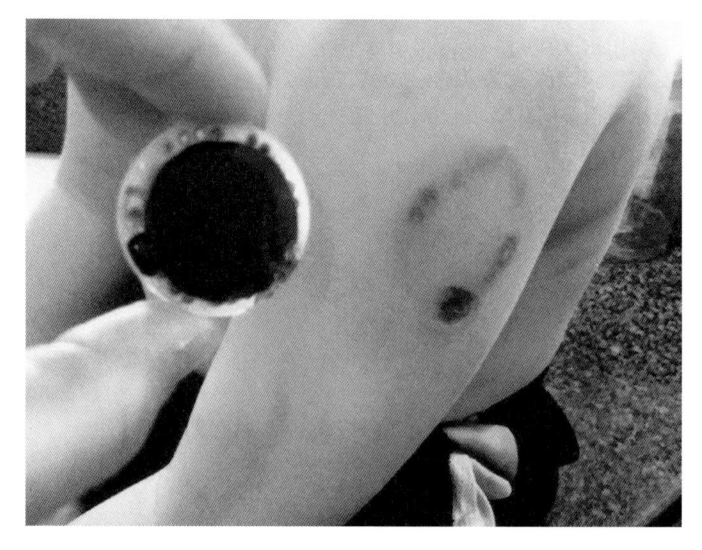

Figure 8.26
Burn injury

The child woke up and went to his parents' bedroom, crying. The parents, thinking that their son had an insect bite on his arm, let him sleep with them. Next morning they realized that it was not an insect bite that had left a round mark on their child's arm, but that it had been caused by the wristwatch (see Fig. 8.27).

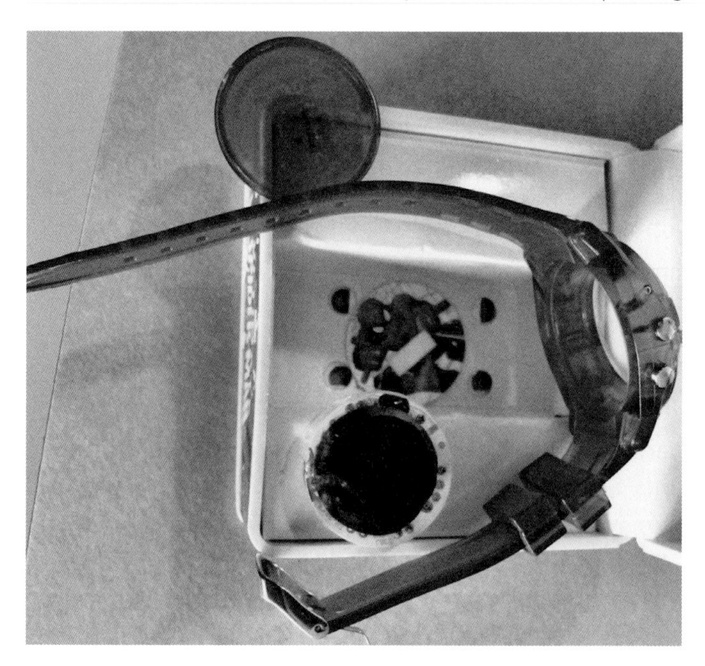

Figure 8.27
Digital wristwatch

The wristwatch case and its back cover were made from a polymer known as PMMA or polymethyl methacrylate, for short, acrylic.

The major design flaw of the wristwatch consists in the fact that the back cover features an annular-type snap fit to mount the cover to the wristwatch case without using an O-ring. Without the O-ring, when a very young child is washing his hands, playing and perspiring, sleeping and perspiring, or spilling water over his hand while wearing the wristwatch, a water drop could easily reach the watch internals causing the battery to overheat and induce an acid burn.

Annular snap fits, even when properly designed and manufactured, cannot seal against water intrusion.

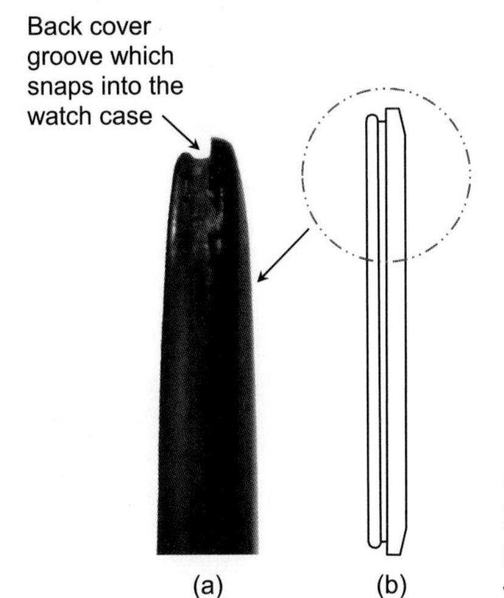

Figure 8.28
Female portion of the annular snap fit, placed on the back cover, employed to secure the cover to the digital watch case: (a) detail; (b) back cover sketch

The injection molding tool used for the manufacture of the digital wristwatch, made by a toolmaker, was also flawed.

The toolmaker used an *electrical discharge machining*, or EDM for short, process to create the annular snap fit groove for the back cover (see Fig. 8.28). This process is also called spark machining, spark eroding, burning, die sinking, wire burning, or wire erosion. It represents a manufacturing process where the desired shape of the steel tool is obtained using electrical discharges or sparks. Material is removed from the tool with a number of current discharges between the two electrodes, which are separated by a dielectric liquid. The process depends on the two electrodes not making an actual physical contact.

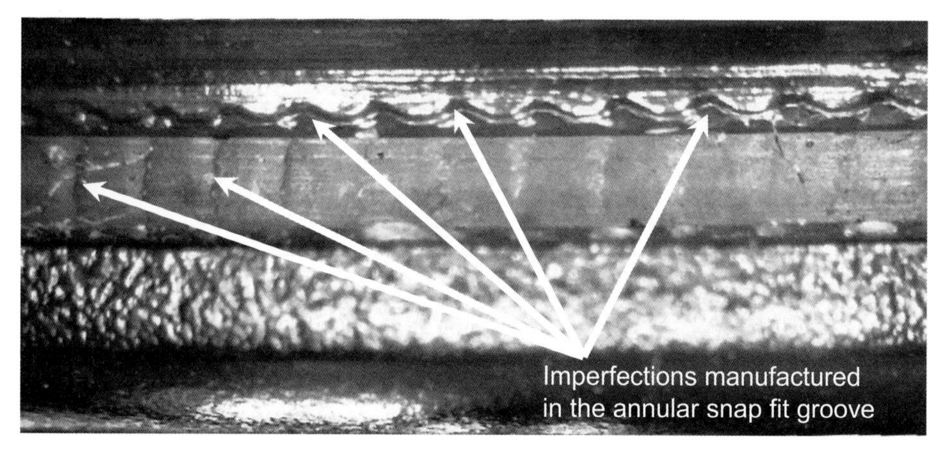

Figure 8.29 Detail of the watch back cover groove showing the manufacturing defects present in the area-magnification factor 500× using a digital microscope

As explained above, the EDM process is the cause of the surface imperfections present in the annular snap fit groove of the digital watch cover, as shown in Fig. 8.29, which displays a 500× magnification image. The tool manufactured to mold the watch back cover is not polished properly.

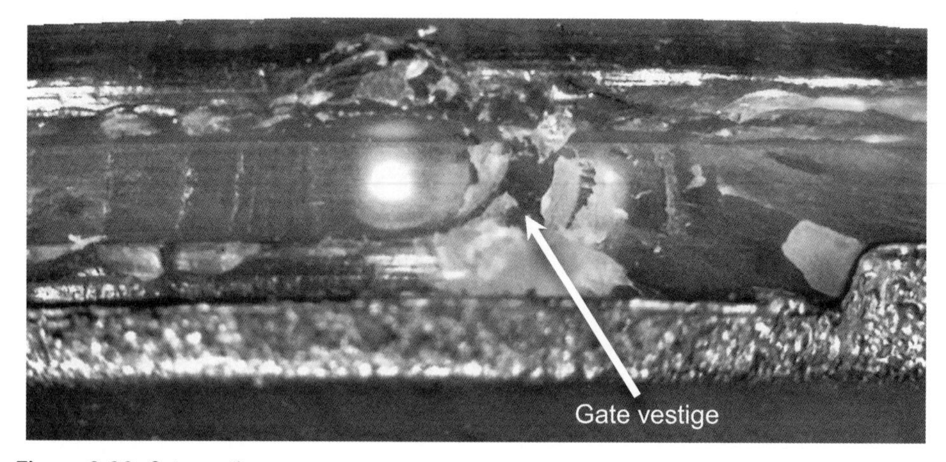

Figure 8.30 Gate vestige

In addition, the toolmaker of the injection molded tool placed the *gate location* (the place where the molten polymer enters the tool cavity) right inside the groove (see gate vestige shown in Fig. 8.30).

When using snap fits to seal against water, a polymeric O-ring or *seal* should always be used.

Around the gate location of the watch back cover, made visible by using a polarizer apparatus, turbulence is created in the PMMA molten polymer entering the injection molded cavity (see Fig. 8.31).

Figure 8.31
Gate location visible when polarized light passes through the back cover and appears on an observation polarizer filter

Light that has been polarized passes through the transparent or translucent polymer component. The light wave components propagate either parallel or normal through the transparent plastic part at various speeds. This behavior is called *retardation*, which is directly proportional to the stress in the molded polymer at that location. Retardation can be made visible using a polarizer. Both parallel and normal waves of the original light beam interfere with one another. They appear as a display of varying colors and intensities, known as *fringes*, where the stress is present.

The color fringes can be used to evaluate residual stress in the molded part or, as in this case, to identify the gate location and the molten flow throughout the mold cavity.

This technique of measuring stress levels is explained in great detail in the Standard Test Method for Photoelastic Measurements of Birefringence and Residual Strains in Transparent or Translucent Plastic Materials, known as Standard D4093-95 (2014) from the American Society for Testing and Materials (or *ASTM* for short).

Table 8.2 Correlation of Observed Sequence of Colors and the Corresponding Retardation Value

Color	Retardation (δ), nm	Retardation, *N* fringes	
Black: zero-order fringe	0	0	0 fringe order
Gray	160	0.28	
White-yellow	260	0.45	
Yellow	350	0.60	
Orange (dark yellow)	460	0.79	
Red	520	0.90	
Indigo-violet: tint of passage #1 (1st-order fringe)	577	1.00	1st fringe order
Blue	620	1.06	
Blue-green	700	1.20	
Green-yellow	800	1.38	
Orange	940	1.62	
Red	1050	1.81	
Indigo-violet: tint of passage #2 (2nd-order fringe)	1150	2.00	2nd fringe order
Green	1300	2.25	
Green-yellow	1400	2.46	
Pink	1500	2.60	
Violet: tint of passage #3 (3rd-order fringe)	1700	3.00	
Green	1750	3.03	

Table 8.2 shows the relationship between the observed color sequence using a polarizer and its corresponding retardation value. Then, using this formula:

$$\sigma = \frac{N \cdot F}{t} = \frac{R \cdot F}{565 \cdot t} \tag{8.78}$$

where:

R = retardation

N = observed fringe order

t = part thickness in millimeters

F = 8.75 MPa,

the edge stress (σ) for any given fringe can be determined.

8.7 Torsional Snap Fits

Analysis types for torsional snap fits are

1. unknown force/deflection (undercut) for a given geometry and material,
2. unknown dimension(s) for a given deflection (undercut)/forces and material, and
3. unknown material for a given force/deflection (undercut) and dimension (geometry).

The methods shown in Section 8.4 can be used to develop formulas for this family of snap fits. The following shows formulas only, without their development, for reasons of space and redundancy.

8.7.1 Notations

l_1 = Distance between the undercut and the axis of the torsional beam

l_2 = Distance between the location of the applied load and the axis of the torsional beam

l_3 = Distance between side of the rotating beam and the fixed end of the torsional beam

d = Width of the torsional beam

f = Undercut

Q = Deflection force

F = Engagement force

α = Engagement angle

W = Polar moment of inertia

I = Moment of inertia

E_S = Secant modulus (tension/compression)

G_S = Secant modulus (torsion)

γ_A = Allowable/permissible/design torsional strain

β = Angle of deflection

ν = Poisson's ratio

μ = Coefficient of friction

ϕ = Angle of friction

F_D = Disengagement force

ε_A = Allowable/permissible/design strain

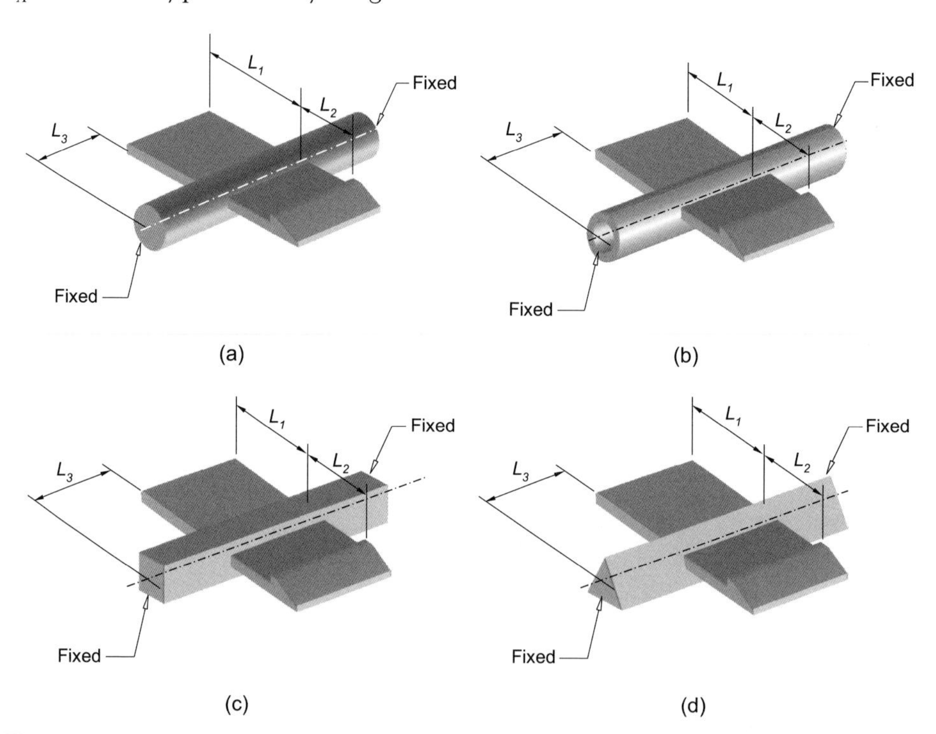

Figure 8.32 Torsional snap fit: (a) round cross section beam; (b) round hollow cross section beam; (c) square cross section beam; (d) triangular cross section beam

8.7.2 Basic Formulae

Polar moment of inertia:

$$W = 0.1667d^3 \tag{8.79}$$

Moment of inertia:

$$I = 0.0834d^4 \tag{8.80}$$

Torsional allowable strain:

$$\gamma_A = (1+\nu)\varepsilon_A \tag{8.81}$$

Angle of deflection:

$$\beta = \frac{W}{I}\gamma_A l_3 \tag{8.82}$$

Undercut (or deflection):

$$f = l_1 \sin\beta \tag{8.83}$$

Torsional secant modulus:

$$G_S = \frac{E_S}{2(1+\nu)} \tag{8.84}$$

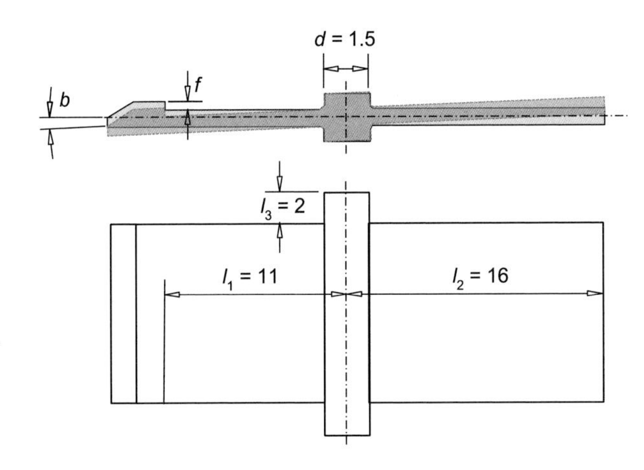

Figure 8.33
Example: torsional snap fit with square cross section beam

Deflection force:

$$P = \frac{WG_S}{l_2}\gamma_A \tag{8.85}$$

Engagement force without friction:

$$F_E = P \tag{8.86}$$

Disengagement force without friction:

$$F_D = P \tag{8.87}$$

Angle of friction:

$$\varphi = a \tan \mu \tag{8.88}$$

Engagement force with friction:

$$F_E = P \tan(a_1 + \varphi) \tag{8.89}$$

Disengagement force with friction:

$$F_D = P \tag{8.90}$$

8.7.3 Material Properties

This application uses a thermoplastic elastomer that has an 82-D Shore durometer hardness. Its properties are

Secant modulus	E_S	= 853.53 MPa
Poisson's ratio	v	= 0.45
Ultimate strain	ε_U	= 17.97%
Allowable strain	ε_A	= 3.52%
Ultimate stress	σ_U	= 38.87 MPa
Allowable stress	σ_A	= 30.03 MPa

8.7.4 Solution

Applying the geometric dimensions shown in Fig. 8.33 and the material properties from Sections 8.7.3 to 8.79 through 8.90, the following results are obtained:

Polar moment of inertia:

$$W = 0.702 \text{ mm}^3 \tag{8.91}$$

Moment of inertia for the torsional beam:

$$I = 0.713 \text{ mm}^4 \tag{8.92}$$

Torsional allowable strain:

$$\gamma_A = 5.1\% \tag{8.93}$$

Angle of deflection of the rotating beam:

$$\beta = 0.1004 \tag{8.94}$$

Undercut (height of the hook or catch):

$$f = 1.104 \text{ mm} \tag{8.95}$$

Torsional secant modulus:

$$G_S = 294.32 \text{ MPa} \tag{8.96}$$

Deflection force, or thumb force necessary to open the snap:

$$P = 0.957 \text{ N } (0.215 \text{ lbf}) \tag{8.97}$$

Engagement force without friction:

$$F_E = 0.957 \text{ N } (0.215 \text{ lbf}) \tag{8.98}$$

Disengagement force without friction:

$$F_D = 0.957 \text{ N } (0.215 \text{ lbf}) \tag{8.99}$$

Angle of friction:

$$\varphi = 16.69^\circ \tag{8.100}$$

Engagement force with friction (force needed to close the snap without pressing on it):

$$F_E = 1.02 \text{ N } (0.22 \text{ lbf}) \tag{8.101}$$

Disengagement force with friction:

$$F_D = 0.957 \text{ N } (0.215 \text{ lbf}) \tag{8.102}$$

This is the same as the deflection force. The snap fit can be opened only by hand or by tool operation.

8.8 Case History: Injection Blow Molded Bottle Assembly

The injection blow molded bottle assembly is a package that can be used for dispensing liquids or powders in a variety of industries such as medical, pharmaceutical, cosmetics, and others.

The assembly features an injection blow molded bottle, an injection-molded cap and an injection-molded dispensing cup. The bottle neck consists of annular snaps designed to provide interference fit for the primary and secondary retention points of the cap. There is also a larger annular snap to fit the dispensing cup (Figs. 8.34 and 8.35).

The cap has two seals: a primary seal to promote secure assembly and a secondary seal to prevent rotation of the cap (Fig. 8.36). A tamper-evident band is situated between the snap rings on the cap. When the seal is broken and removed, a remaining length of the band serves as a living hinge, allowing multiple operations without loss of the cap.

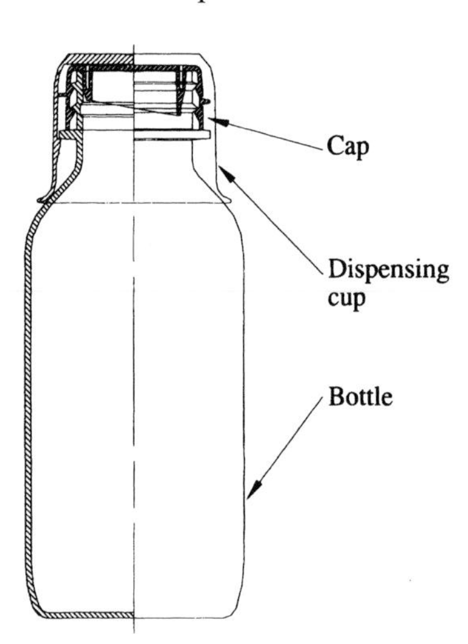

Figure 8.34
Injection blow molded bottle assembly
(Courtesy of Molding Technology Corporation)

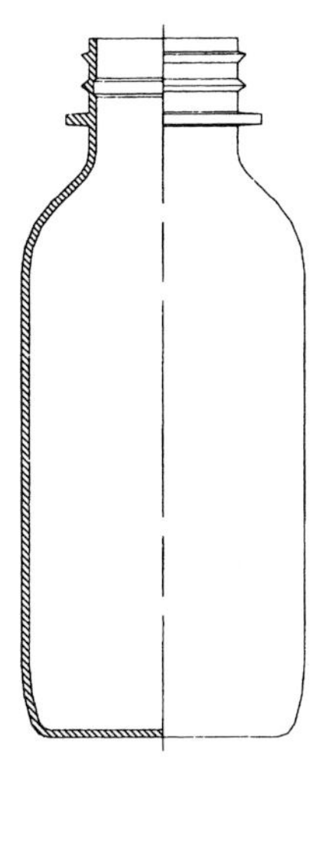

Figure 8.35
Bottle *(Courtesy of Molding Technology Corporation)*

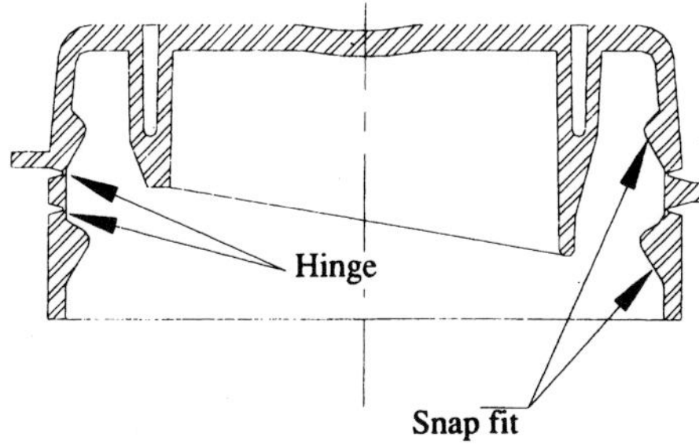

Figure 8.36
Cap *(Courtesy of Molding Technology Corporation)*

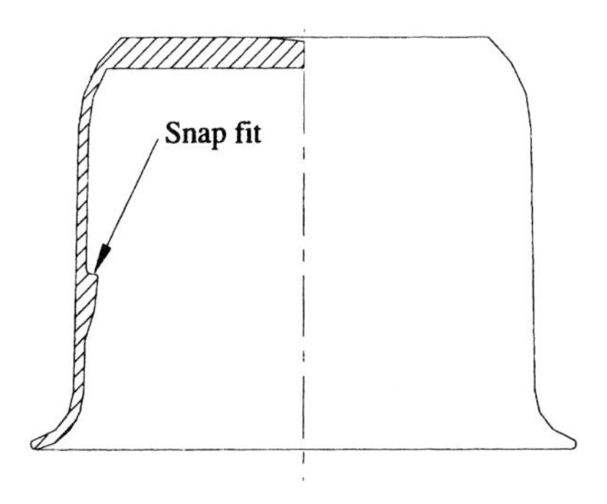

Figure 8.37
Dispensing cup *(Courtesy of Molding Technology Corporation)*

The dispensing cup has retaining lugs that fit over the corresponding collar on the bottle neck, holding it in place. The cup is designed to have a low center of gravity when dispensing the contents of the bottle (Fig. 8.37).

This product is an excellent example showing practical applications for some of the many assembly methods described in this book.

8.9 Tooling

At the preprototype or prototype stage, fine-tuning of design, tool, and processing windows will be needed. Subtle variations in the joint design such as inconsistent wall thickness, varying rib patterns, or any other features will make hand calculations difficult.

Fine-tuning the tool includes matters such as finding the ideal gate location and proper vent locations as well as ensuring that the tool is balanced, having runners of equal length connecting the sprue and the gate of each part.

Fine-tuning the processing window is a matter of finding the optimal point on the processing controls that yield the best results. These parameters include crystallization for crystalline materials, holding time, injection speed, cooling, and others.

Although precision snap fits are difficult to achieve, they are in great demand for consumer products such as computers, photo cameras, video cameras, and toys as well as health care instrumentation and devices and automotive interiors and exteriors. One should not try to make the design too dependent on critical dimensions, especially where tight tolerances are required. Tooling always accounts for material shrinkage, and substituting a different material will result in a different degree of shrinkage, resulting in parts that do not comply with required dimensions. The best precision snap fits are achieved through a combination of design, tooling, and processing.

Snap fit fasteners often have undercuts (the height of the hook on the lock arm, or the depth of the groove in the mating part) that require special slides and cam-actuated tool mechanisms. Such features will increase tooling costs due to their complexity of design and necessitate more frequent tool maintenance. It is always beneficial to strive for designs that minimize the use of mechanisms, to simplify procedures and keep tooling costs low.

Tooling for parts that incorporate snap fits usually involves much closer tolerances than for other applications. For high-quality parts such as those mentioned above, the specifications for tooling and processing are as important as those of the parts themselves. The importance of tooling and processing cannot be overemphasized. No matter how good the design, without proper tooling and processing, a poor quality part will result. For that reason, these requirements should be considered very carefully, especially when the parts are molded by an outside molder. Using preprototype and prototype tooling when time and budget allow can greatly improve the quality of the final part.

The effective use of tooling should allow for *metal-safe* design prototypes. This means designing a tool to which modifications would be made by removing metal, which is less expensive than adding metal by welding and then machining it.

Minimizing or eliminating the need for undercuts, which would otherwise require special side pulls or lifter pins, makes a more robust tool possible. Using side pulls or lifter pins would increase the tooling costs significantly while reducing the tool robustness or reliability of the part.

Productivity and quality are greatly enhanced if the part drawing includes the gate location and type, the parting line location, and complete tolerances for the part design. This also helps ensure that the toolmaker builds the tool to the engineer's specification.

8.10 Case History: Snap Fits That Kill

In 1947, at the Western Reserve University in Cleveland, Ohio, a surgeon named Claude Beck was performing an operation on a 14-year-old boy whose heart arrested. Heart arrest, also known as cardiac arrest, cardiopulmonary arrest, or circulatory arrest, is the cessation of normal circulation of the blood due to the failure of the heart to contract effectively. Arrested blood circulation prevents delivery of oxygen to the body. The doctor performed a cardiac massage and gave the kid various medication. None worked. He tried an experimental defibrillator from his laboratory. By administering an electric shock, such a device can save the patient's life by resuscitating a heart that had stopped beating. This was the first time this device was used on a human being–and the boy survived.

Ten years later, in 1957, a German couple moved from Munich, Germany, to the state of Washington, USA. They had two sons. The young one, only two when moved to the United States, grew up to become a general superintendent and member of a local union, got married, and had five children. In 2002, while he was working in California, he had an accident and was brought to a local hospital. While being treated he had a cardiac arrest. The nurse reached out and grabbed an *automated external defibrillator* (AED), the FirstSave® 9000 model, made by Survivalink Corporation of Minnetonka, Minnesota (Fig. 8.38). However, the AED did not work. The reason why the AED failed was that the snap fit latch holding the battery in the AED was not locked. The snap fit did not snap, so no electrical connection was made. The current was not received by the AED, and the patient died. He was just 46 years old.

Figure 8.38 Automated external defibrillator *(Courtesy of ETS Inc.)*

When activated by the operator (in this case the nurse), the AED releases an electric shock, which is required to resuscitate the patient [180]. To provide electricity for the AED devices, lithium batteries are used (see Fig. 8.39). Once charged, the battery can hold a charge for one to five years. The batteries are enclosed in a polycarbonate acrylonitrile butadiene styrene alloy (PC/ABS) case made of Cycoloy® C2950 polymer, initially manufactured and sold by GE Plastics and, since 2007, manufactured and sold by SABIC *(Saudi Basic Industries Corp.)*, which that year acquired the GE Plastics business for almost $12 billion.

Figure 8.39 Snap fit latch design feature of the AED lithium batteries in two colors: warm red C and dark blue *(Courtesy of ETS Inc.)*

Figure 8.40 Side view of snap fit latch for the AED lithium batteries in two colors: warm red C and dark blue *(Courtesy of ETS Inc.)*

The battery case was made in two halves ultrasonically welded together. One of the halves had an *energy director* feature (similar to the design shown in Section 5.1.4.3, Fig. 5.16) along the joint area that allows the amorphous polymer to be ultrasonically welded, while the other half incorporated a snap fit latch. The battery was then assembled to the AED device employing a latch snap fit.

The product development conducted by the company used the polymer mechanical properties, such as stress/strain curves, made available by the resin vendor, in this case GE Plastics. All stress/strain curves for thermoplastic polymers, which are made available by the material suppliers to designers and engineers worldwide, are based on only one color: natural color, or NC for short. For all other colors, users have to test and generate their own data, such as stress/strain curves. In this case, the *stress/strain curve* for the PC/ABS alloy having the trade name Cycoloy C2950, natural color, is shown in Fig. 8.41. The polymer stress at yield is just below 60 MPa.

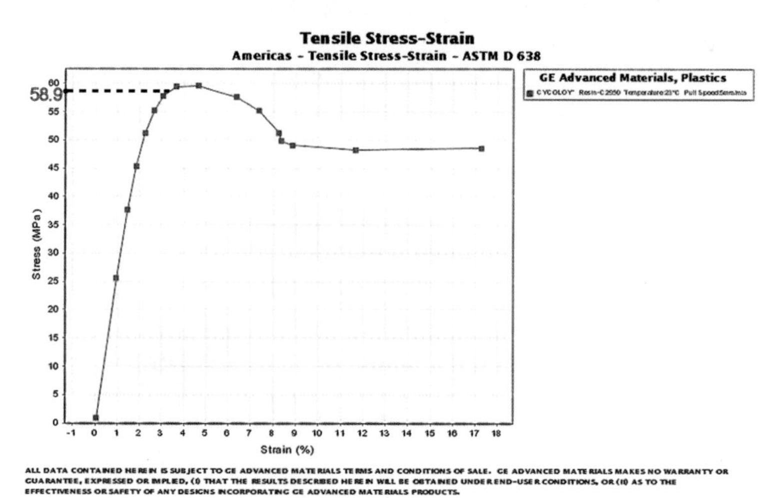

Figure 8.41 The stress/strain curve for the Cycoloy® C2950 polycarbonate acrylonitrile butadiene styrene (PC/ABS) alloy at room temperature as provided by the supplier *(Courtesy of General Electric website)*

However, when the AED product, FirstSave® 9000 series, was released for sale in the marketplace, its color was no longer natural; it was matched to a painted sheet-metal chip called "warm red C." The amount of pigment used to match the polymer's color to the painted metal chip exceeded the maximum level of 2% by weight–almost doubling it. The stress level for polymer colored to match a painted "warm red C" sample employing close to 4% pigment can drop by as much as 50%.

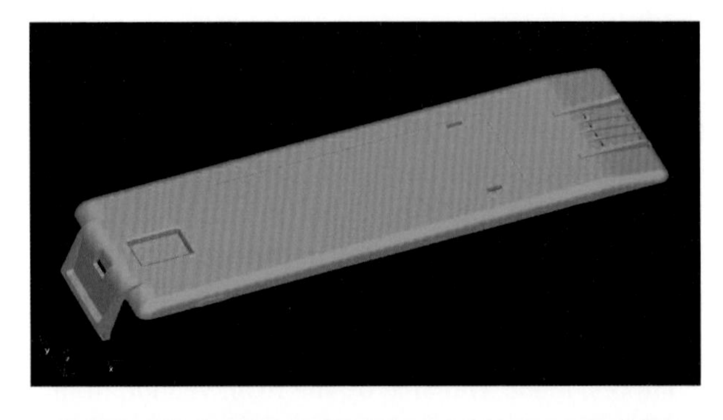

Figure 8.42
3-D model of the upper portion of the case holding the lithium battery including the snap fit latch *(Courtesy of ETS Inc.)*

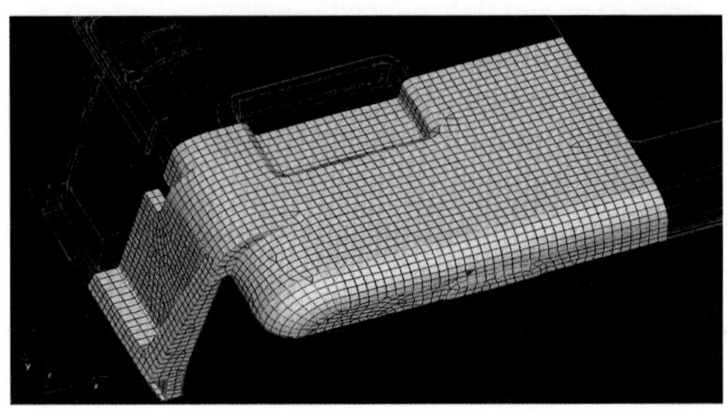

Figure 8.43
Snap fit latch mesh created for the nonlinear finite element analysis *(Courtesy of ETS Inc.)*

For the legal case brought against the manufacturer by the patient's surviving wife and children, a nonlinear material properties (Fig. 8.41) finite element analysis (FEA) was conducted (Figs. 8.42, 8.43, and 8.44). The results show that the von Mises stress level in the latch, when deflected almost 3 mm to be secured in the AED body and made of a natural color Cycoloy C2950, is 58.1 MPa or 8,424 psi: very close to the material yield stress level. During the molding process, any small variation of parameters set on the injection-molding press may result in lower mechanical properties of the polymer shown in Fig. 8.40.

To perform well when in use, a snap fit must always be in the preyielding strength range until the end of life for that given product. Once the yield stress levels are reached, the latch will deform when pressed but will not snap back to its original position because the polymer will yield and stay deformed–thus not snapping back to secure the battery into the AED–as was the case here when a man died because of it.

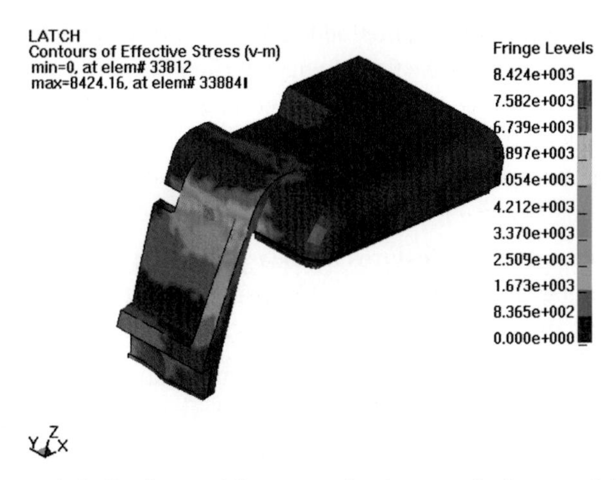

Figure 8.44 Contours of effective von Mises stress level present in the snap fit latch when deflected 3 mm to engage into the AED device for a battery case made of natural color Cycoloy® C2950 *(Courtesy of ETS Inc.)*

New batteries made since this accident are of different colors, like the dark blue, which holds better polymer mechanical properties than does red [180].

At the end of 2013 the U.S. Food and Drug Administrations (FDA) website showed 107 recalls of medical devices, including AED, made since it started keeping records 10 years earlier. Most manufacturers are on the FDA recall list for various reasons, including plastic part design.

8.11 Assembly Procedures

Although one of the first items to be determined by the designer is whether the part will be assembled manually or automatically, it is best to design all components as if they will be assembled by robots. This approach encourages design choices that will help make manual assembly easier.

The assembly procedure for plastics components is a 3-D movement required to position the parts and snap them into place. Think of the assembly process in terms of spatial movements the hand or robot arm must go through to have the components assembled. In practice, five movements are common. They are push, slide, tip, spin, and pivot.

The presence of certain features in the part design improves the assembly process. Some of the design features that could be used are guides, back-ups, symbols, locators, and stops.

Primary guides are design features added to the part design for the purpose of providing guidance. They are the first to engage when the parts are brought together. The guides need to engage readily, and once engaged, there will be no need for further alignment by the operator or robot. The primary guides are designed, in certain cases, as part of the holding fixture or nest, rather than the components themselves.

Secondary guides are design features already present in the part. They are bevels, radii, and clearances incorporated into the part design to ensure that component edges do not hang up on each other as the parts are brought together.

Other design features that permit easy disassembly can also be added. Extending the lock arm beyond the lock-in undercut provides access and makes the snap fit easier to release. Tool access provides an easy disassembly of an otherwise hidden and difficult lock. A good example is a slot in the part allowing a screwdriver to reach and deflect the lock arm.

Back-up design features that allow reassembly of components if the lock arm breaks can also be included into the final part design. To prevent scrapping the damaged part, space can be provided to repair the joint using assembly methods other than the snap fit itself.

Locators are pins or protrusions of various cross sections, structurally strong, that help align or nest one part to the other. They can be molded into the component itself, such as walls, or they may be part of the fixtures. For manual assemblies, it is recommended that locators be incorporated into the part; for automatic assemblies, fixture-based pins are recommended.

Locators restrict part movement in a certain direction relative to its mating counterpart. Theoretically, 12 possible movements in the assembly process should be considered when designing a snap fit. They are

- two translations in X direction
- two translations in Y direction
- two translations in Z direction
- two rotations about X direction
- two rotations about Y direction
- two rotations about Z direction

These possible movements are also known as *degrees of freedom*. Nesting provides a way of mechanically interlocking the parts through the use of grooves, notches, tabs, and so on. These restricting design features, which should not be flexible, help prevent any motion in a restricted direction. While they provide nesting, they also help in positioning the part and provide assistance in carrying loads. The nesting features are present on one or both parts.

It is best to use features that have one degree of freedom. A degree of freedom (DOF) is one of the 12 directions in which a part can move in a 3-D space. On each axis there are two translations and two possible rotations about the axis. A snap fit can restrict movement of the part it belongs to in up to 11 directions. Examples of design features having multiple degrees of freedom are a pin, which provides good location for a hole or a slot in the mating part (two DOFs); a lug, which is a nondeflecting beam and can provide location for a surface or a wall (three DOFs); a track, which is used as a guide for sliding (two DOFs); a wedge, which fits into a slot (two DOFs); a lock arm, which restricts movement only in the disassembly direction (one DOF), deflecting during assembly and then locking in place in the desired nested position.

There are a number of lock types, each with different degrees of freedom. The hook, for example, uses the tensile strength of its material to resist the removal. The catch is a protrusion on a wall or a surface whose material shear strength resists the removal. The annular ridge engages in an annular slot and resists removal in both tension or compression and shear. The torsional snap has a hook mounted on a torsional beam to resist removal in tension.

In order to engage, the lock arm must deflect, providing from one to three degrees of freedom. In most cases the lock arm is an integral component of the part. Otherwise it is designed as a separate part, which locks two components in place during assembly.

During disassembly, loads in the removal direction should be low or ideally non-existent. The axis of installation is dictated by packaging restraints, the amount and direction of the loads, and the position the parts have with respect to each other.

The lock arms prevent movement in the direction of removal. They should be flexible, to allow deformation to the new position and provide locking. They should also be able to sustain the amount of load they were designed for.

Certain design features can be added to enhance quality control in both manual and automatic assembly lines. These features provide a way of indicating to the operator or to sensors that a good assembly has been achieved. For example, a definite "click" of a lock arm returning to its locked position may be heard by an operator or sensed by a robot to indicate a good joint. A visual position indicator that only appears for good joints should be used for manual assemblies. Tactile features are another quality control technique. Both auditory and tactile features can be applied in such a way that the part suddenly releases energy, which is detected by the assembly robot. A good assembly feel should allow for strong lock arm return, allowing the alignment of the show surfaces. Poor assemblies should then be easily recognizable.

In some applications, such as the one shown in Fig. 8.14, snap fits are used as temporary holders until a different assembly method is applied. In this case it is ultrasonic welding. Snap fits are used as temporary holders for other assembly methods such as adhesive bonding or welding.

Based on the nesting features for a specific design, a number of concerns may arise. The snap (beam) may not nest properly in its location after it has been assembled. There may be incompatibility between the direction of installation, the movement needed to assemble, and the actual design of the attachment. The hook can become overconstrained as a result of design features that negate each other or by temperature variance during service, which causes thermal expansion and contraction of mating parts, affecting the fit. If under-constrained, the hook will be required to carry additional loads or even higher loads than intended. This can also occur when the wrong type of hook is used to carry those loads. When the part is designed for tensile strength and the part is actually loaded in shear, it will fail.

8.12 Issues with Snap Fitting

After design, tooling, and processing have been considered, problems may occur in the manufacturing process. The following are common problems in snap fit assembly, probable causes, and recommended solutions.

The deflection is too high. This is usually indicated by plastic deformation in the lock arm. This can be corrected by reducing the undercut height, lowering the pull-out angle, or tapering the lock arm vertically (if it is already tapered, increase the amount of taper). The plastic deformation will become elastic, allowing the lock arm to return to its original position. Other alternatives include increasing the lock arm length or substituting a polymer that has lower strain values.

The retention force is too low and is therefore not holding the assembly. Possible solutions include increasing the lock arm thickness or width, which provides a higher value of the moment of inertia. Other solutions include decreasing its length, using a higher pull-out angle, or changing to a stiffer material.

Lock arm length is too small, preventing deflection. Increasing the lock arm length or decreasing the lock arm height may be the solution. Substituting a material with higher strain values (more elasticity) might help as well.

Interference between the two mating parts is too small. This can result in rattles and squeaks. Possible solutions include reducing the pull-out angle, reducing the lock arm height, and increasing the undercut. Part tolerances may also be lowered.

A poor lock arm is obvious when adjustment is required after the parts have been assembled. It is also evident when high retention force is needed, when insufficient nesting is provided in the mating part. These problems occur often when the material properties are unknown.

8.13 Serviceability

Serviceability is the ability of a snap fit to be taken apart and reassembled over the predicted design life of the product.

The part should be designed so that breakage of the snap fit does not damage the rest of the part. One must also consider designing molded-in bosses for reinstallation with mechanical fasteners when the snap fits are broken.

It should be noted that, when designing a component with integrated snap fits, the part will probably be disassembled for maintenance or other reasons during its useful life. The designer should consider the viewpoint of a service person and try to take the assembly apart. This will help determine whether special tools or fixtures will be needed and ultimately enhance the serviceability of the final product.

An easy method of solving many serviceability or recyclability problems is to have symbols engraved into the hook, groove, or on the actual component or assembly. The symbols could be as simple as arrows or as complex as letters or words if space allows. The symbols, letters, or words should be common throughout the product, product line, company that produces these products, or sometimes even for a whole industry. They should also be referenced in the service manual.

One should ensure during product subassembly testing that the snap fit would not break during handling and shipping to a dealer, service person, or the end user.

Figure 8.45 Recyclability symbols for various polymer families *(Courtesy of Society of Plastics Industry)*

8.14 Exercise

Figure 8.46 shows a rectangular cross section continuous cantilever beam that has a load applied at room temperature (23°C) onto the free end. First, calculate the deflection of the cantilever beam immediately after the load is applied at room temperature. Next, calculate the total deflection of the cantilever beam after three days or 72 hours of being exposed to an operating temperature of 80°C.

The cantilever beam is made of an amorphous polycarbonate polymer, Lexan® EM3110, manufactured and sold by SABIC (Saudi Arabia Basic Industries Corp.).

Figs. 8.47 and 8.48 show the creep curves at two distinct temperatures, 60°C and 99°C, for Lexan EM3110 at three stress levels, 3 MPa, 6 MPa, and 8 MPa.

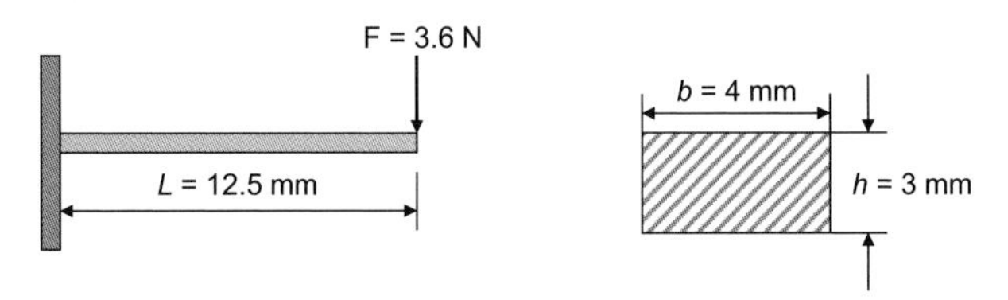

Figure 8.46 Rectangular cross section continuous cantilever beam loaded with a force of 3.6 Newton at the free end

The modulus at room temperature for Lexan EM3110 polycarbonate is 2,034 megapascals (MPa). The formulas employed to solve this exercise are shown below.

Moment of inertia:

$$I = \frac{bh^3}{12} \tag{8.103}$$

Section modulus:

$$z = \frac{I}{c} = \frac{bh^2}{6} \tag{8.104}$$

Stress level in the beam:

$$\sigma = \frac{FL}{z} \tag{8.105}$$

Initial deformation in the beam after the load was applied at room temperature:

$$y_{t=0} = \frac{FL^3}{3EI} \tag{8.106}$$

Creep modulus:

$$E_{\text{Creep}} = \frac{\sigma_{\text{Constant}}}{\varepsilon_{t=72}} \tag{8.107}$$

Final deformation of the cantilever beam after three days of being exposed to a continuous operating temperature of 80°C:

$$y_{t=72} = \frac{FL^3}{3E_{\text{Creep}}\text{I}} \tag{8.108}$$

Deformation amount due to creep when the beam is exposed to 3.6 N load continuously for three days at 80°C:

$$y_{\text{Creep}} = y_{t=72} - y_{t=0} \tag{8.109}$$

8.14.1 Solution

First the moment of inertia for the rectangular cross section of the beam is calculated:

$$I = \frac{4 \cdot 3^3}{12} = 9\ \text{mm}^4 \qquad (8.110)$$

Then, the section modulus of the beam is

$$z = \frac{4 \cdot 3^2}{6} = 6\ \text{mm}^3 \qquad (8.111)$$

followed by the stress generated in the beam by the load applied at the free end

$$\sigma = \frac{3.6 \cdot 12.5}{6} = 7.5\ \text{MPa} \qquad (8.112)$$

Now, the initial deformation in the beam, when a load of 3.6 N is applied at room temperature, can be calculated:

$$y_{t=0} = \frac{3.6 \cdot 12.5^3}{3 \cdot 2{,}137 \cdot 9} = 0.122\ \text{mm} \qquad (8.113)$$

To find the deflection in the beam after three days or 72 hours of being exposed to a continuous load of 3.6 N while the operating temperature is 80°C, the creep curves for the polymer are necessary. The creep data available from the vendor are shown in Figs. 8.47 and 8.48. Constant stress of 7.5 MPa generated by the 3.6 N load is not available in the creep data provided. By using the graphical interpolation technique used earlier (see the step-by-step technique in Chapter 6), the 7.5 MPa constant stress level is generated, as shown in Figs. 8.49 and 8.50 in a heavy black line.

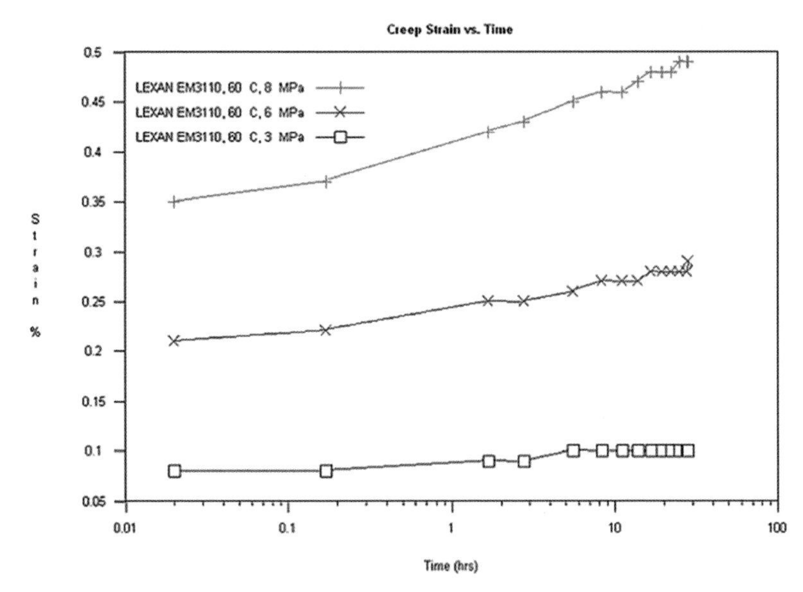

Figure 8.47 Creep curves for Lexan® EM3110 at 60 °C at 3 MPa, 6 MPa, and 8 MPa constant stress levels

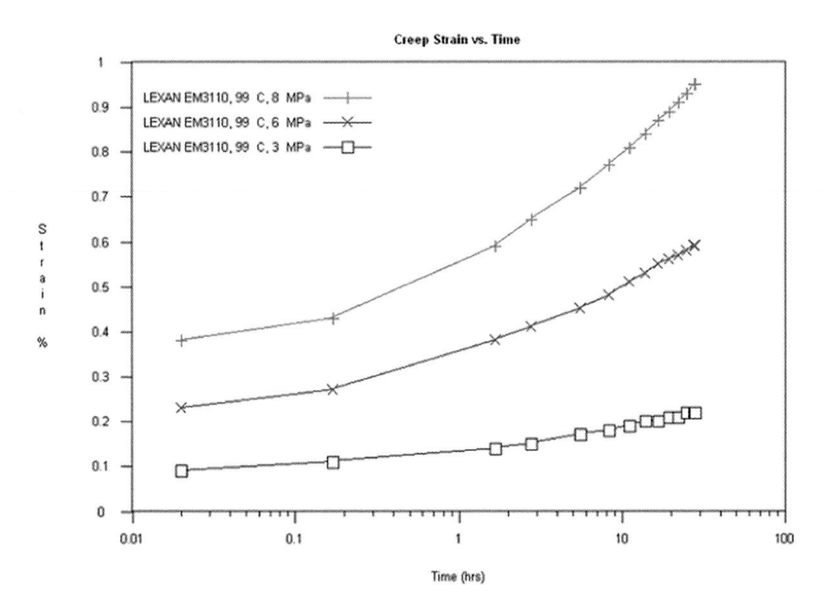

Figure 8.48 Creep curves for Lexan® EM3110 at 99 °C at 3 MPa, 6 MPa, and 8 MPa constant stress levels

To extract the Lexan EM3110 strain level after 72 hours of being exposed to a constant stress of 7.5 MPa, first identify the time on the horizontal axis of the graph. The time of 72 hours is close to the 100-hour point because it is a logarithmic graph showing time in increments of power of 10. Once the 72-hour point is properly identified, we move vertically until we intersect the 7.5 MPa constant stress level, which was graphically interpolated previously. Then, projecting horizontally and intersecting the strain axis, we find the maximum strain in the beam after 72 hours ($t = 72$) at 60 °C as 0.456%.

Now, using the graph shown in Fig. 8.48, which represents creep data at 99 °C for the same polymer Lexan EM3110 at three constant stress levels of 3 MPa, 6 MPa, and 8 MPa (by graphic interpolation as done previously), the constant stress level of 7.5 MPa is generated. It is also located below the constant stress level of 8 MPa. Based on this graph we obtain the maximum strain as 0.985% at 99 °C and $t = 72$ (time = end).

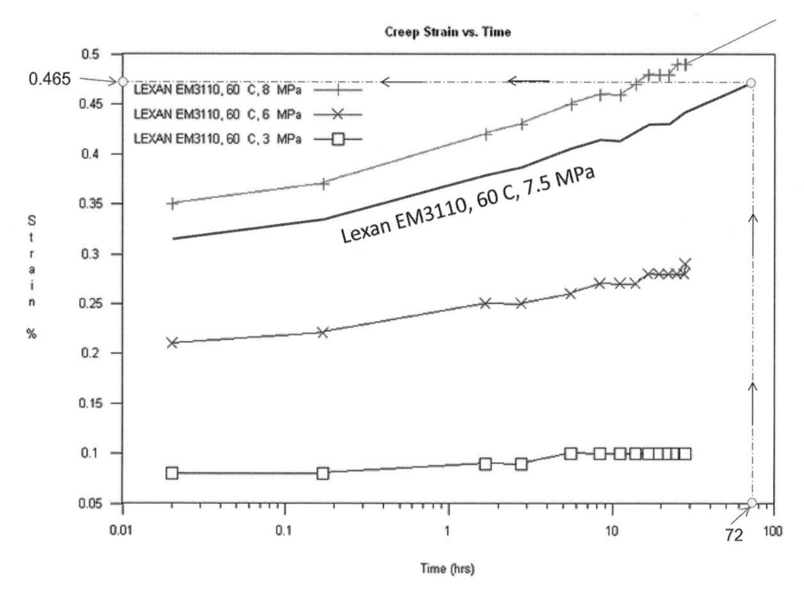

Figure 8.49 Creep curve at a constant stress 7.5 MPa for Lexan® EM3110 at 60 °C

The strain after three days at 80 °C is required. To obtain this value we average the strains at 60 °C (0.465%) and 99 °C (0.985%) to find the strain at 80 °C as 0.725%:

$$\varepsilon_{t=72} = \frac{0.465 + 0.985}{2} = 0.725\% \tag{8.114}$$

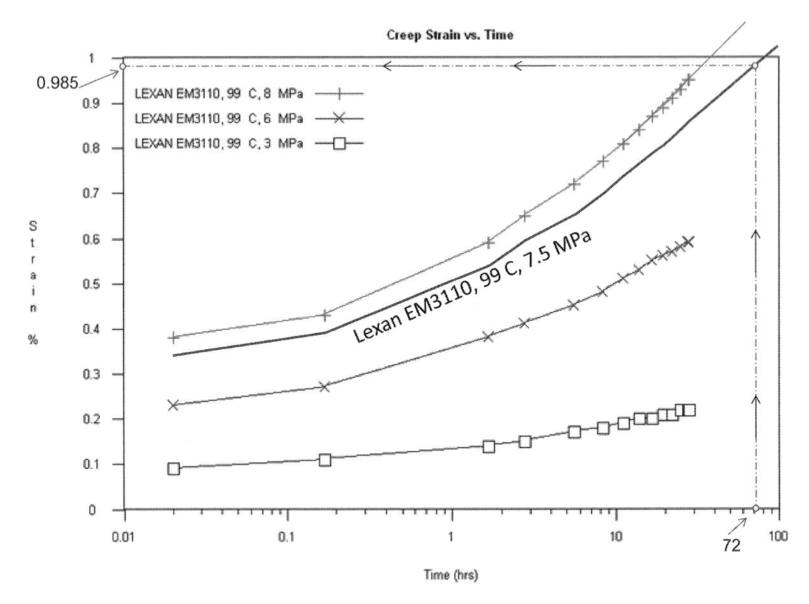

Figure 8.50 Creep curve at a constant stress 7.5 MPa for Lexan® EM3110 at 99°C

Then the creep modulus at 80°C is

$$E_{\text{Creep}} = \frac{7.5}{0.00725} = 1{,}034 \text{ MPa} \tag{8.115}$$

So, the deformation in the rectangular cross section cantilever beam loaded for three days continuously with the 3.6 N force while at a temperature of 80°C is

$$y_{t=72} = \frac{3.6 \cdot 12.5^3}{3 \cdot 1034 \cdot 9} = 0.252 \text{ mm} \tag{8.116}$$

Therefore, for this rectangular cross section cantilever beam design, when a load of 3.6 N is placed at the free end for 72 hours at 80°C, the amount of creep generated is

$$y_{\text{Creep}} = 0.252 - 0.122 = 0.13 \text{ mm} \tag{8.117}$$

Please observe that only after three days does the beam deform a great deal: 0.13 mm. This occurs because the amorphous thermoplastic polymer was used. Compared to crystalline polymers, which have a precise melting point, the amorphous resins exhibit a large melting temperature gradient, which explains the loss of properties at elevated temperatures.

8.15 Conclusions

Most calculations assume overall infinite rigidity in mating (grooved) parts and features. All initial designs and calculations should be based on such assumptions. This is a good first step, which can also improve engineering productivity.

Sometimes the easiest and simplest method of designing snap fits that perform properly is to overdesign them. By playing it safe, one can still save money over conventional methods of fastening. The way to overdesign snap fits is to make them larger or thicker by using more material. The material itself might be expensive, but adding extra thickness to a snap fit shouldn't increase the unit cost by much.

When approaching a snap fit design project, it is advisable to be as flexible as possible. It is not beneficial to commit to one particular design, especially if it is an unfamiliar one. Successful engineers will seek input from within their design work group, as well as in marketing, manufacturing, sales, and other departments within the organization when considering a new design.

It is always good to think broadly and challenge the ways in which products are traditionally designed and produced. A useful approach is to conceive a couple of completely different designs and evaluate each on its merits: assemblability, manufacturability, processability, novelty, quality, and other characteristics that may be important to the product.

If outside molders or toolmakers will be involved in the process, they, as well as material suppliers, should be consulted early in the program.

9 Bonding

There are bonding techniques employing two types of consumables: adhesives and solvents.

Adhesive bonding is an assembly process by which two parts are held together through surface attraction, also known as mechanical interlocking. The adhesive itself is a substance capable of adhering to the surface of the parts being bonded, developing strength after it has been applied and, afterwards, remaining stable. It can be described as a specific interfacial tension phenomenon. A useful way to classify adhesives is by the way they react chemically after they have been applied to the surfaces to be joined.

The principle of *solvent* bonding, on the other hand, consists of applying a liquid solvent that dissolves the surfaces of the joint area. Once the solvent evaporates, the parts are assembled by applying a small pressure in the joint area.

There are many adhesives and solvents available. The task of selecting the best adhesive or solvent for any given bonding application is a difficult one.

■ 9.1 Failure Theories

There are two types of failure for bonded joints: adhesive failure and cohesive failure.

Adhesive failure refers to a bonded assembly that fails because the adhesive peels away from one or both of the surfaces it was supposed to hold together (Fig. 9.1(b)). In most bonded applications it is an unacceptable failure. It can also be defined the same way when both the substrate and the bonded joint would fail at the same time.

Cohesive failure represents the breakage of the adhesive used for the bonded joint. On both components, the surface of the bonded joint after failure would have adhesive remnants visible. Cohesive failure also refers to the dual failure of the polymer being bonded and the adhesive–both failures occuring at the same time (Fig. 9.1 (a)). While adhesive failure is considered unacceptable, cohesive failure is the preferred failure.

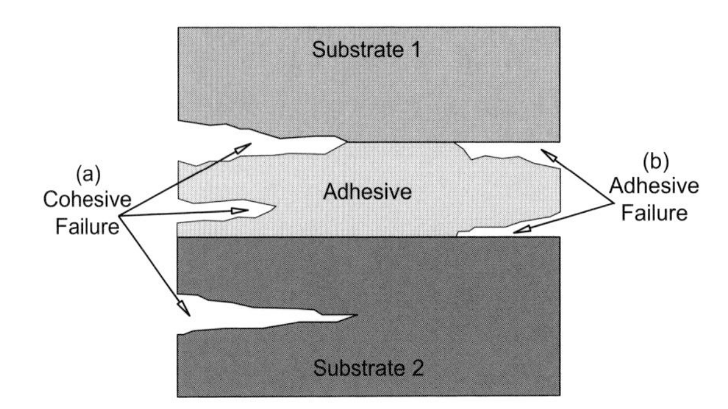

Figure 9.1
Failure theories: (a) cohesive, (b) adhesive

9.2 Surface Energy

A characteristic called *wetting* is an important element in both the adhesive and solvent bonding processes. Wetting refers to the intimate contact between the liquid solvent or adhesive consumables and the surface of the polymer. Good wetting occurs when there is strong attraction between the parts and the liquid.

Table 9.1 Surface Energy of Various Materials (Expressed in dynes/cm or 0.001 N/m)

SOLIDS (dynes/cm)	LIQUIDS	SURFACE TENSION (dynes/cm)
100+ (metals, glass, ceramics)		100
80	Water	72
	Glycerol	63
	Formamide	58
47	Phenolic	
46 (PA)		
43 (PET)		
42 (PC, ABS)		
39 (PVC)		
38 (PMMA)		
37 (PVA)		
36 (POM)		
33 (PS, EVA)		

SOLIDS (dynes/cm)	LIQUIDS	SURFACE TENSION (dynes/cm)
31 (PE)	Cellosolve	30
29 (PP)	Toluene	29
	N-Butanol	25
	Alcohol	22
18	Fluoropolymer	

Polymers in general are difficult to bond. As Table 9.1 shows, metals, ceramics, and glass are the materials that are easiest to bond succsssfully. Plastics, on the other hand, have surface energies required for bonding much lower than metals, glass, and ceramics. Thermoplastic nylon, also known as polyamide (PA), exhibits the highest *surface energy,* of 46 dynes/cm. Polypropylene (PP) is the worst thermoplastic polymer to bond–nothing sticks to it. Its surface energy is less than 30 dynes/cm.

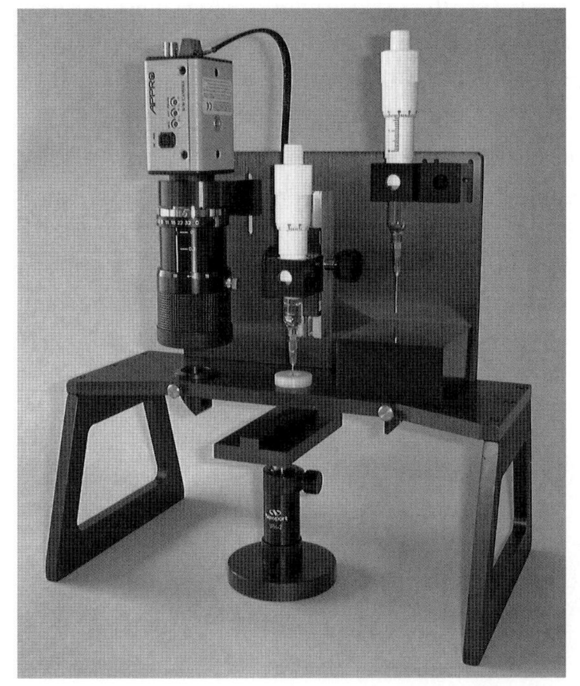

Figure 9.2
Contact angle goniometer instrument

Reinforcements, fillers, additives, pigments, dyes, and flame retardants all play a role in modifying the polymer surface energy. To properly measure the exact surface energy of the polymer, an instrument called the "contact angle *goniometer*" (see Fig. 9.2) can be used. The device measures the angle between the tangent to surface of the thermoplastic material and the tangent to the curvature of a drop of distillated water placed on the polymer surface. If the angle between the two tangents is less than 60°, then the surface energy of the thermoplastic to be bonded is excellent (see Fig. 9.3(b)). However, if the same angle is greater than 90°, that means that the surface of the polymer being bonded has extremely low surface energy and additional steps to improve it are required, as will be discussed in Section 9.3.

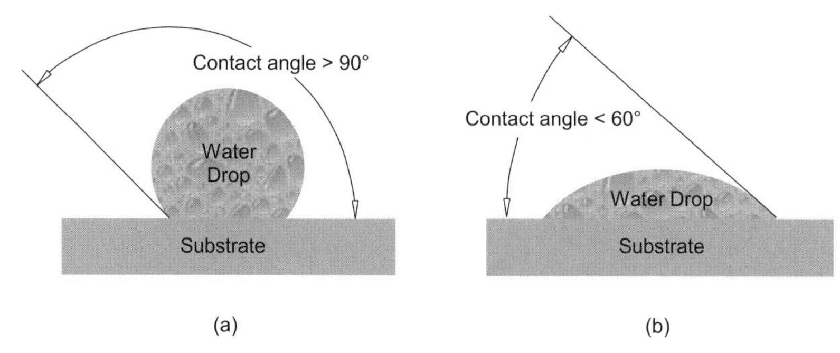

Figure 9.3 Surface energy: (a) poor surface energy (wettability), and (b) good wettability

Another technique for determining precisely the surface energy of the thermoplastic polymer, when a contact angle *goniometer* instrument is unavailable, involves a liquid ink set. The test liquids come in sets from as little as six to as many as 24 small bottles, each with a specific dynes per centimeter label designation.

Fluid from the bottle is applied to the surface of the thermoplastic material to be bonded, with the small brush mounted inside the cap (Fig. 9.4). If the test liquid draws back into droplets in less than 1 second after it has been applied, then the surface energy of the polymer substrate is lower than that of the fluid itself. The exact surface energy *(dyne level)* is determined by subsequently applying a range of increasing or decreasing values of dyne test inks until the fluid spreads like a film over the plastic surface. The label designation on the bottle will indicate the polymer surface energy level.

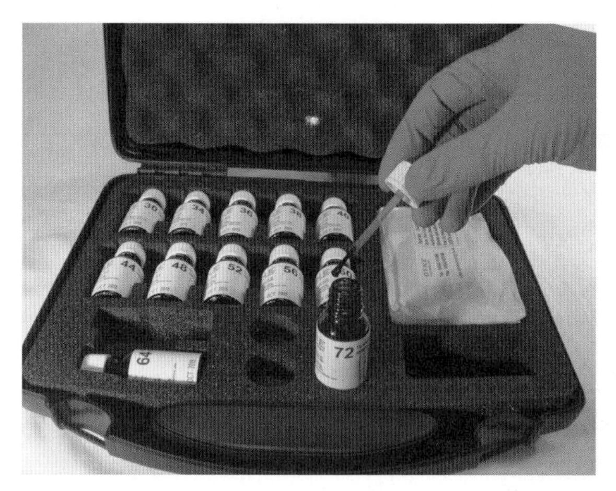

Figure 9.4 Set of dyne test liquids containing 12 fluid bottles ranging from 30 to 72 dynes/cm

Dyne pens provide another method of determining the surface energy of a thermoplastic polymer. Similar to the fluid bottle sets, the pens in the set are of various dyne designations. However, they have a lower shelf life compared to the test liquids. Figure 9.5 shows a set of 16 pens with designations between 30 and 60 dyne/cm.

Figure 9.5 Dyne test pen set

In the last few years a new test device was brought to market by Brighton Technologies Group of Cincinnati, Ohio. The instrument is called *Surface Analyst*™ (Fig. 9.6). A small droplet of deionized water is created on the thermoplastic material surface from a pulsed stream of microdrops. These droplets, once placed onto the polymer surface, grow together on the surface to form a single, slightly larger drop. Then, the average contact angle of the resulting drop is calculated, within a measurement cycle of just three seconds, based on the volume and the average diameter of the newly formed drop. This method involves the use of a single button and requires no operator input.

Once the surface energy of the component being bonded has been identified using one of the techniques described above, a number of procedures can be employed to increase the surface energy of the polymer, which we will discuss in the next section.

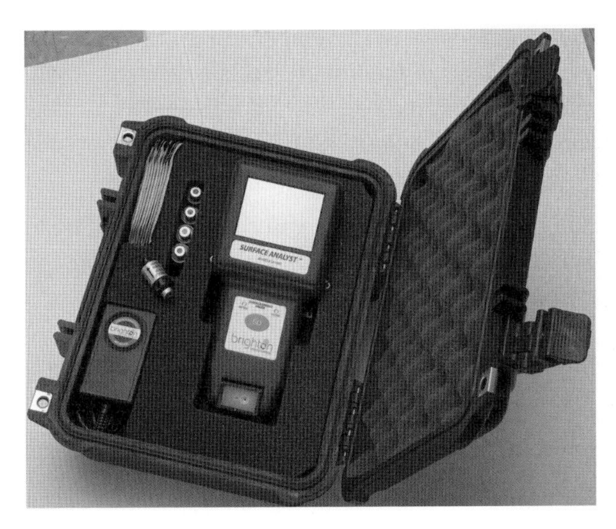

Figure 9.6
Surface Analyst™ set

Figure 9.7
Surface Analyst™ handheld instrument

■ 9.3 Surface Treatment

There are a number of techniques employed in the industry to increase the surface energy of thermoplastic polymers. They are corona, flame, and plasma treatments.

The *corona* treatment methodology was invented in the early 1950s by Verner Eisby, a Danish engineer–it is also called open air plasma. It uses two electrodes connected to a power source. When the electric current is turned on, a spark is created between the electrodes. When the spark touches the thermoplastic surface of the joint, it treats the surface, thus allowing up to 50% improvements in surface energy. Figure 9.8 shows the schematic of a portable corona unit. It has two electrodes connected to a power source. When the power is turned on, the fan rotates. The air flow generated pushes the spark toward the thermoplastic substrate, which is treated instantly.

Extruded film treated with a corona process, for example, can reach speeds of up to 800 meters per minute, at the same time providing various intensity levels for the treatment measured in watts per minute per square meter.

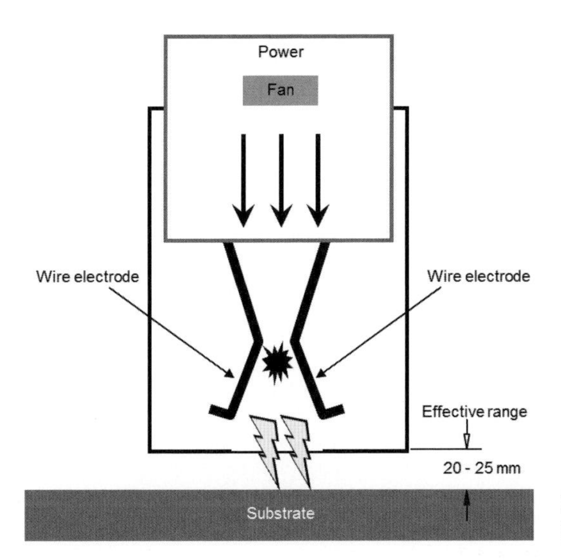

Figure 9.8
Portable corona unit

For example, to calculate the amount of power required to treat one side of a thermoplastic film, Equation 9.1 can be used:

$$P = kvw \quad (9.1)$$

where

k = process constant

v = line speed (m/min)

w = width (m)

P = power required (watt)

The process constant for extrusion varies between 8 and 15, for coating it is about 20, for cast film it is also about 20, and so on.

If, for example, there is a thermoplastic film that has to be treated on both sides (surfaces), then Equation 9.2 is employed:

$$P = 2kvw \quad (9.2)$$

Another technique to improve the polymer's surface energy is *flame treatment*. This method uses a natural gas burner, similar to those used for preparing food, which is usually placed in a vertical position. Thermoplastic components are mounted, placed, or hooked on a conveyor, which brings the components in front of the flame one at a time, thus treating the polymer surface.

Figure 9.9
Corona process

A more expensive technique is *plasma treatment*, which is performed using a closed chamber. This technique is used mostly for medical applications, because most plastic parts in that sector are of small volume. For other industries, like automotive in which the components tend to be rather large, the open-air plasma, also called corona, treatment is employed.

It is important to note that the surface treatment performed should be done just before adhesive is dispensed onto the joint area. There are a number of thermoplastic polymers that are rather sensitive to the amount of time elapsed between the treatment and the application of the adhesive. Some polymers may lose fairly quickly the treatment applied to increase the surface energy. For some thermoplastics it could occur within one day. A good rule when *troubleshooting* bonding joints is to check to see how much time has passed between the manufacturing stage at which the treatment was performed and when the actual bonding is performed.

Figure 9.10 shows that certain thermoplastic polymers, within the first 25 hours after they were actually treated, have completely lost the treatment benefits induced. If bonded, they will not be up to a normal standard of the joint strength.

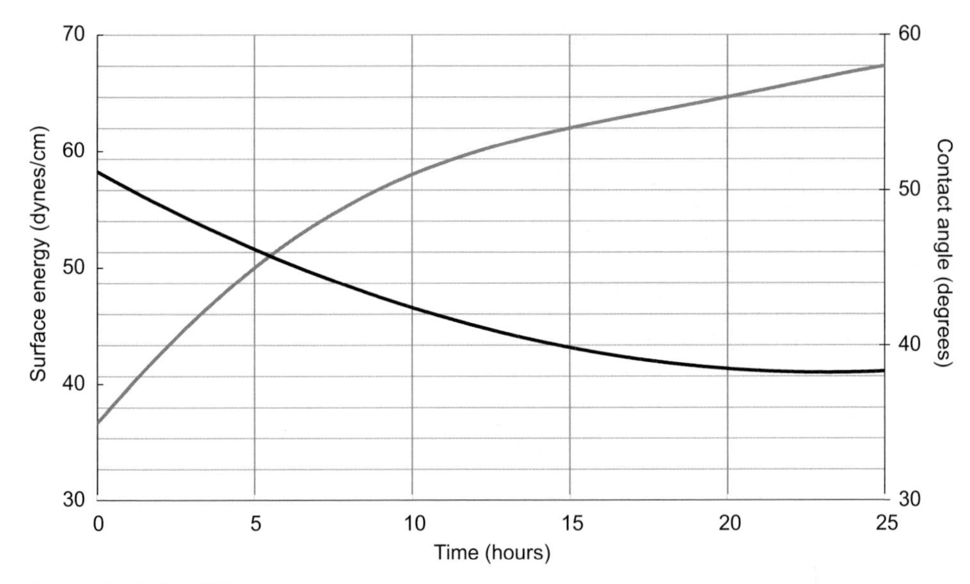

Figure 9.10 Shelf life

9.4 Types of Adhesives

There are many different types of adhesives, each with different properties. *Heat-cured* adhesives require the presence of heat in order to promote curing. *Holding* adhesives hold parts together for limited periods of time, not permanently. Masking tape is an example of a holding adhesive. Hot-melt adhesives are applied in a molten state and then allowed to harden as they cool. *Instant* adhesives cure within seconds. *Structural* adhesives can sustain loads and stresses of at least 50% of the original stress level of the polymer the components are made of.

Adhesives contain a variety of substances, each added to enhance a certain property. For example, activators, accelerators, and curing agents may be used to speed up the solidification process.

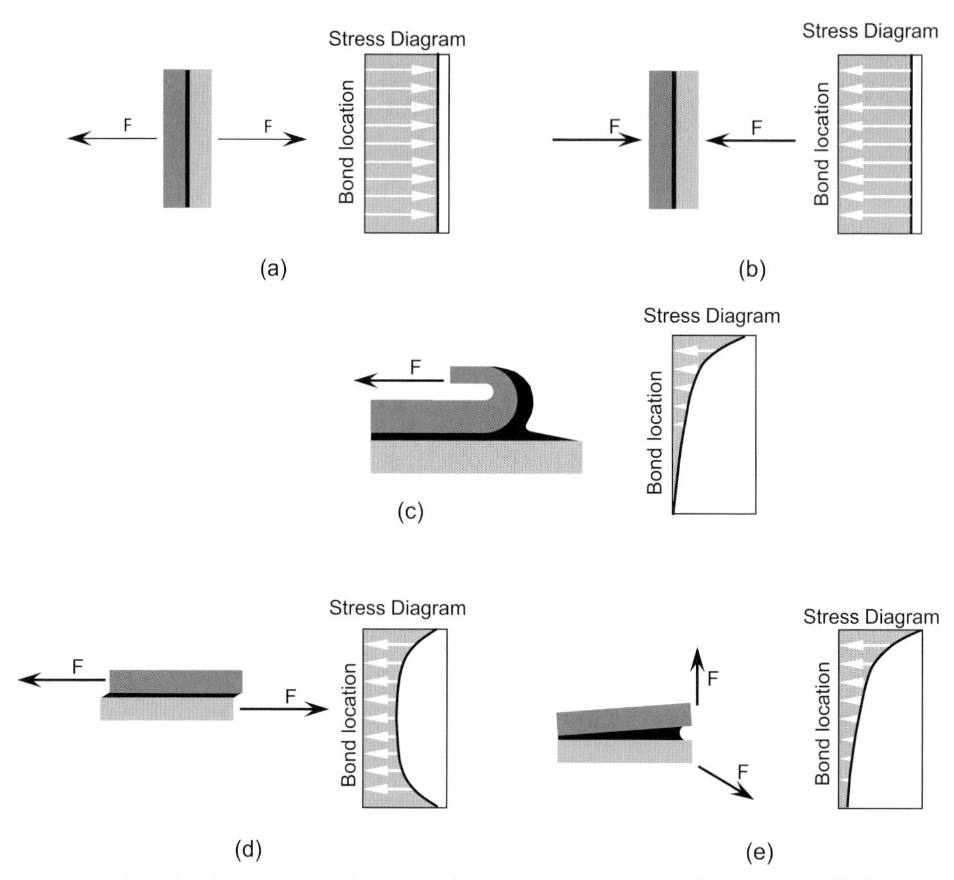

Figure 9.11 Stresses present in a bonded assembly: (a) tensile stress; (b) compressive stress; (c) peeling stress; (d) shear stress; (e) cleavage stress

Figure 9.11 shows the five stresses that must be considered when selecting an adhesive. When two mating parts assembled through adhesive bonding are pulled apart in a horizontal plane, the loading is *tensile stress* (Fig. 9.11(a)). If components are pressed against each other, the loading that develops at their interface is *compressive stress* (Fig. 9.11(b)). When the bonding agent is bent away from the bond line in a flexible joint, the stress is called *peeling stress* (Fig. 9.11(c)). When the components are loaded in the same plane or parallel planes, *shear stress* develops (Fig. 9.11(d)). Finally, when the two parts are pulled apart at the end of a lap joint, the assembly is loaded with *cleavage stress* (Fig. 9.11(e)).

As explained earlier, there are two major modes of failure in adhesive bonding: *adhesive failure*, which occurs when the adhesive peels away from the part surface; and *cohesive failure*, which occurs when the bonding agent adheres to both surfaces but rips itself apart.

Delamination is a third mode of failure that occurs when the bonding agent sticks to both components but pulls a thin layer of the thermoplastic polymer off one or both parts. This happens when the internal stresses between the bonding agent and the material exceed the strength of the thermoplastic polymer being bonded.

9.5 Advantages and Limitations of Adhesives

There are many adhesives available to accommodate a wide range of applications. Each has its advantages and limitations.

Hot melts are fairly good at filling the gap that in most cases occurs due to warpage between the components. Hot melts form both rigid and flexible bonds and set up quickly, requiring rather short curing times. However, hot melts exhibit poor wetting capabilities and can only be used in cases in which components can be assembled quickly before the melt solidifies. Hot melts are potentially dangerous to work with and can burn workers.

Silicone adhesives make excellent sealants in low-stress applications. They offer good water resistance and perform well in temperatures as high as 200 to 260°C (400 to 500°F). Silicones are relatively flexible for structural designs. However, silicone applications are limited due to low resistance to solvents. It should also be noted that silicones are corrosive and difficult to clean.

Solvent cements offer good wetting and penetration. Because a wide variety are available, they can be used on many different polymers. Solvent cements can be easily applied to large areas, and they can be applied by spraying. These adhesives work with low clamping pressures, offer good storage life, and require no special equipment.

Limitations of solvent cements include low strength, which is below that of the base polymer. They also tend to shrink up to 70% and require long curing times. These adhesives must be applied to both components in most cases, and they can attack certain thermoplastic polymers. Another inconvenience is that solvent cements are flammable chemicals, which limits their application, and special storage and dispensing procedures are required.

Knowledge of the material of the components to be bonded is essential in determining the best adhesive or solvent for the application. Most of the difficulties encountered when using adhesive bonding procedures arise from poor adhesive compatibility with the bonded thermoplastics or poor surface wetting.

9.6 Stress Cracking in Bonded Joints of Adhesives

Stress cracking occurs when adhesives are applied to the material surface of an already stressed part. The prestressed condition could be the result of improperly executed injection molding, sharp corners in the part, or sudden changes in the wall stock. Internal stresses are greater in the parts that have metal inserts molded into them. These stress concentrations create microcracks in the part. When an adhesive or solvent is applied, the liquid can penetrate the part, further weakening the thermoplastic polymer.

Polymers that are prone to stress cracking include acrylics, polycarbonate, polystyrene, SAN, polysulfone, ABS, and PPO.

Materials with good stress cracking resistance are thermosets, acetals, PET/PBT, rigid PVC, polyamide, polyolefins, and PPS.

The following procedure can help control problems associated with stress cracking. The surface of the thermoplastic component being bonded is abraded with sandpaper and then cleaned with *isopropyl alcohol*. The parts are assembled immediately after the solvent adhesive has been applied. Once assembled, the parts are clamped together under moderate pressure, which is held constant for a short amount of time, typically anywhere from 30 to 60 seconds.

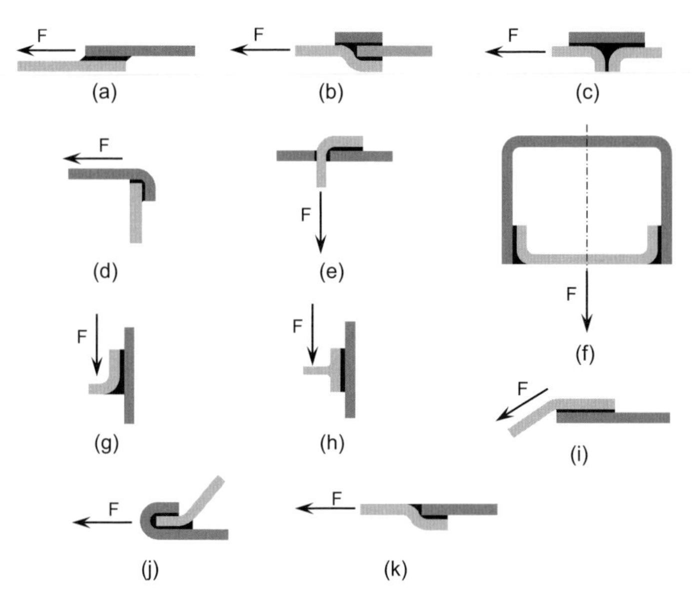

Figure 9.12 Bonded joints designed to avoid cracking caused by various stresses: (a) shear; (b) tensile; (c) tensile and shear; (d) compressive and shear; (e) compressive; (f) circumference shear; (g) cleavage; (h) peeling; (i) variation of shear; (j) variation of tensile; (k) variation of tensile and shear

9.7 Joint Design

Certain allowances made during the design stage can help facilitate optimum adhesive bonding. Joints that create compression loading rather than tensile loading increase the life expectancy of the bond. Also, *shear joints*, rather than peel or cleavage joints, often result in stronger adhesive bonding. It should also be noted that the *joint width* is more important in adhesive joint design than the *overlap length* of the parts.

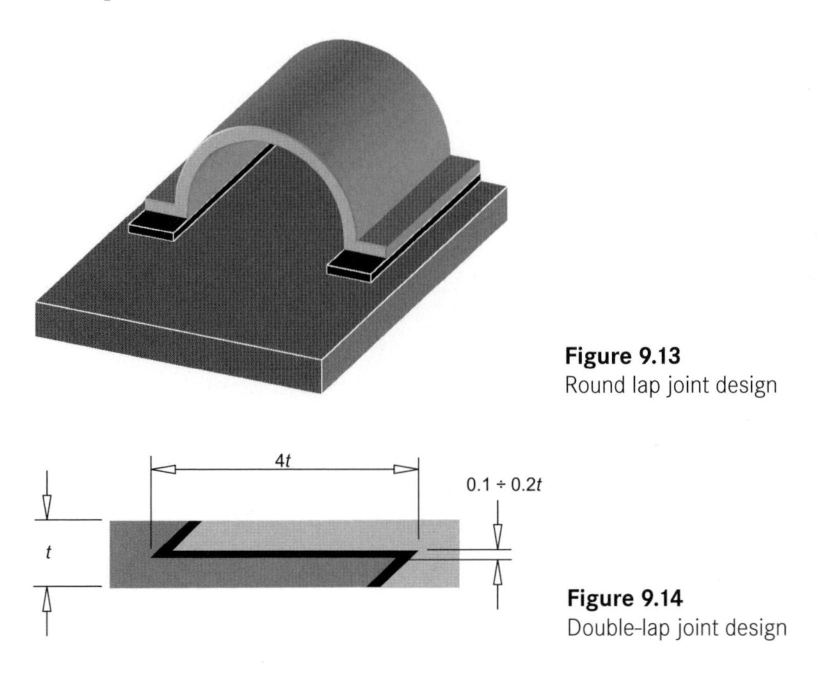

Figure 9.13
Round lap joint design

Figure 9.14
Double-lap joint design

Fiat Chrysler Automobile Group developed the first plastic-body car in North America, the Chrysler Concept Vehicle (nicknamed CCV). The CCV's outer body, inner body, and cage (typically a tubular structure that protects the driver and passengers by collapsing like an accordion in a crash situation), were all made of thermoplastic resins. Figure 9.15 shows one vehicle among the one or two dozen assembled, which was tested on the proving grounds during the development stage.

The vehicle body is made of four large components, each molded using a process called "gas-assisted injection molding," employing multiple gates and multiple gas nozzles and using a very large press. The 9000-ton press, belonging to Cascade Engineering, a molder from Grand Rapids, Michigan, was at that time the largest injection-molding machine in North America. The joint design used for the CCV was the double-lap joint (see Fig. 9.14) because it has good performance when peel and cleavage loads are encountered when the vehicle is in use.

Figure 9.15
CCV, Chrysler Concept Vehicle

Figure 9.16 shows the robot arm removing the driver-side outer body part from the tool when the injection-molding cycle time is finished and placing the component made of thermoplastic polymer into a fixture to further cool, preventing any post-mold warpage from taking place [181].

Figure 9.16
A pick-and-place robot arm removes one of the four body parts from the mold once the injection-molding cycle is over

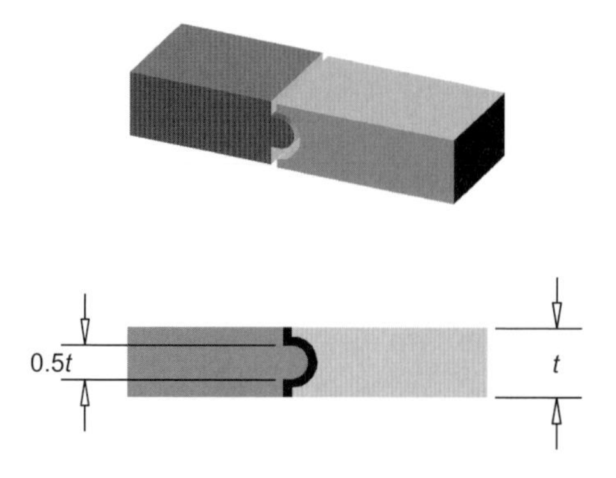

Figure 9.17 Tongue-and-groove joint design

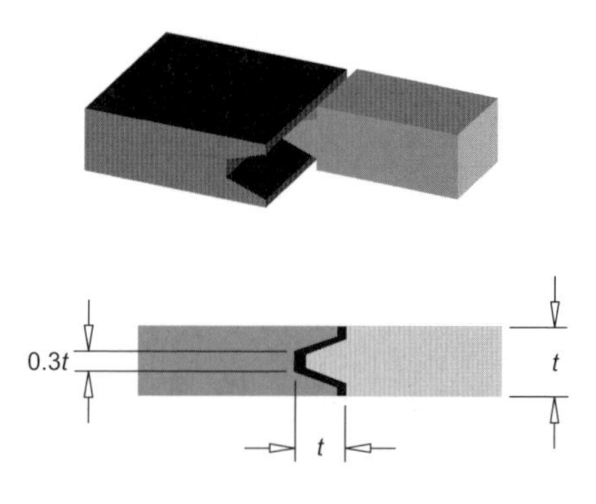

Figure 9.18 Scarf-joint design

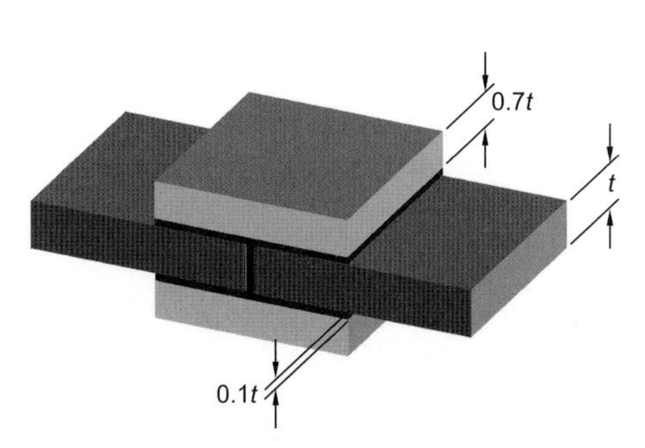

Figure 9.19 H-joint design

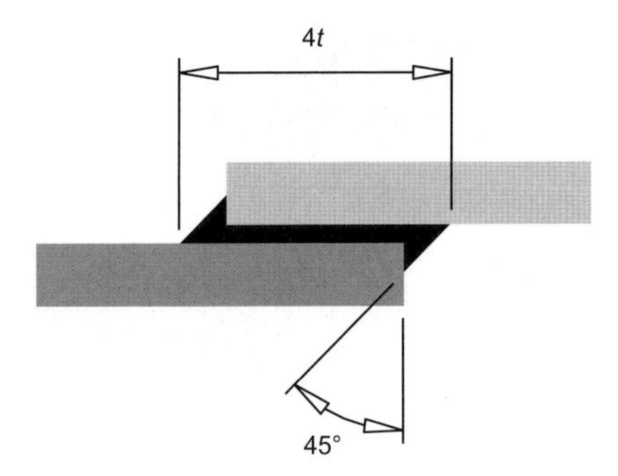

Figure 9.20 Single-lap joint design

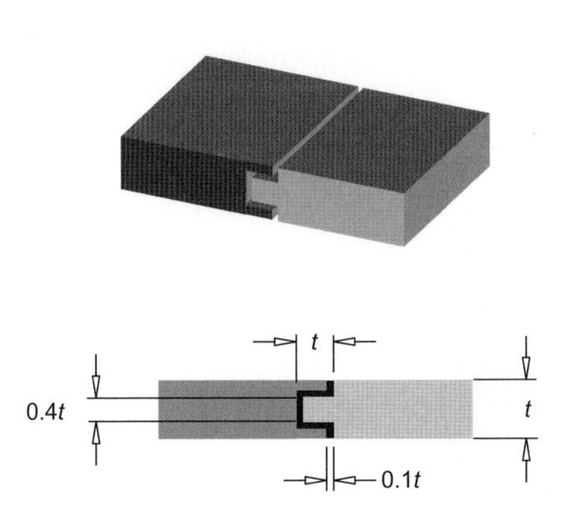

Figure 9.21 Tongue-and-groove joint design variation

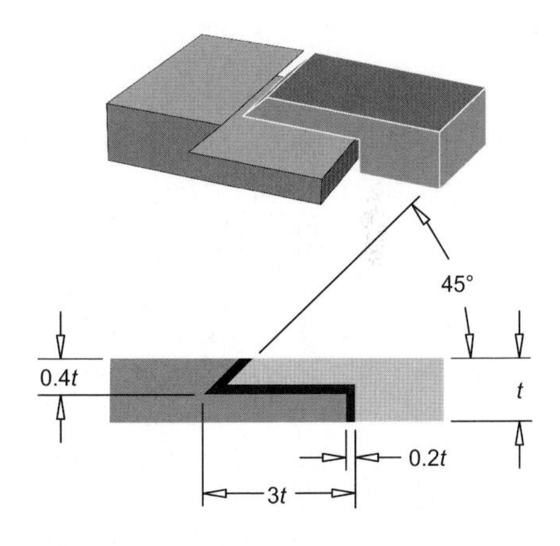

Figure 9.22 Butt-scarf lap joint design

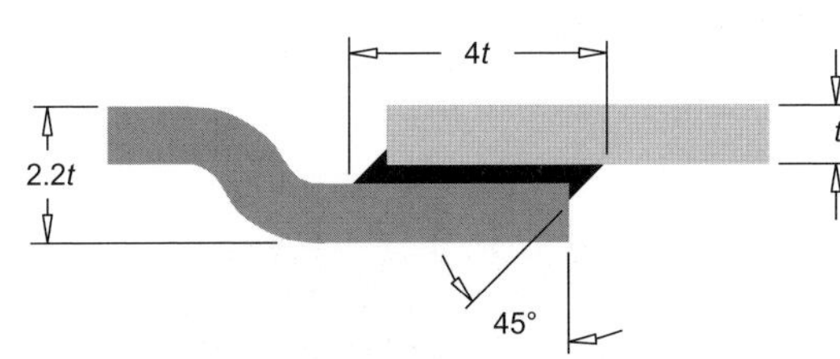

Figure 9.23
Joggle lap joint design

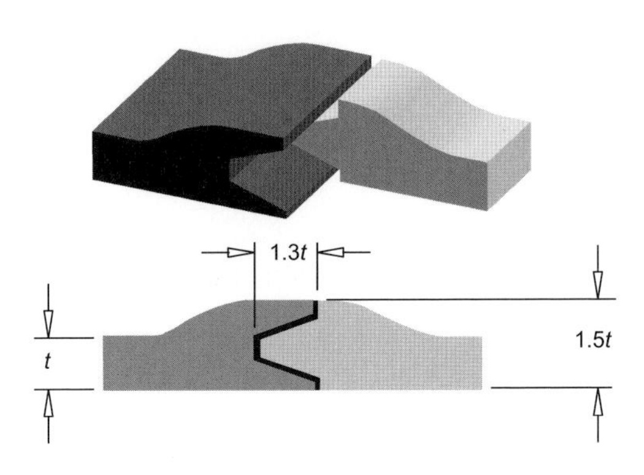

Figure 9.24
Tapered tongue-and-groove joint design

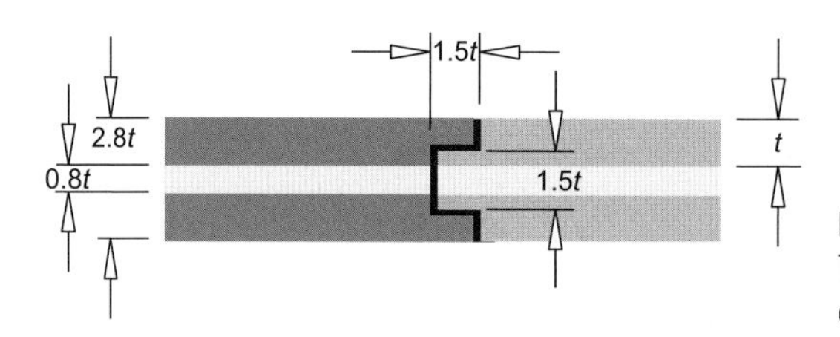

Figure 9.25
Three-piece tongue-and-groove joint design variation

9.8 Conclusion

Depending on the application, materials, design, and conditions under which the product will be used, it is possible to assemble plastic parts effectively.

Bonding allows the assembly of components made of similar thermoplastic polymers, but it also permits the assembly of dissimilar materials such as thermoplastic amorphous polymers with semicrystalline thermoplastics or even, in certain cases, with thermoset polymers.

Other considerations should also be taken into account when selecting an assembly procedure, such as cost and experience with any given bonding method.

10 In-Mold Assembly

The rather new injection-molding techniques known as "overmolding" and "in-mold assembly" (IMA) have seen exponential growth lately.

In the overmolding injection-molding process, a number of different polymers or the same polymer in various colors are molded in the same tool or mold in subsequent steps. This process is also called *multipolymer* injection molding" or *multimaterial* injection molding," terms that refer to thermoplastic resins. Overmolding is a common way of manufacturing products that require a certain decorative look and certain tactile characteristics that no one thermoplastic polymer is able to provide. A second, third, fourth, or even fifth polymer can be overmolded to provide added visual appeal, a softer texture, functionality, or marketability over traditional designs. Various examples of overmolding components include automotive turn signals, tail lights, and high mounts, handles for toothbrushes, and various seals and gaskets. There are no standards or best practices within the industry to ensure manufacturing success because each application for overmolding multimaterial parts is unique. The development stages for each multimaterial have to be adjusted so that the end product meets the end-user requirements.

In use for the last few years, IMA, also called "*multicomponent* injection molding," completely eliminates the need for additional postmolding assembly. It reduces inventories and variable costs, the most expensive of which is labor. Greatly reducing or even eliminating the number of assembly steps provides the significant advantage of combining classic assembly processes with injection molding. To be successful, the IMA injection-molding technique requires that all thermoplastic polymers have different glass transition temperatures (T_g) and dissimilar shrinkage rates. The IMA process allows the manufacturability of articulated 2-D or even 3-D joints that can move with respect to each other in a plane or a 3-D space. However, to achieve the simplicity of an articulated joint, the mold or the tool and the injection-molding press used to make the multicomponent parts will be more complex.

10.1 Overmolding

The main requirement for manufacturing components successfully using the overmolding or multimaterial injection-molding process is that the polymers have very close if not identical shrinkage. Chemical bonds achieved between various polymers used in the molding process should be considered a bonus.

A chemical bond takes place between different thermoplastic polymers or parts made of the same resin but of different colors in the overmolding process. In the vast majority of multipolymer applications, chemical bonds between the polymers employed do not take place. To overcome the shortcoming, it is best to rely upon mechanical interlocking features designed into the components themselves, as shown in Fig. 10.1.

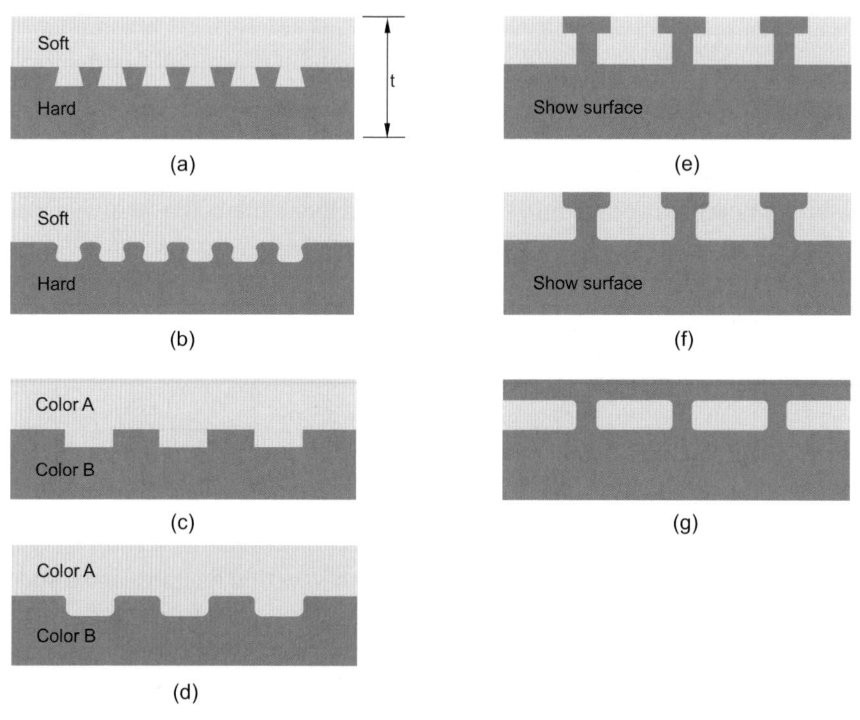

Figure 10.1 Mechanical interlocking features: (a) dovetail design for an extruded hard thermoplastic component overmolded with a soft polymer; (b) dovetail design for an extruded hard thermoplastic component overmolded with a soft polymer having no sharp corners; (c) zigzag design for molding different color polymers; (d) zigzag design for molding different color polymers with no sharp corners; (e) design for show surface; (f) design for show surface having no sharp corners; (g) first polymer completely embedded

Many other design options that achieve mechanical interlocking between various polymers are possible. It should be noted that by employing fillets, the peel strength improved on average by 5 to 8% versus sharp-corner designs (Figs. 10.1(a), (c), and (e)).

10.2 In-Mold Assembly

Similar to the overmolding process, the in-mold assembly (IMA) injection-molding technique requires multicavity molds as well. This technique could use as few as two cavities to as many as four or five. Once the first cavity of the mold is completely filled and packed with the first polymer, the tool opens, and the cavity rotates around the horizontal axis or in a direction perpendicular to the horizontal axis of the injection-molding machine. This step allows the mold to close using the second cavity of the tool. Then the second thermoplastic polymer completely fills and packs the remaining volume of the second cavity. Finally, the tool opens and the process repeats itself for subsequent cavities if more than two are used.

The IMA technique provides several advantages over traditional methods, which involve molding various components separately, followed next by the assembly step:

- fewer parts required for assembly and less time required to assemble the components;
- no need for work-in-progress inventories;
- no need for additional injection-molding machines and tools, thus reducing floor space requirements;
- reduce labor content required by purchasing, inspecting, and warehousing functions;
- lower capital requirements and piece costs of a product;
- improved part alignment in assembly for highly cosmetic parts;
- less time required to mold each part separately and assemble them later;
- ability to produce components that otherwise would be impossible to produce using conventional manufacturing techniques;
- ability to produce better-quality products because the components are aligned perfectly and material interfaces have tighter tolerances.

The key to successful IMA injection-molded components relies upon the ability to select thermoplastic polymers that will not achieve a chemical bond when molded one after the other in the same tool, as shown in Fig. 10.2. It is also very important that, besides not melting each other, the thermoplastic polymers employed in the IMA process have distinct shrinkage rates. Key points in the prevention of adhesion between polymers for IMA are

- resin shrinkages create separation forces at the joint interface between various polymers;
- disallowing bonding at the molecular level prevents polymer cross-linking;
- using lower injection-molding pressures prevents bonding between resins;
- lower temperature gradients prevent adhesion;

- antistatic friction additives are preferable because static friction is overcome (also known as "stiction") to enable relative motion of stationary objects in contact;
- low tool polishing prevents mechanical interlocking between resins.

The best way to limit adhesion is to select materials that are chemically incompatible and hence do not promote cross-linking of polymers.

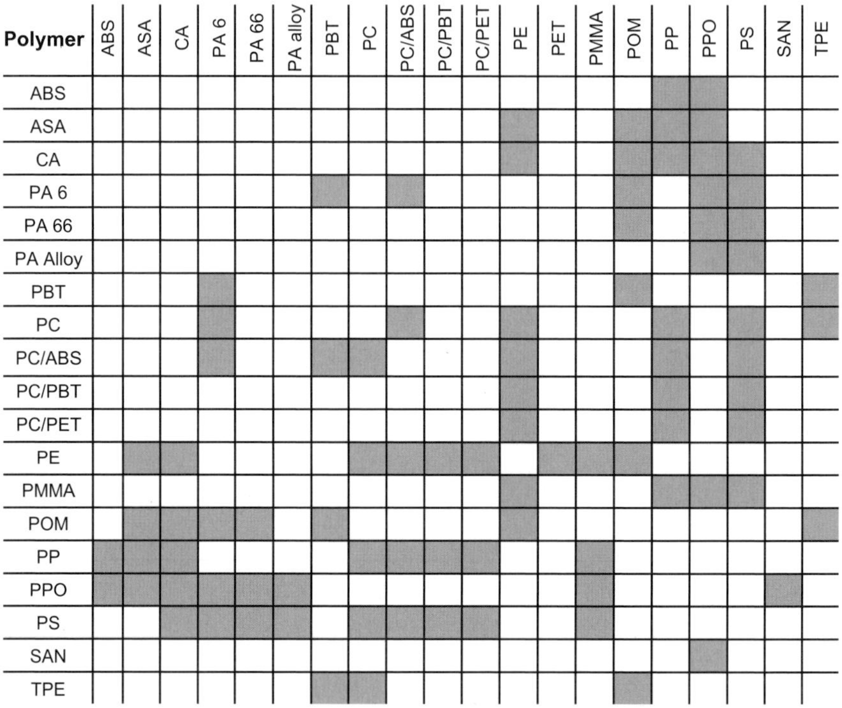

Polymer	ABS	ASA	CA	PA 6	PA 66	PA alloy	PBT	PC	PC/ABS	PC/PBT	PC/PET	PE	PET	PMMA	POM	PP	PPO	PS	SAN	TPE
ABS																■	■			
ASA												■			■	■	■			
CA												■			■	■	■	■		
PA 6							■		■						■		■	■		
PA 66															■		■	■		
PA Alloy																	■	■		
PBT				■											■					■
PC				■					■			■				■		■		■
PC/ABS				■			■	■				■				■		■		
PC/PBT												■				■		■		
PC/PET												■				■		■		
PE		■	■					■	■	■	■		■	■	■					
PMMA												■				■	■	■		
POM		■	■	■	■		■					■								■
PP	■	■	■					■	■	■	■			■						
PPO	■	■	■	■	■	■								■					■	
PS			■	■	■	■		■	■	■	■			■						
SAN																	■			
TPE							■	■							■					

Figure 10.2 Material compatibility for in-mold assembly process

■ 10.3 Joint Design

There are three ways of creating a joint design for IMA.

The simplest is the axial joint, which allows only translation movement. As the name implies, this type of joint provides relative motion between the two components in only one direction. A joint design that allows motion between the parts in one plane is known as a rotation joint. Finally, the joint design that allows motion between parts to take place in 3-D space defines a spherical motion. There are also options to combine various joint approaches in a single design.

Shrinkage differentiation between polymers is the best way to create a joint design. Shrinkage diferentiation is defined as the dimensional difference between the molded part, smaller at room temperature than the tool cavity, and the mold at the same temperature. If the shrinkage calculations are not performed properly, the chances of either sink marks or voids in the thickest area of the molded component increase.

For example, the Pulsafe FitLogic™ safety glasses (Fig. 10.3), manufactured by Uvex, then part of the French company Bacou-Dalloz and now part of Honeywell, use all three joint designs described above.

Figure 10.3
FitLogic™ safety glasses

The temple arms rotate 90° from the stored position into a fully open position when in use. Articulation in one plane is achieved within the molding cycle by employing two different polymers (see Fig. 10.4).

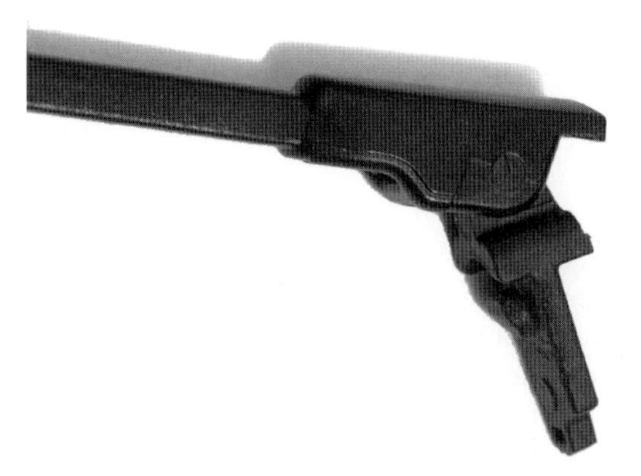

Figure 10.4
Detail of the one-plane (2-D) articulation, which acts as the temple piece hinge

Furthermore, when the mold opens, the plastic assembly is not ejected from the tool, but instead the cavity rotates. When the tool closes again, the third component of the assembly, the soft temple component, is molded (see Fig. 10.5).

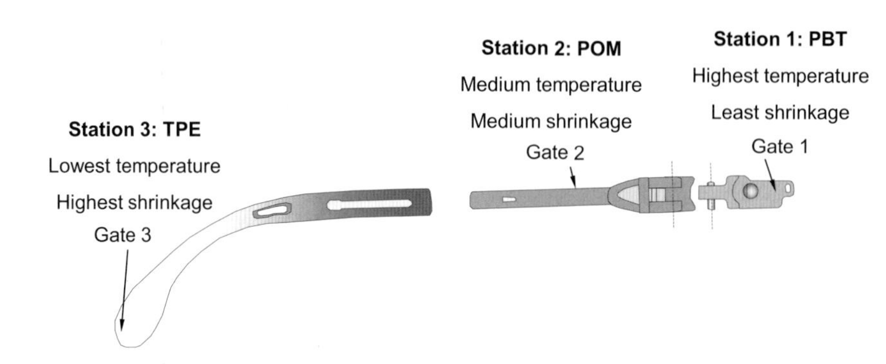

Figure 10.5 Temple arm

The critical dimension to articulate the temple piece in a 2-D plane is the pinhole interface, which has the nominal diameter of 2 mm (see Fig. 10.6).

Figure 10.6 Pinhole detail of the temple arm hinge

Table 10.1 Shrinkage Rates

Polymers	Shrinkage (%)	Melting temperature, °F (°C)
PBT	1.4	482 (250)
POM	2.2	390 (195)
TPE	2.5	350 (175)

Shrinkage calculation for the pin is given by Eq. 10.1:

$$S_{\text{Pin}} = \xi_{\text{PBT}} D_{\text{Pin}} = 0.014 \cdot 2 = 0.028 \text{ mm} \tag{10.1}$$

Similarly, the hole shrinkage is

$$S_{\text{Hole}} = \xi_{\text{POM}} D_{\text{Hole}} = 0.022 \cdot 2 = 0.044 \text{ mm} \tag{10.2}$$

By adding the shrinkage amount to the diameter of the hole, made of POM resin, and by subtracting the shrinkage amount from the pin diameter, made of PBT polymer, the clearance value of the 2-D articulated joint is

$$D_{\text{Hole}} = 2 \text{ mm} + 0.044 \text{ mm} = 2.044 \text{ mm}$$

$$D_{\text{Pin}} = 2 \text{ mm} - 0.028 \text{ mm} = 1.972 \text{ mm}$$

$$C = D_{\text{Hole}} - D_{\text{Pin}} = 0.072 \text{ mm} \tag{10.3}$$

Due to the variation of the processing setting adjustments made to mold a given batch of resin, the wear of the tool, and other variables, a tolerance should be included in the above calculations, as shown by Eq. 10.4 for maximum tolerance and by Eq. 10.5 for minimum tolerance.

$$T_{\text{Max}} = D_{\text{Hole}}^{\text{Max}} - D_{\text{Pin}}^{\text{Min}} \tag{10.4}$$

$$T_{\text{Min}} = D_{\text{Hole}}^{\text{Min}} - D_{\text{Pin}}^{\text{Max}} \tag{10.5}$$

The highest melt temperature thermoplastic resin with the lowest shrinkage rate should always be injected first. Then, for the second shot, the thermoplastic that has lower melt temperature and slightly higher shrinkage rate should be injected, and so on.

Similarly, the nose piece of the safety glasses is assembled inside the tool as well. The first resin molded is clear polycarbonate (PC). After the tool opens again, the part is not ejected, but instead the cavity rotates and the tool closes again. The thermoplastic polyolefin elastomer is now injected into the tool. Its shrinkage rate is higher than that of polycarbonate. A number of processing parameters must be adjusted so that the motion of the soft nose pad molded over the polycarbonate moves to a position desired by the user and will remain in that position. There is a narrow path to adjust the motion of the nose piece (see Fig. 10.7) and to keep it in that position once adjusted.

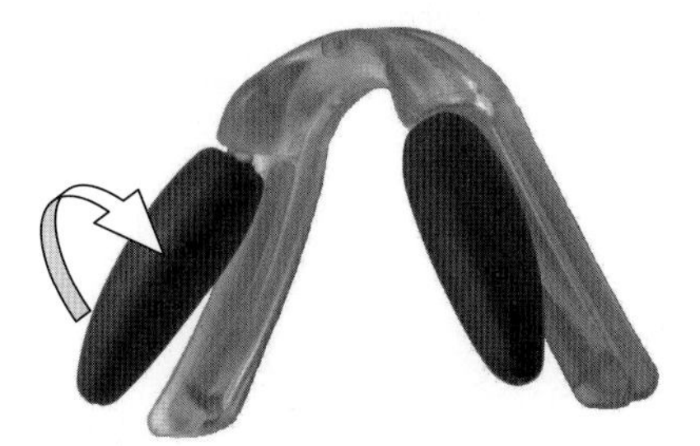

Figure 10.7
Nose piece

10.4 Tool Design

There are two distinct approaches to designing a tool for IMA.

The first approach depicted in Fig. 10.8 shows a tool using a core retraction mechanism, in this case a movable slide inside the mold. The highest melt temperature polymer, which also has the lowest shrinkage rate, is injected first by using one of the injection-molding press general-purpose screws (it could also be a barrier screw if, for example, the polymer is colored in the injection-molding press). Once the high melt temperature polymer is properly packed, the mold does not open. Instead, activated electrically, pneumatically, or even hydraulically, the core retracts, and the second polymer having a lower melt temperature and higher shrinkage rate is then injected.

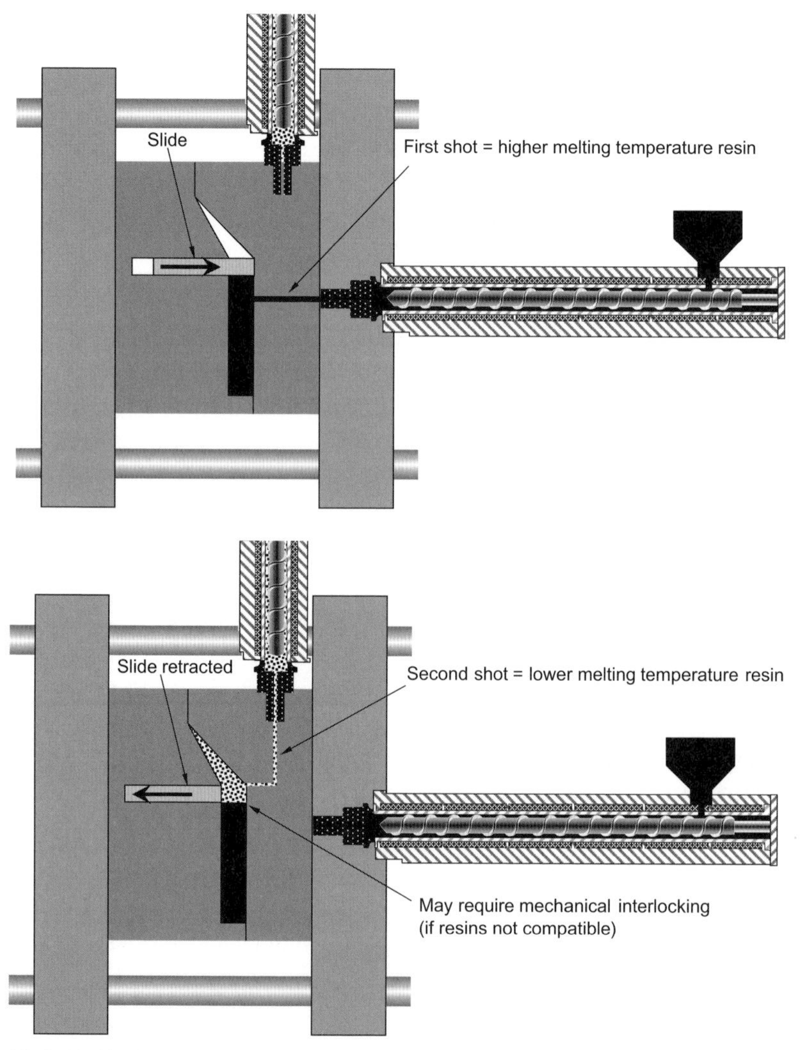

Figure 10.8 Mold with core retraction: (a) injection of the highest melt temperature resin, (b) injection of the second polymer having a lower melt temperature

In Fig. 10.8(a), the core retraction is shown as a movable slide inside the tool. Manufacturing tools in this manner provides lower-cost options. However, there are limitations to the type of components that can be produced. Figure 10.9 shows other types of core retraction options.

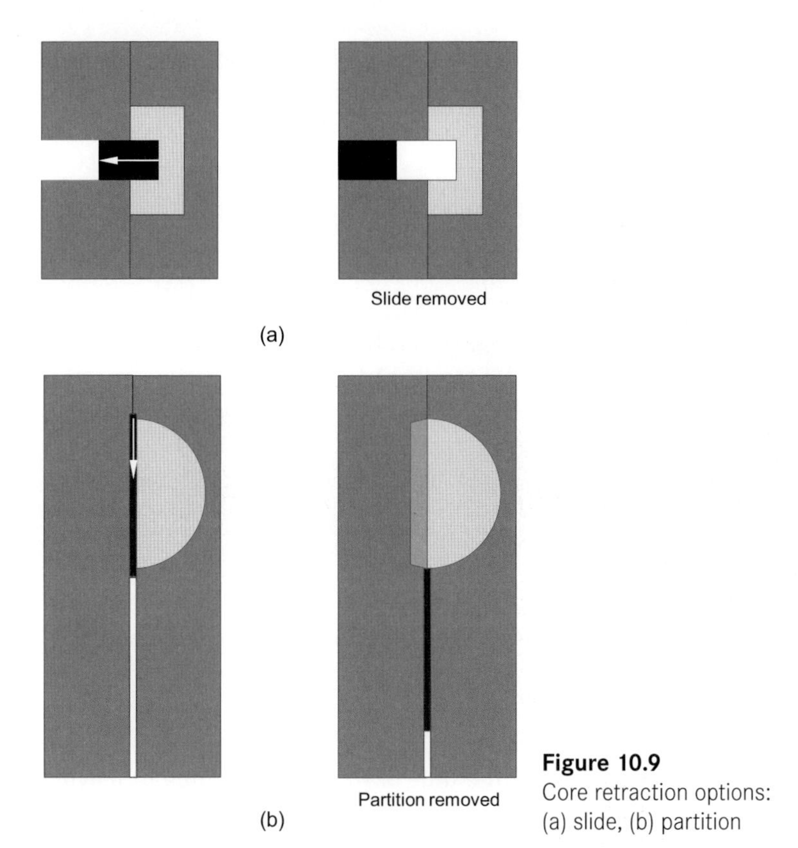

Figure 10.9
Core retraction options:
(a) slide, (b) partition

The second manufacturing approach to making IMA molds is to use a vertical or horizontal injection-molding press that has a movable platen, which, when rotating, allows a new cavity to become available to the same core (Fig. 10.10). This is achieved by using multiple-parting-line stack molds in traditional linear or newer rotary-table designs. The so-called single-face molds use rotating or sliding plates to change cavity configurations between shots. Both single-face and stack tools have become more versatile recently by employing one or even more turning center turrets. Each turret can have up to four faces, allowing each face to perform different functions.

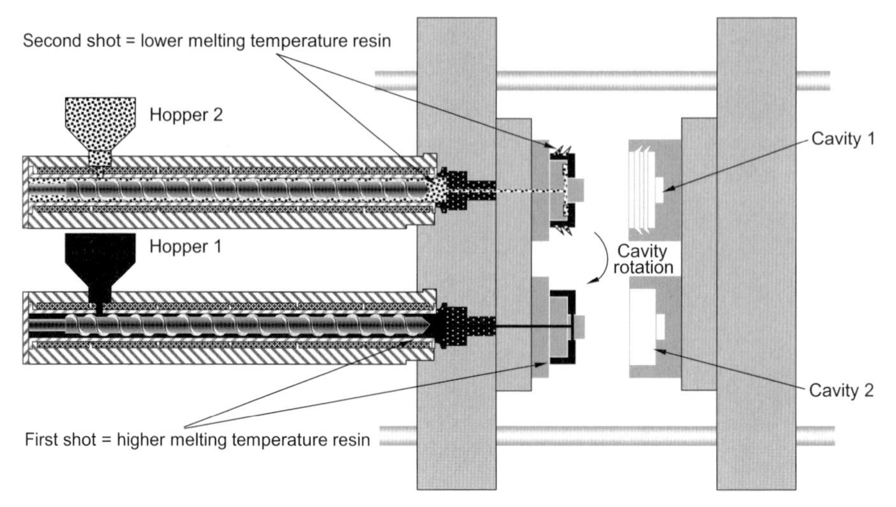

Figure 10.10 Indexing, also called rotating platen, tool

The rotation can be achieved in a number of ways:

- entire movable platen rotation;
- one section of the movable platen rotating at time;
- rotation of center plate for stack molds;
- sliding or moving one or more cores at the time of rotation, or sliding or moving one or several cores at one time or subsequent times;

The IMA technique is used in many industries, such as eyewear, automotive, appliance, consumer products, toys, and tools. Figure 10.11 shows a number of toys, for example, molded using the IMA process.

Figure 10.11 Various children's toys: elephant, chimpanzee, panda, and gorilla

The chimpanzee is fully assembled inside the tool (see Fig. 10.12). When the tool finally opens at the end of the complete molding cycle time and the assembled component is ejected from the mold, the toy has rotatable arms and legs. Furthermore, the head can not only rotate; it can also move up and down. In order to achieve such articulation joints in a plane and in 3-D space, the mold has three distinct cavities in which three different resins are injected.

Figure 10.12 From left to right: chimpanzee, panda, and gorilla

The molding cycle takes place as follows: when the tool is first closed, the first resin, yellow polybutylene terephthalate or PBT for short, is injected (see Fig. 10.13, Station 1). This yellow polymer is used to create the chimpanzee face. Once the face is created, the tool opens again and the part is not ejected; instead, the cavity rotates. The second resin, polyamide 6 (PA 6) or nylon 6, is injected into the new space created inside the tool, forming the chimpanzee body (see Fig. 10.13, Station 2).

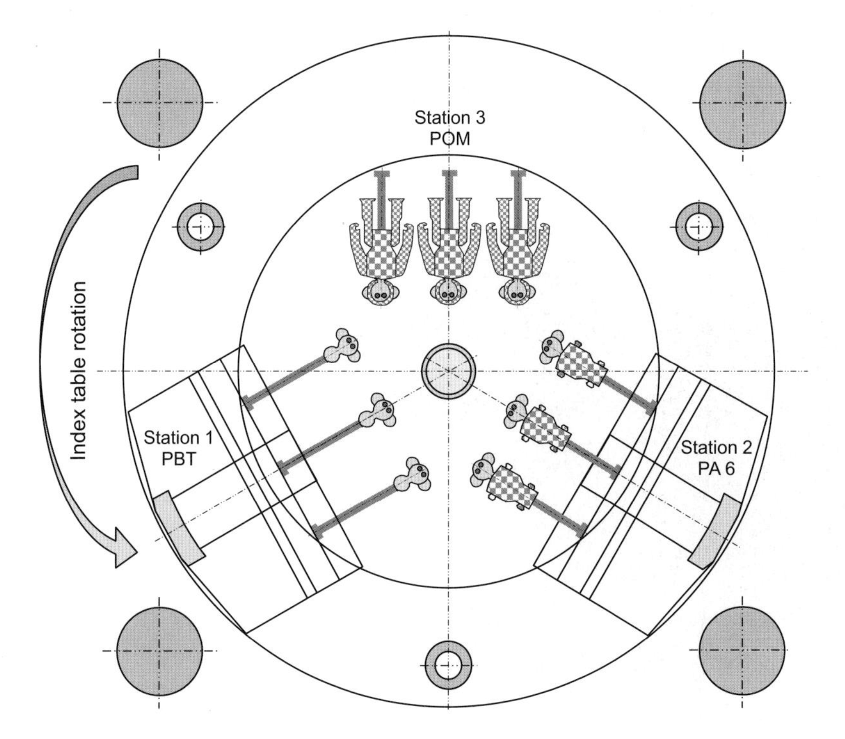

Figure 10.13 Tool with rotating cavity, three cavities in total

Next, the tool opens once more and the part is still not ejected from the mold. Again, the cavity rotates. In the new cavity, the chimpanzee's arms, legs, and head are now molded using the third polymer, polyoxymethylene (POM) or acetal. When the mold opens for the last time, the fully assembled toy is ejected. No secondary operations are necessary.

Table 10.2 Shrinkages and Melting Temperatures for the Three Polymers Used for the Chimpanzee

Polymer	Shrinkage (%)	Melting temperature, °C (°F)
PBT	1.2	250 (482)
PA 6	1.5	220 (428)
POM	1.9	195 (390)

The IMA technique requires that the first polymer injected in the tool be the resin that has the highest melting temperature and the lowest shrinkage. That will be followed by the resin with the second-highest melt temperature and the higher shrinkage, and so on. Differentiation of polymer melting temperatures together with shrinkage differentiation ensures that the polymers do not chemically bond with each other. When the melting temperature gradients are not large enough to prevent polymer bonding after ejection from the tool, a quick rotation for articulated joints or a quick pull for translating joints should be performed.

Figure 10.14 Gate vestiges for telephone key pads

Telephone keys are molded in a tool with a rotating cavity (Fig. 10.14) using the same acrylonitrile butadiene styrene (ABS) polymer in two distinct colors: light gray and dark gray.

10.5 Case Histories: Automotive IMA

The first case history highlights an under-hood application while the second deals with an automotive interior assembly, both achieved using IMA.

A mixture of air and gasoline vapor is burned in the internal-combustion engine. The gasoline is delivered to the engine from the fuel tank, and the air it mixes with is delivered by the intake manifold. Before the fuel-injection technology was in use, the carburetor would mix fuel vapor and air, and the mixture was delivered to each cylinder by the intake manifold.

More efficient computer-controlled fuel-injection systems have long since replaced the carburetor. Now, the intake manifold delivers only air. The fuel-injection system delivers gasoline much closer to the cylinder–at the base of the intake manifold next to the intake valves of the cylinder head–where it mixes with air from the intake manifold to become an extremely precise mixture of air and fuel vapor. The air flows from the air intake through the air cleaner and into the intake manifold. This is where the throttle valve adjusts and measures the total volume of air. When maximum power is required, the valve is fully open. At idle, the valve is closed, and a tiny amount of air ported around the throttle valve keeps the vehicle from stalling.

Once drawn into the manifold, the air mixes in a central chamber called a plenum. The air is directed from there to the cylinders through individual intake runners. The engine has an exhaust gas recirculation (EGR) valve. When the vehicle is driven under certain conditions, the EGR valve introduces a small amount of exhaust gas into the air/fuel vapor mixture to lower the temperature gradient of the combustion process and at the same time to lower exhaust emissions.

Each runner must be exactly the same length and have very smooth turns to provide for an even flow of air to the cylinders. The tumble valves are necessary to increase performance and reduce fuel consumption. As an added feature, some air intake manifolds "tumble" the incoming air by design. The tumble generator valves are positioned directly above the fuel injectors. When the engine is at idle, tumble valves are closed and the incoming air flow bypasses through a passage to increase the swirl of the air/fuel vapor mixture, thus achieving a more clean operation for the engine. A sketch of the air intake manifold, including the tumble generator valves above the fuel injectors, is shown in Fig. 10.15.

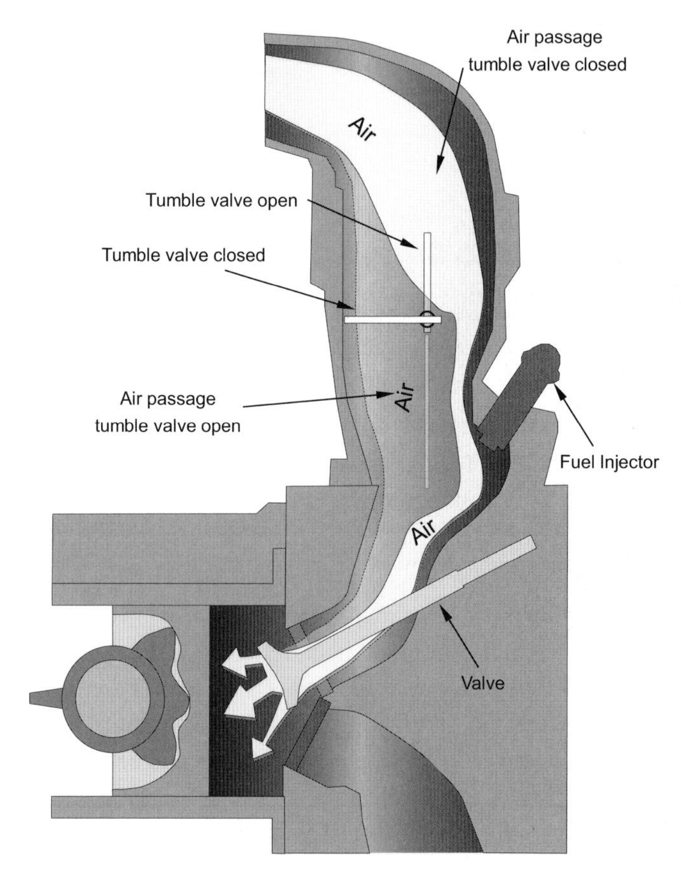

Figure 10.15 Schematic of the role of tumble valve in engine design

Figure 10.16 shows the bearings of the tumble valve feature employed in the air intake manifold by Mercedes for the C-Class series of vehicles, made using the IMA process.

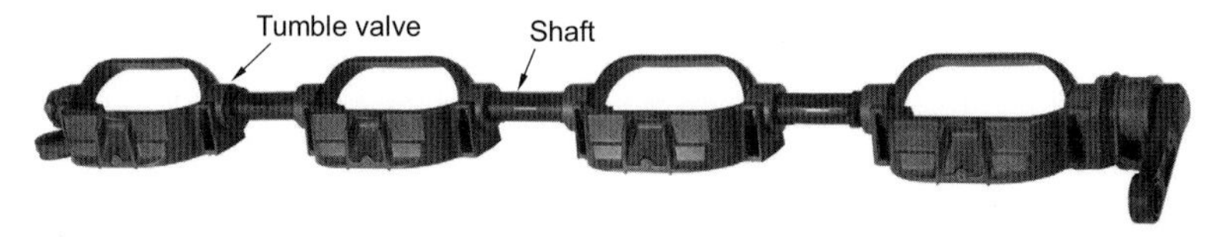

Figure 10.16 Tumble valve in the air intake manifold used by Mercedes in the C-Class engines, made by IMA process

The shaft is made of polyamide 46 having 2% shrinkage and a melting temperature of 295°C, and the tumble valves are made of polyphenylene sulfide (PPS) having 0.7% shrinkage and melting temperature of 277°C. Because the shrinkage coefficients are so differentiated, they allow free movement after the molding operation is accomplished and while in use under the engine's operating conditions and temperature gradients of 150°C.

The second case history is an automotive dashboard air vent, also referred to as an automotive louver (see Fig. 10.17).

Figure 10.17
Automotive louver made using IMA process

The IMA process is also used to manufacture the assembled louver. For this component, the first resin injected into the tool is a polyamide (PA 6,6) having the melting temperature of 290°C. Next, the second polymer, polyester (PBT), which has a melting temperature of 264°C, is injected in the second cavity. When the mold opens next, the cavity rotates once more and when closed, the third or last resin is injected: an alloy of polycarbonate and acrylonitrile butadiene styrene (PC/ABS). So, when the tool opens for the last time, a complete louver is ejected from the mold.

Shrinkage variance and the difference in melt temperature ensure proper functionality of the louver across its operation temperature range of –40°C to 120°C.

10.6 Conclusion

It is important to note that each component manufactured with the IMA process is custom. The IMA allows substantial reduction of variable manufacturing expenses while at the same time significantly increasing the capital in-flow in specialized injection-molding machines and highly specialized tooling.

Although the IMA techniques have been commercial for a few years, the technology is still in its infancy.

11 Fasteners

There are many types of fasteners which could be used to assemble plastic parts: screws, bolts, metal inserts, clips, spring washers, etc. The difference between screws and bolts is that bolts necessary to assemble two unthreaded components require a nut and/or a washer, while screws form an internal thread during the installation either by thread forming or thread cutting.

In this section we will address two types of screws: *thread forming* and *thread cutting*. Thread forming screws, as the name implies, deform the polymeric material of the boss or hole in which they are driven. On the other hand, thread cutting screws actually remove material from the boss or hole in which they are driven to form the thread. Thread forming screws are preferred if repeated disassembly and re-assemblies are required.

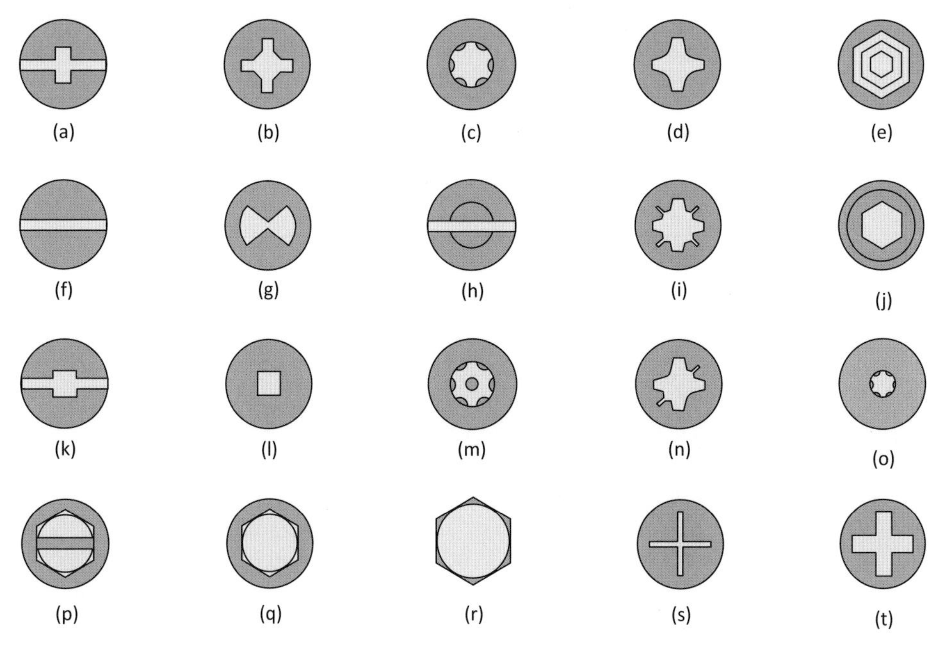

Figure 11.1 Types of screw driving head: (a) Phillips slotted; (b) Quadrex; (c) Torx; (d) Phillips; (e) Uni-screw; (f) slotted; (g) one way; (h) Torx slotted; (i) Pozidriv; (j) hex socket; (k) combo squared; (l) squared; (m) tempered Torx; (n) Supadriv; (o) Torx plus; (p) slot hex washer; (q) hex washer; (r) hex; (s) Frearson; (t) ACR Phillips

Thread forming screws should be installed by using automated fastening systems or manual power drives with torque limiters, also called transducers, which limit the rotation to a maximum of 800 rotations per minute. To start the thread forming process, the screw has to be slightly pressed into the *pilot hole* or boss. Screws can be removed, for servicing and repair work, and re-installed, a few times, as necessary.

There are many types of screw heads used in various industries. Figure 11.1 shows a variety of screw heads, some very common, like the hex screw, while others are fairly new, like the *Uni-screw* head developed in Britain a few years ago.

11.1 Thread Forming

As already stated, thread forming screws, when driven in the plastic part, deform the polymer, thus forming the thread. Type AB, Type B, and Type C screws (see Fig. 11.2) are employed to assemble metal components. None of them are recommended for plastic components because their thread flight angle of 60° may crack the plastic boss or the plastic hole in which they are fastened, typically at or near the weld line. However, Types AB and B are used in highly reinforced polymers–those exceeding 35% fiber reinforcement. The Type C screw is probably the worst type to be used for plastic components because of the very small thread pitch (see Fig. 11.2 (c)).

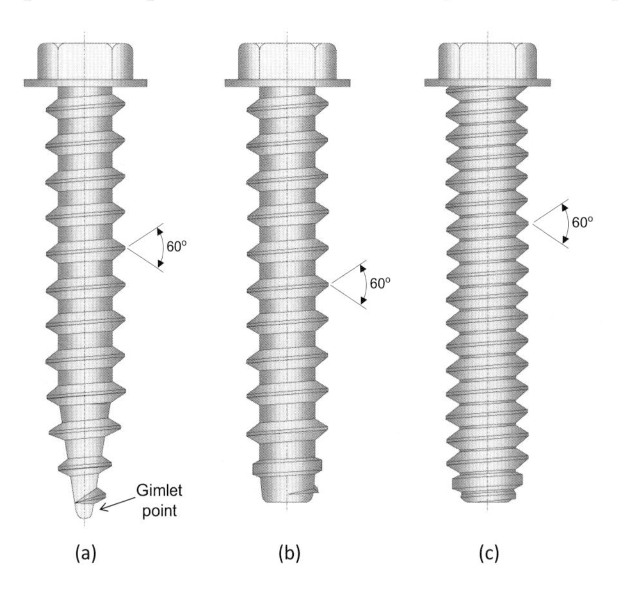

Figure 11.2
Thread forming screws for metal components: (a) Type AB; (b) Type B; (c) Type C

There are a number of screws specifically designed for assembling plastic parts. *HiLo*® is a registered name belonging to the Illinois Tool Works company. This type of screw, HiLo® (see Fig. 11.3 (a)), combines both 30° and 60° thread flights providing good retention capability for low to medium pull-out loads. The pull-out force is defined as the tensile load necessary to pull and remove the self-tapping screw from the mating boss. The HiLo® screw reduces radial stress exercised against the mating boss and at the same time has a tendency to prevent boss cracking. They require low driving torque for installation. A similar screw, from the Stanley Engineering Fasteners company, called *DST* (dual-spaced threads) also offers both 30° and 60° thread flights (see Fig. 11.3 (b)).

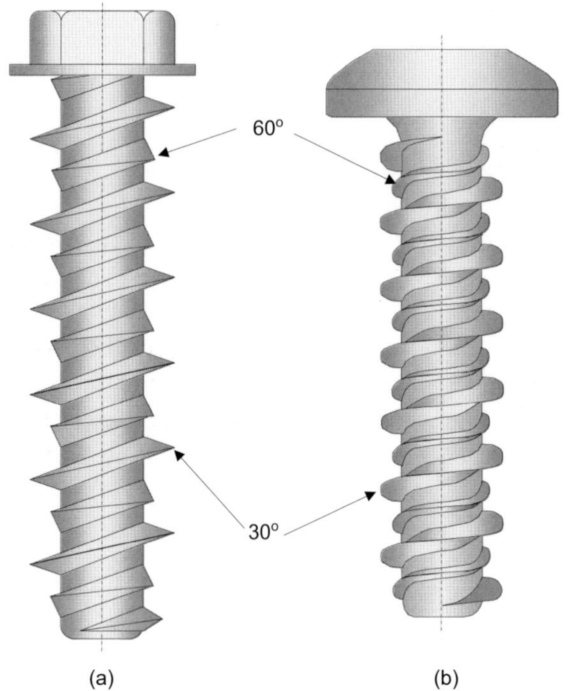

Figure 11.3
(a) HiLo® thread forming screw;
(b) DST thread forming screw

Another thread forming screw is *Plastite*®, a name registered to the Research Engineering & Manufacturing company, which exhibits a rather unusual shank, *trilobular* in shape (resembling a rounded triangle–Section A-A in Fig. 11.4). With this shape of the shank the installation torque necessary to drive in the screw is reduced when compared with other types of screws. At the same time, the screw design reduces the hoop stress and the root interference friction, which in general could cause *cracking*, especially for bosses having thin walls. Another feature of this type of screw is the *thread angle*, which in this case is 45° or 48°. The smaller thread angle, when compared to metal thread forming screws, which have a thread angle of 60°, penetrates deeper into the plastic boss, allowing for a higher pull-out load. The screw is available as single flight or double flight (see Fig. 11.5).

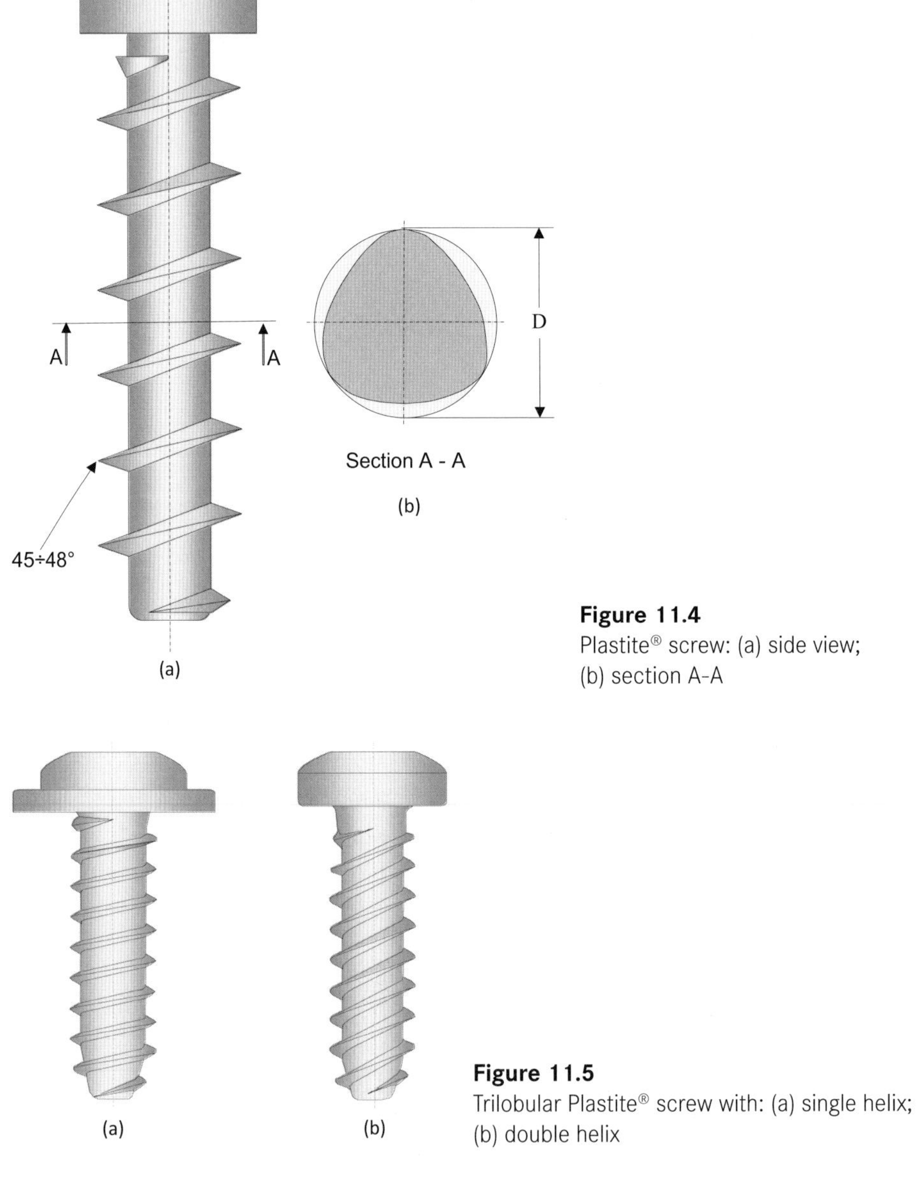

Figure 11.4
Plastite® screw: (a) side view; (b) section A-A

Figure 11.5
Trilobular Plastite® screw with: (a) single helix; (b) double helix

Simplex Corporation also registered a thread forming screw for plastics under the name of *Polyfast*® (see Fig. 11.6). An asymmetrical thread flight design which has a total angle of 45°, of which the leading angle represents 35°, while the remaining angle of 10° acts as the *trailing angle*, are this screw major features. The Polyfast® flight design requires low driving torque from around 2.4 in.·lb (0.016 MPa) to as much 14 in.·lb (0.1 MPa). It is most suitable for unreinforced and ductile thermoplastics.

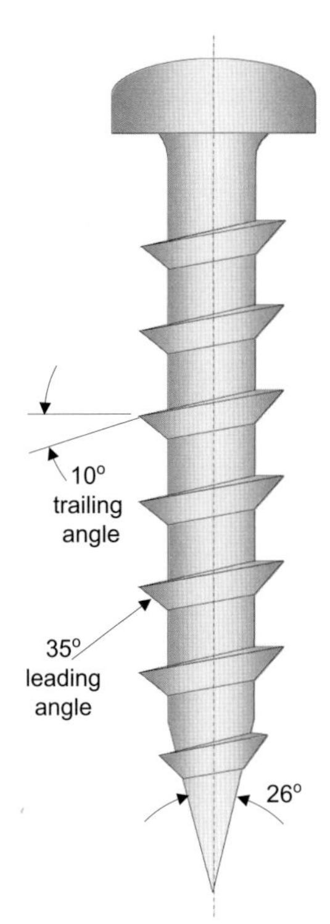

Figure 11.6
Polyfast® screw

Probably the best thread forming screw for plastics is the screw known as *PT* or *Plastic Thread* (see Fig. 11.7). The thread flight of this type of screw is 30° and it is available in single or double flight. Two of the registered designs are available from the German company EJOT Verbindungstechnik GmbH & Co. KG, under the name of *Delta PT*® and from the Turkish company Keba Fastenings, under the name of *K-PT*®. Both products are available from many resellers worldwide.

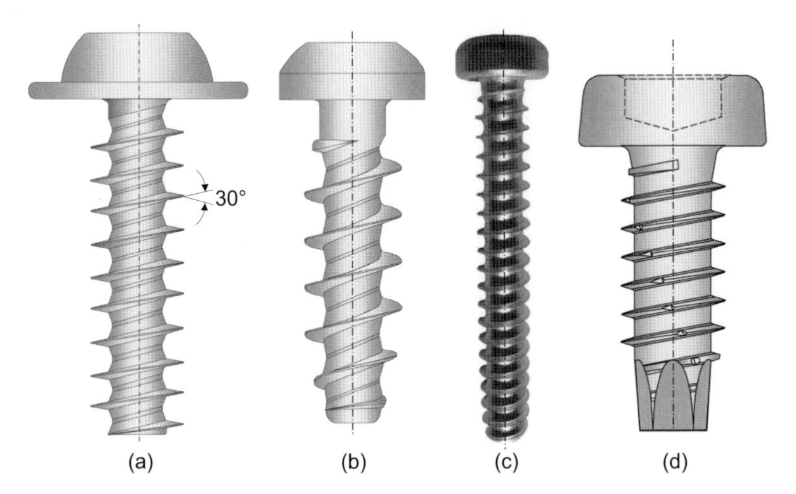

Figure 11.7 Plastic Thread screw: (a) PT screw with integrated washer; (b) PT screw; (c) Delta PT® screw; (d) RS Duroplast

The key feature of the PT screw is the 30° flight angle (see Fig. 11.8). When compared with a regular thread forming screw having 60° thread flight angle, used to assemble metal components and some highly glass or aramid fiber reinforced polymers, the PT design allows for a significant reduction of the radial stress, while at the same time increasing somewhat the pull-out load. Figure 11.9 shows that the PT screw design penetrates deeper into the plastic boss or hole, thus reducing the hoop stress and improving the retention load of the assembly.

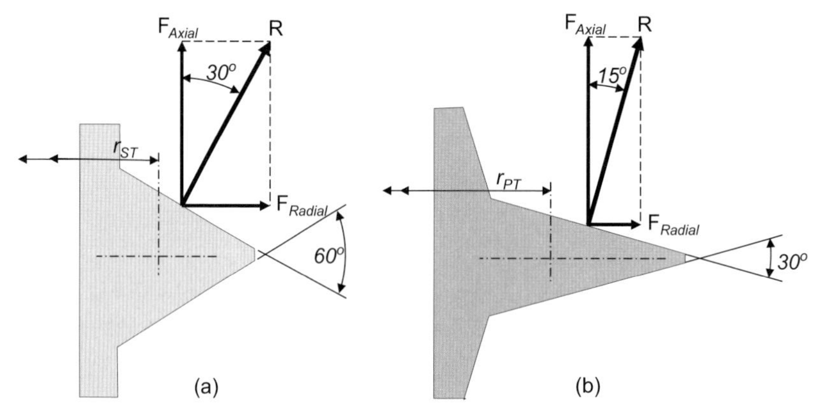

Figure 11.8 PT thread flight detail compared with the standard thread

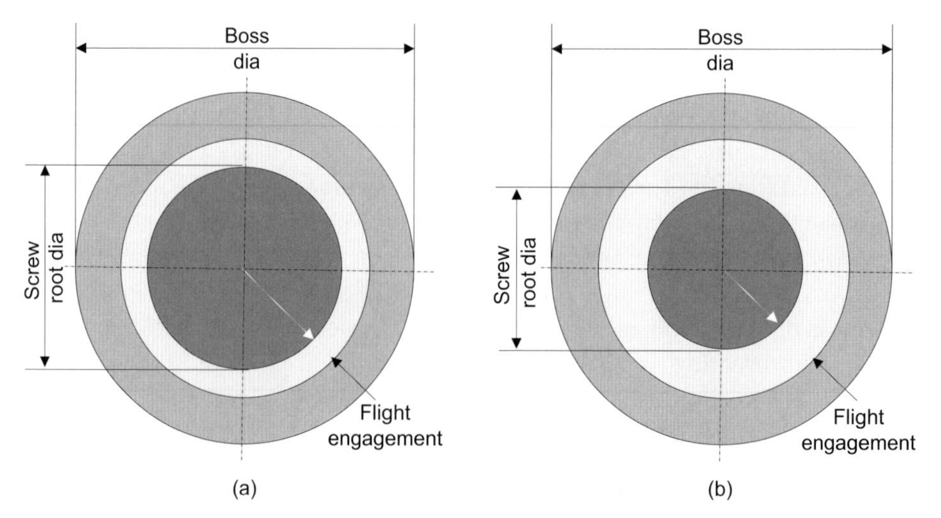

Figure 11.9 (a) Standard thread flight penetration detail; (b) plastic thread flight penetration detail

From the free body diagram shown in Fig. 11.8 the *radial force* for standard thread of 60° is

$$F_{Radial} = 0.5R \tag{11.1}$$

while the radial force for a plastic thread having a thread flight angle of 30° is

$$F_{Radial} = 0.26R \tag{11.2}$$

The plastic thread reduces the radial force by 48%, lowering significantly the load which could crack the plastic component.

From the same Fig. 11.8 we can calculate the axial load capability for the 60° flight:

$$F_{Axial} = 0.87R \tag{11.3}$$

and for the plastic thread the axial force is:

$$F_{Axial} = 0.97R \tag{11.4}$$

which represents a 12% improvement in pull-out load capability.

Another important feature of the PT screw is the *shank design* (see Fig. 11.10). The cylindrical portion of the shank is surrounded by two angled planes—easy to observe in a 2D view—having an angle between them of 140°. The stress for each of the thread flights is obtained by dividing the screw pull-out force by the area under each flight. Therefore, first the area under each thread flight, A, is obtained by subtracting the screw minor diameter area from its major diameter area:

$$A = \frac{\pi\left(d_M^2 - d_m^2\right)}{4} \tag{11.5}$$

Then the pull-out stress is:

$$\sigma = \frac{4F}{\pi\left(d_M^2 - d_m^2\right)} \tag{11.6}$$

When the screw is driven and forming the thread in the mating hole or boss, the 140° angle allows for the plastic/elastic deformation of the polymer (see Fig. 11.11). This in turn provides a rather low *radial stress* in the plastic component in which the PT screw is driven, while at the same time allowing for increased load carrying capability and preventing plastic relaxation of the polymer. The boss or hole in which the PT screw is mounted could have a very thin wall stock, around 0.5 to 0.67 of the overall wall thickness of the plastic component.

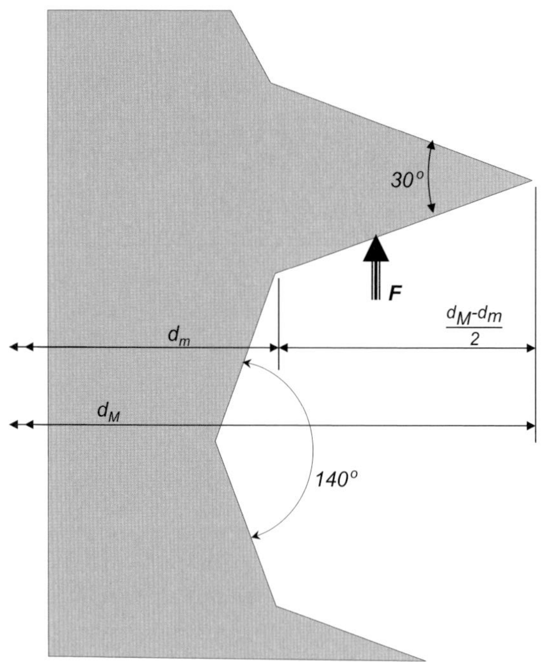

Figure 11.10
Cross-section of the plastic thread flight

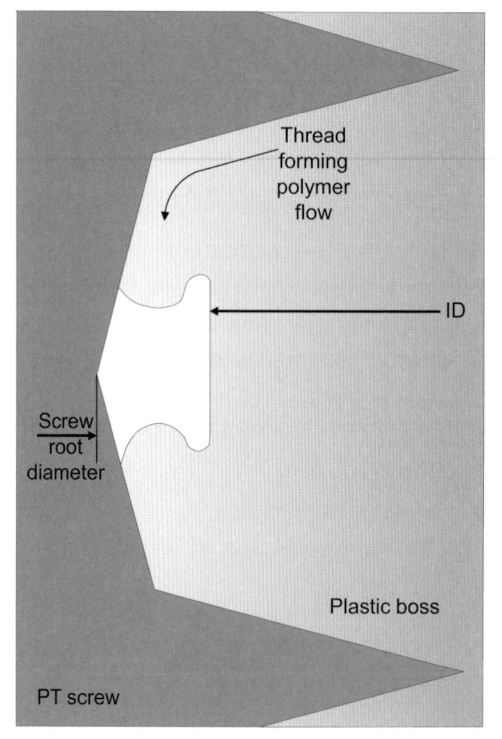

Figure 11.11
Polymer flow during the thread forming process

An important aspect for proper assembly using thread forming screws is the actual plastic boss/hole design (see Fig. 11.12). To properly align the thread forming screw with the mating hole or boss, it is necessary that the boss or hole in which the screw is fastened has a *counter bore* greater than the minor diameter of the screw by at least 0.2 mm or 0.008 in. The depth of the counter bore varies based on the type of polymer used between 30% and as much as 50% of the major diameter of the thread forming screw. The installation depth for the screw also varies depending on the type of resin used (see Table 11.1).

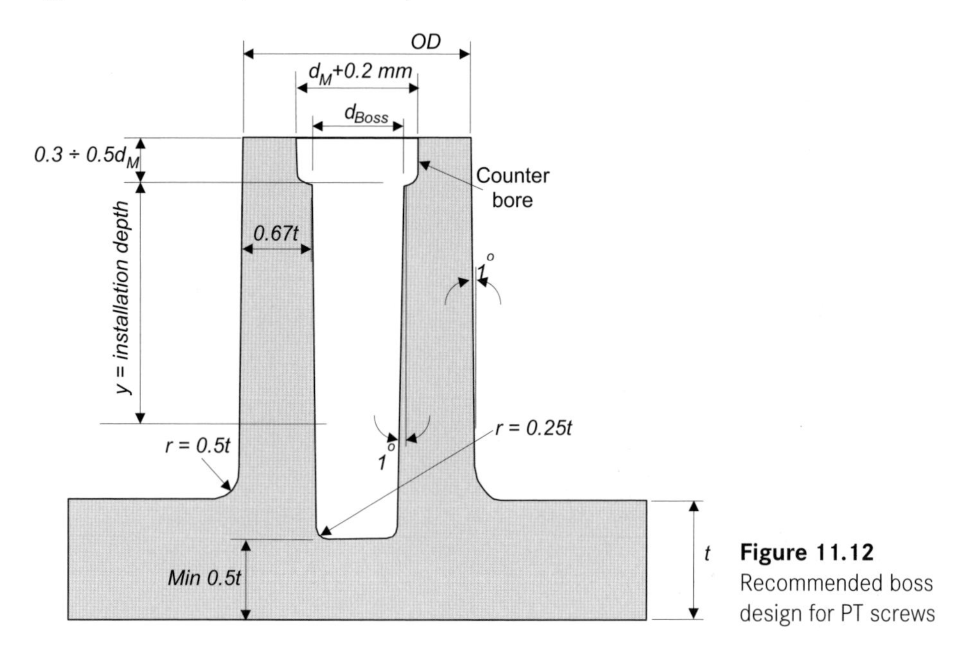

Figure 11.12 Recommended boss design for PT screws

Table 11.1 Recommended Dimensions for Different Types of Polymers Based on d_M (Major Thread Forming Diameter) for the Boss or Hole in Which the Screw Is Driven

Polymer	Inside diameter	Outside diameter	Minimum installation depth
ABS	$0.8 \cdot d_M$	$2 \cdot d_M$	$2 \cdot d_M$
ABS/PC	$0.8 \cdot d_M$	$2 \cdot d_M$	$2 \cdot d_M$
ASA	$0.8 \cdot d_M$	$2 \cdot d_M$	$2 \cdot d_M$
HDPE	$0.75 \cdot d_M$	$1.8 \cdot d_M$	$1.8 \cdot d_M$
LDPE	$0.7 \cdot d_M$	$2 \cdot d_M$	$2 \cdot d_M$

(continuation next page)

Table 11.1 Recommended Dimensions for Different Types of Polymers Based on d_M (Major Thread Forming Diameter) for the Boss or Hole in Which the Screw Is Driven *(continuation)*

Polymer	Inside diameter	Outside diameter	Minimum installation depth
PA 4.6	$0.7 \cdot d_M$	$1.85 \cdot d_M$	$1.8 \cdot d_M$
PA 6	$0.75 \cdot d_M$	$1.85 \cdot d_M$	$1.7 \cdot d_M$
PA 6 GR	$0.8 \cdot d_M$	$2 \cdot d_M$	$1.9 \cdot d_M$
PA 66	$0.75 \cdot d_M$	$1.85 \cdot d_M$	$1.7 \cdot d_M$
PA 66 GR	$0.8 \cdot d_M$	$2 \cdot d_M$	$1.8 \cdot d_M$
PBT	$0.75 \cdot d_M$	$1.85 \cdot d_M$	$1.7 \cdot d_M$
PBT GR	$0.8 \cdot d_M$	$1.8 \cdot d_M$	$1.7 \cdot d_M$
PC	$0.9 \cdot d_M$	$2.5 \cdot d_M$	$2.2 \cdot d_M$
PC GR	$0.9 \cdot d_M$	$2.2 \cdot d_M$	$2 \cdot d_M$
PET	$0.75 \cdot d_M$	$1.85 \cdot d_M$	$1.7 \cdot d_M$
PET GR	$0.8 \cdot d_M$	$1.8 \cdot d_M$	$1.7 \cdot d_M$
POM	$0.75 \cdot d_M$	$1.95 \cdot d_M$	$2 \cdot d_M$
POM GR	$0.8 \cdot d_M$	$1.95 \cdot d_M$	$2 \cdot d_M$
PP	$0.7 \cdot d_M$	$2 \cdot d_M$	$2 \cdot d_M$
PP GR	$0.72 \cdot d_M$	$2 \cdot d_M$	$2 \cdot d_M$
PP TR	$0.72 \cdot d_M$	$2 \cdot d_M$	$2 \cdot d_M$
PPO	$0.85 \cdot d_M$	$2.5 \cdot d_M$	$2.2 \cdot d_M$
PS	$0.8 \cdot d_M$	$2 \cdot d_M$	$2 \cdot d_M$
PVC	$0.8 \cdot d_M$	$2 \cdot d_M$	$2 \cdot d_M$
SAN	$0.77 \cdot d_M$	$2 \cdot d_M$	$1.9 \cdot d_M$

It should be noted that when employing self-tapping screws for an assembly, it is very important to use a torque limiter for the driving tool which holds the screw socket. As shown in Fig. 11.13, the stripping zone can very easily be reached without a proper torque limiter. Keeping the screw driving tool speeds between 300 rotations per minute (rpm) at the low end and 800 rpm at the high end ensures that the plastic material is not damaged, thus providing reliable installations. The heat generated during the fastening process can break down the plastic by melting it, thus reducing the required torque levels stipulated by the specific application. Speeds should be kept below 800 rpm so that the polymer does not melt, thereby destroying the plastic component. To reduce or even eliminate the eventuality of screw stripping, the ratio between stripping torque and driving torque should be minimized.

Driving guns activated pneumatically have large torque variance even when a torque controller is employed. Tolerance for such *torque limiters* varies between ±15% at the low end and as much as ±20% at the high end of the required set torque. Electronic tools, activated electrically, provide much greater accuracy in the range of ±2–4%. Fully automated electric tools should have a drive to strip ratio of 1 to 5. For hand tools the ratio is lower, from 1 to 2.

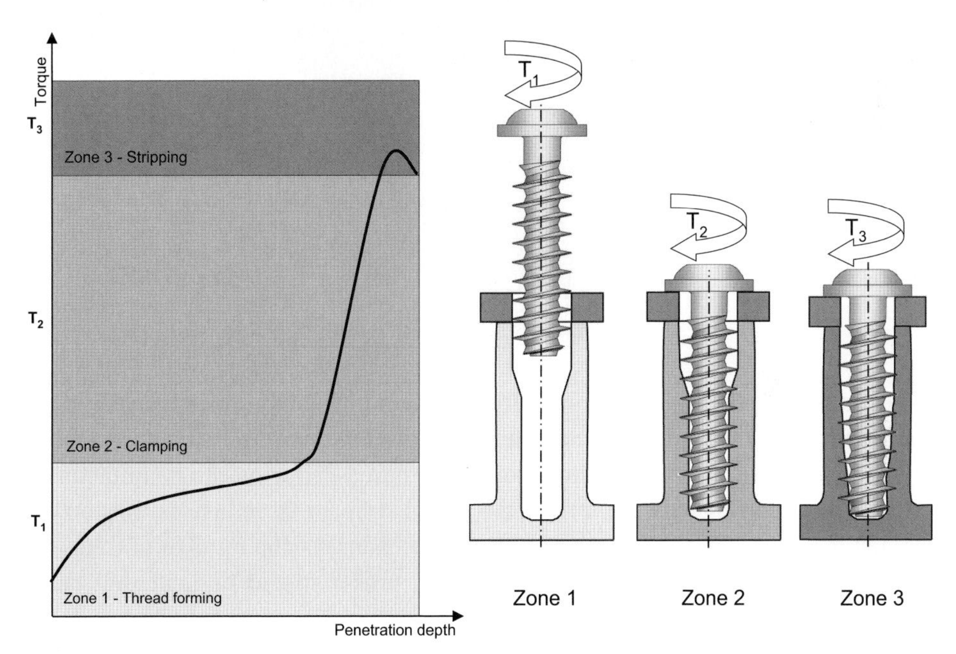

Figure 11.13 Screw stripping zone

As a rule of thumb, the longer the screw, the higher the pull-out forces and strip torques are. However, the improvement in load bearing is not directly proportional to the length of the screw utilized.

Thread forming screws are inexpensive and quite reliable, allowing assembly and disassembly of dissimilar polymers up to 10 times. At the same time, components assembled with thread forming screws are prone to stress concentrations at the joined location. When employing screws to assemble components the increased labor costs could become a major issue.

■ 11.2 Case History: Automotive Undercarriage Splash Shield

European original equipment manufacturers (OEMs) had to use splash shields (see Fig. 11.14) under their vehicles in order to improve the aerodynamics and at the same time to increase the gas mileage, the main reason being the high cost of gasoline and diesel fuels when compared with North America. The automobiles with splash shields under the carriage appeared in the North American marketplace in the 1980s. Then, a few years later, other car manufacturers, from Asia, started using splash shields to improve the gas mileage for their vehicles as well.

Figure 11.14
Typical undercarriage splash shield

Almost all splash shields use commodity polymers like polypropylene or polyethylene. However, in most cases, the polymers employed are recycled resins. Depending on how many heat histories the polymer was exposed to during the manufacturing process, the resin mechanical properties could drop by as much as 50%. When recycled polymers are used, the number of heat histories (the number of times the polymer passes through the injection molding press) should be kept well below seven.

An OEM from Asia had used a recycled unreinforced polypropylene resin to manufacture an undercarriage splash shield for one of their vehicles. To assemble the shield to the chassis of the vehicle a metric hex bolt without washer was used (see Fig. 11.15 (a)). Many vehicles developed shield failures during the warranty period. They were breaking just underneath the bolt head, holding the shield to the chassis, at the weld line.

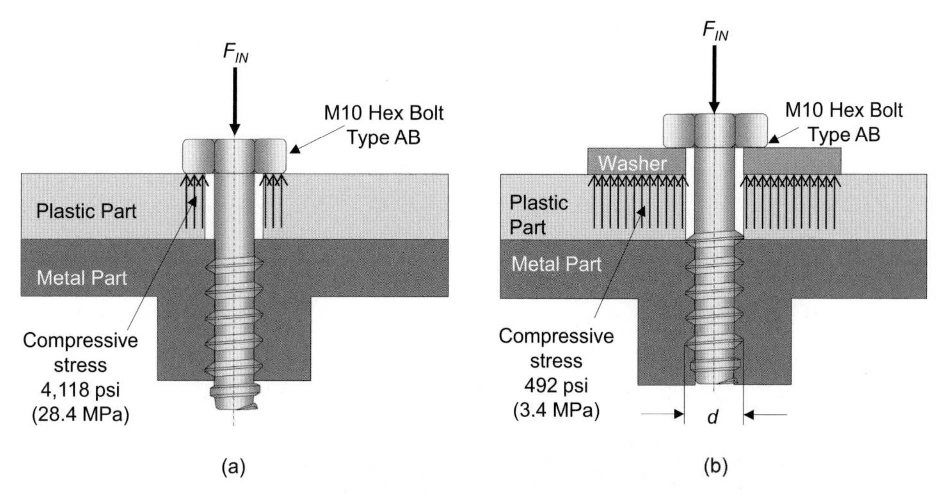

Figure 11.15 Cross-section of a splash shield held to the chassis with an M10 hex bolt: (a) without washer; (b) with washer

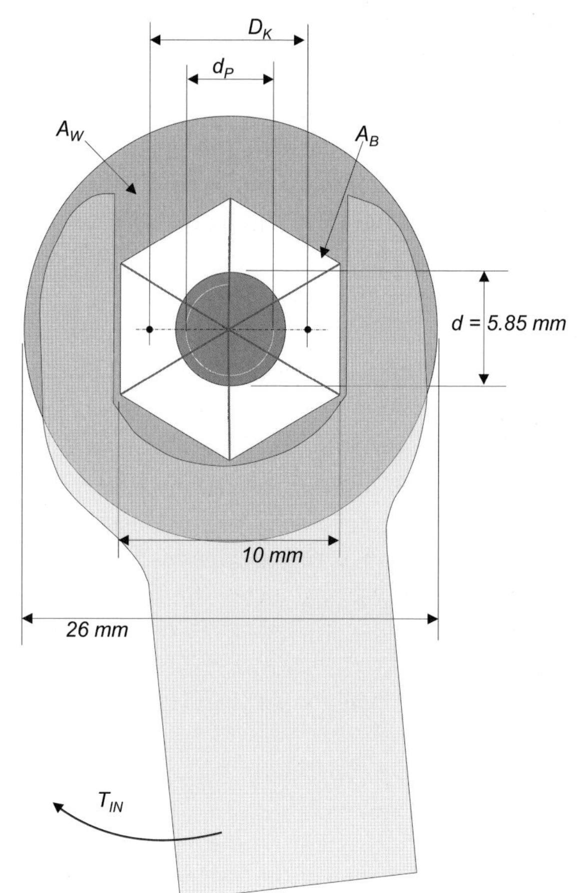

Figure 11.16
Torque load applied when assembling the bolt to the chassis

The bolt head shape is a hexagon (see Fig. 11.16). The area under the bolt head represents the area of the hexagon from which the area of the bolt shank is subtracted. The hexagon is made of six equilateral triangles. The side of each triangle is:

$$a = \frac{2h}{\sqrt{3}} = \frac{10}{\sqrt{3}} = 5.77\,\text{mm} \tag{11.7}$$

The area of each triangle is:

$$A_T = \frac{a^2}{4}\sqrt{3} = 14.42\ \text{mm}^2 \tag{11.8}$$

Then the hexagon area is six times greater:

$$A_H = 6A_T = 86.5\ \text{mm}^2 \tag{11.9}$$

This is how the area under the bolt head is obtained:

$$A_B = A_H - \frac{\pi d_S^2}{4} = 86.5 - \frac{\pi \cdot 5.85^2}{4} = 60\ \text{mm}^2 \tag{11.10}$$

Screw tension is generated when the screw elongates during tightening, producing the clamp load that prevents movement, rattles, and squeaks between the polypropylene aerodynamic splash shield and the chassis of the vehicle. The equation defining the relationship between the applied torque exercised by the wrench tightening the screw and the compression exercised against the PP shield by the screw is described by the *k-method* relationship:

$$T_{IN} = \mathrm{F}_C \cdot k \cdot d_M \tag{11.11}$$

where T_{IN} = torque applied by the wrench, k = friction factor, d_M = bolt major diameter, and F_C = bolt tension generated during tightening.

The friction factor, k, consolidates all factors that affect the clamp load. Most of them are determined by conducting proper mechanical testing. The *friction factor*, also called *nut factor*, is not obtained from engineering calculations, but it is determined through proper testing.

The coefficients which are part of the friction factor, k, are:

μ_k = Coefficient of friction under the screw head

μ_G = Coefficient of friction in the threads

d_P = Pitch diameter

P = Thread pitch

d_M = Bolt major diameter

d_k = Acting diameter for determining the friction torque at the fastener head

Thus, the value of k is obtained from:

$$k = \frac{0.36P + 1.16\mu_G d_P + \mu_k d_k}{2d_M} \tag{11.12}$$

and consequently the compression load under the bolt head, F_C, is:

$$F_C = \frac{T_{IN}}{kd_M} = \frac{2\,\text{Nmm}}{0.2 \cdot 5.85\,\text{mm}} = 1,709\,\text{N} = 384\,\text{lbf} \tag{11.13}$$

This, in turn, generates the compressive stress, σ_C, against the polypropylene shield, when a bolt without washer is used:

$$\sigma_C = \frac{F}{A_B} = 28.5 \text{ MPa} = 4,134 \text{ psi} \tag{11.14}$$

By using a washer having a 26 mm outside diameter, the compressive stress exercised against the shield is greatly reduced.

The area under the washer is:

$$A_W = A_{OD} - A_{ID} = 504 \text{ mm}^2 \tag{11.15}$$

The compressive load, being the same as before, generates a much lower compressive stress:

$$\sigma_C = \frac{F}{A_W} = 3.4 \text{ MPa} = 492 \text{ psi} \tag{11.16}$$

The compressive stress generated by the torque applied to clamp the shield to the chassis, when using a washer under the bolt head, lowers the load by over 88%. Probably the best unreinforced polypropylene has a stress capability of around 5,000 psi or about 34.5 MPa. The undercarriage splash shield, in this case, was made of recycled polypropylene resin having a much lower stress carrying capability—about half or around 2,500 psi or 17 MPa. Furthermore, around each molded-in hole, for assembly screws, in the shield exists a weld or knit line, created by the advancing flows of resin during the injection molding process, which could reduce the stress capability by another 50% to around 1,250 psi or 8 MPa (see Fig. 11.17 (a)).

A weld line forms when frozen layers at the front of each advancing flow front meet, melt, and then re-freeze again within the component being molded. The molecular chains within the polymer orient themselves perpendicular to the flow front at the weld line. It is the sharp difference of the molecular chain orientation at the weld which causes the significant decrease in strength. If the gate location can be changed, one possibility to improve the weld line strength would be to generate a weld line followed by a meld line during the injection molding cycle time (see Fig. 11.17 (b)).

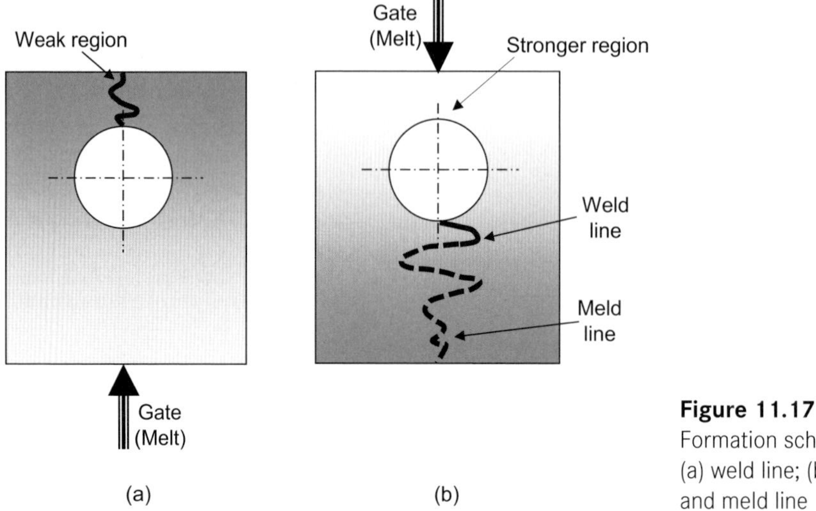

Figure 11.17
Formation schematics of: (a) weld line; (b) weld line and meld line

As a result of the vibration created by the vehicle when driven, the splash shield made of recycled PP was failing because the stress exercised by the M10 bolt without washer was much greater than the resin capability. The Asian OEM started using a washer which translated into reducing the compressive stress load under the washer and M10 hex bolt by a factor of nine with no warranty concerns.

There are injection molding processing steps, tooling changes, and plastic part design changes which can be applied to improve the weld line strength as mentioned above. A *meld line* occurs when two advancing molten resin flow fronts of the molten polymer inside the mold during the injection stage of the molding process blend together at an angle greater than 135°. Figure 11.18 shows a component where a meld line forms.

Meld lines are stronger than weld lines and offer fewer defects that are visible of the surface of the plastic component.

Consistent meld lines are challenging to obtain; however, it is possible to attain them by changing the polymer injection location, also called the *gate* location, or by modifying the wall thicknesses and achieving different fill times. This can be achieved by keeping records of the injection molding process parameters and tool modifications as per Appendix C and D, and by reviewing them. Other modifications of the injection molding process consist in increasing the polymer melt temperature and/or the tool temperature, thus allowing the advancing flow fronts to integrate better. Optimizing the runner system of the mold could prove beneficial as well in achieving consistent meld lines.

The use of *additives*, *fillers*, or *reinforcements* in the polymer results in a tendency to highlight both weld and meld lines. Especially glass fibers, as well as metallic pigments, are good examples of reinforcements and additives which make them more visible.

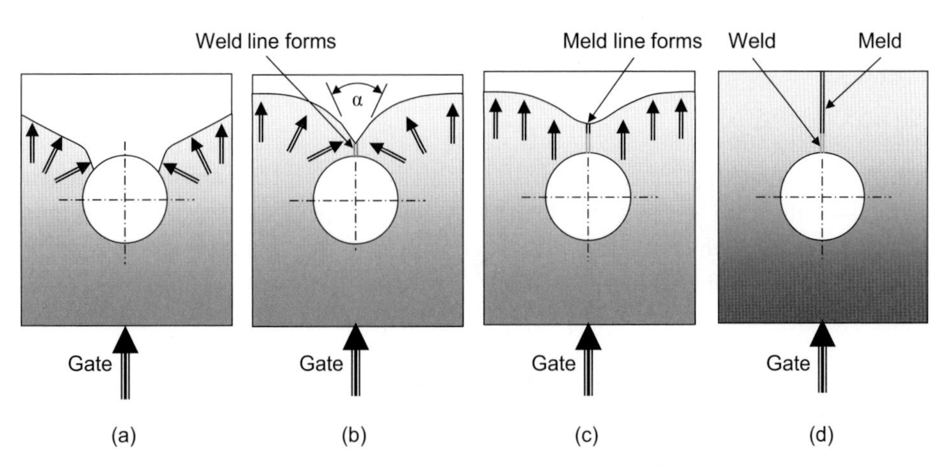

Figure 11.18 Creating a meld line: (a) the advancing flow fronts of the molten resin go around the metal pin forming the hole in the plastic component; (b) the angle α between the flow fronts which "kiss" behind the metal pin, forming a classic weld line, is greater than 135°; (c) adjustment of the injection pressure of the injection molding press such that the molecular chains move perpendicular to the flow direction; (d) during the packing and holding stage of the injection molding process, the meld line forms

The precise strength of both weld lines and meld lines varies between 15 and 85% of the molded polymer. In general, calculations indicate that the resin strength at the weld line is decreased by 50% while for the meld line by just 30%.

One of the newest screws on the market to be used when a plastic product is exposed to vibration in applications such as in the automobile industry is *BosScrew™*. The design was patented by Illinois Tool Works. The distinct features for this type of screw are the small indentations present on the 60° thread flights (see Fig. 11.19). When it is driven into a plastic boss, the polymer tendency to creep under constant load prevents the screw from becoming loose by interlocking it with the polymer.

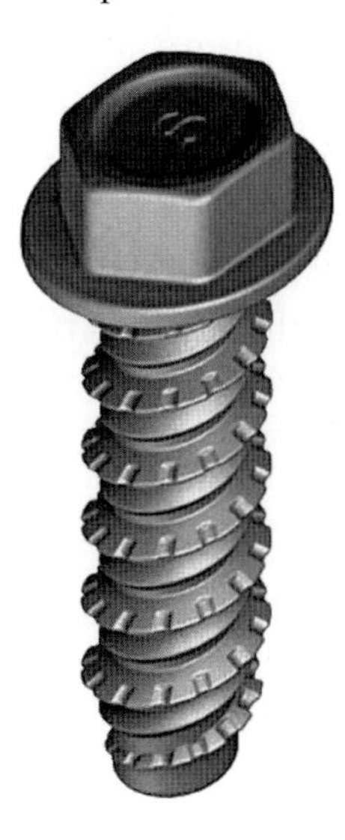

Figure 11.19
Anti-vibration screw: BosScrew™

11.3 Thread Cutting

Thread cutting screws create threads in the mating boss or hole by either having cutting slots or having some of their flights slotted. When they are driven into the plastic part they cut the polymer, creating "chips" or plastic debris. The bottom of the boss or hole in which thread cutting screws are employed acts as a debris cavity, capturing all the falling chips.

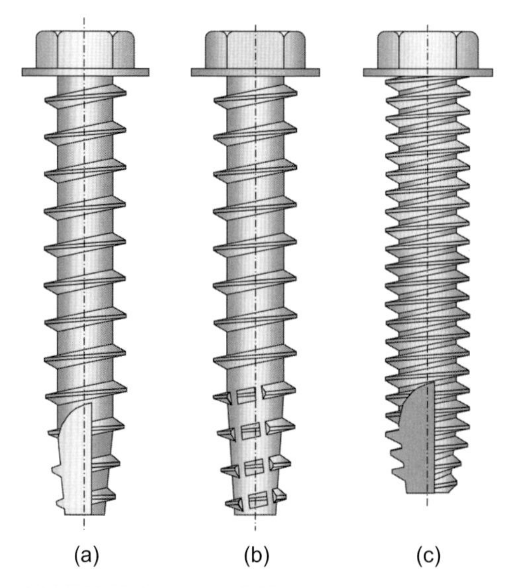

Figure 11.20
Standard thread cutting screws: (a) Type BT; (b) Type BF; (c) Type T

In Fig. 11.20 are shown standard thread cutting screws having 60° flight angles. Type BT (Fig. 11.20 (a)) is probably the most common because has wide thread spacing and a generous cutting slot. They were also previously called Type 25 screws. The next screw type, Type BF (Fig. 11.20 (b)), also displays wide thread spacing, but features slashed cutting flights which have clogging tendency when working with softer polymers around 200,000 psi (1,400 MPa).

In Fig. 11.20 (c) is displayed screw Type T, also called Type 23. This type of screw in general is used only for highly reinforced polymers having fiber content beyond 35% or modulus higher than 1,000,000 psi (7,000 MPa). The threads that cut into the polymer chip the resin making it difficult if not impossible to re-assemble the components. The engagement length for all thread cutting screws has to be at least double than thread forming screws (see Table 11.1).

For softer, unreinforced polymers, thread cutting screws having 30° flights, also known as PT or plastic thread, are recommended. As explained in the thread forming section, these flights bite deeper into the polymer, making them better for sustaining higher pull-out loads.

Figure 11.21 shows the thread cutting screw with 30° flights named *Duro-PT®*. It is mostly recommended for thermoset materials like polyesters, and the name is a registered trademark belonging to the EJOT company. The screw has a cutting slot which cuts the threads into the mating component, and a 30° flight angle, which allows for rather low installation torque levels, at the same time exhibiting high stripping loads, especially when used in thermoset polymers. Duro-PT® screws are used in a variety of industries: consumer products (telephones, coffee makers, printers, toys, etc.), appliances (washers, copiers, etc.), automotive, and others.

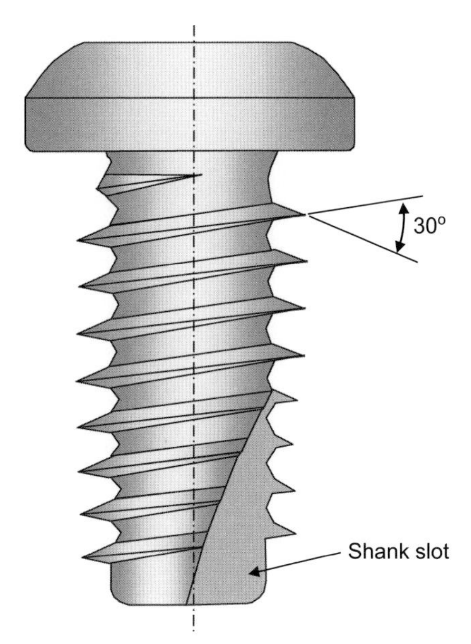

Figure 11.21
Duro-PT® plastic thread cutting screw

11.4 Conclusion

In general, thread forming (a.k.a. thread-rolling) screws are used to assemble components that would need to be disassembled for servicing or during the warranty period of the given product.

On the other hand, thread cutting screws are intended for products that do not require servicing—mostly for one-time use products—because once the threads are cut into the plastic components of the assembly, it is very unlikely that when re-assembling the components the screw flights would follow the cut threads.

Appendix A: Enforced Displacement

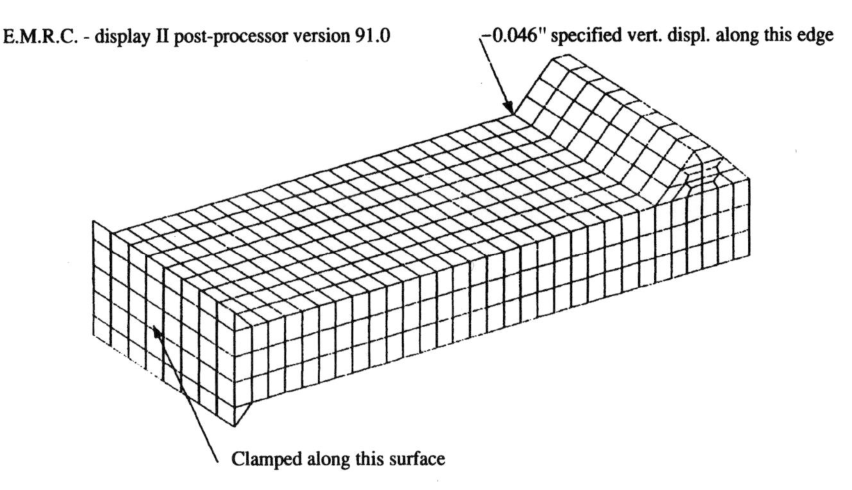

Figure A.1 Finite element model for one-way rectangular cross section continuous beam case history (brick elements with eight nodes per element) showing where the enforced displacement boundaries are applied

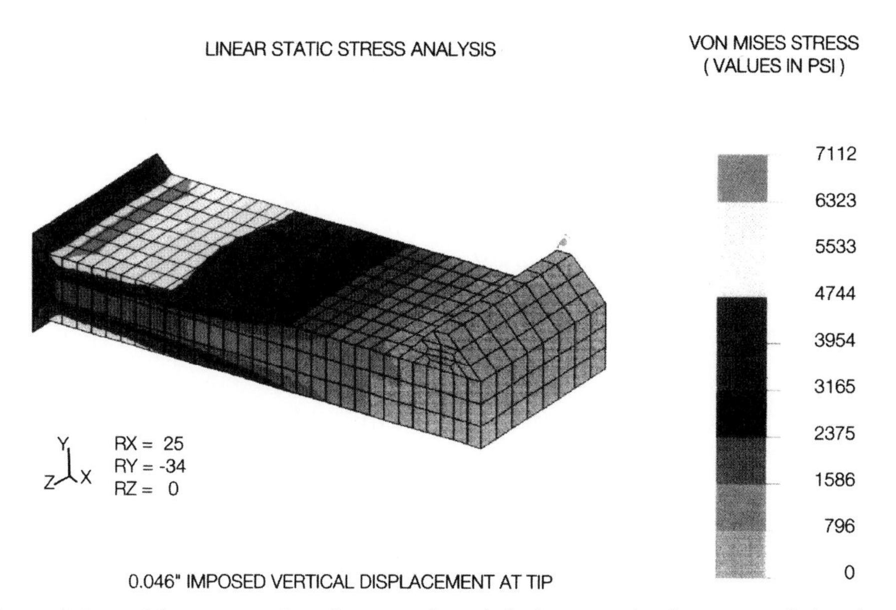

Figure A.2 von Mises stress plot, when an enforced displacement loading was applied to the model, of the linear static stress analysis conducted for the one-way rectangular cross section continuous beam case history

FEA Results Output File

LISTING OF EXECUTIVE COMMANDS

LINE 1 **EXECUTIVE
LINE 2 ANAL = STAT
LINE 3 SAVE = 26,27
LINE 4 FILE = SET-DISP
LINE 5 *ELTYPE
1 *** E M R C N I S A *** VERSION 91.0(06/17/91)

NISA COMPUTER PROGRAM RELEASE NO. 91.0

STATIC ANALYSIS

1 *** E M R C N I S A *** ——— VERSION 91.0 (06/17/91)
4:8

SELECTION OF ELEMENT TYPES FROM THE NISA ELEMENT LIBRARY (*ELTYPE DATA GROUP)

NSRL	NKTP	NORDR	NODES/EL	DOF/NODE
1	4	1	8	3
2	4	10	6	3

1 *** E M R C N I S A *** ——— VERSION 91.0 (06/17/91)
5:52
MATERIAL PROPERTY TABLE (*MATERIAL DATA GROUP)

MATERIAL INDEX 1

EX 1 0 3.1200000E+05 0.0000000E+00
0.0000000E+00 0.0000000E+00 0.0000000E+00
NUXY 1 0 4.0000000E-01 0.0000000E+00
0.0000000E+00 0.0000000E+00 0.0000000E+00

1 *** E M R C N I S A *** ——— VERSION 91.0 (06/17/91)
5:53

DEFINITION OF SETS (*SETS DATA GROUP)

SET-ID LABEL ——— MEMBERS (S=SINGLE, R=RANGE, E=EXCLUDED) ———

1 R 104, 936, 104
1 *** E M R C N I S A *** ——— VERSION 91.0 (06/17/91) LOAD CASE ID NO. 1
5:35

```
OUTPUT CONTROL FOR LOAD CASE ID NO.   1

INTERNAL FORCE AND STRAIN ENERGY KEY.... (KELFR) =  0
REACTION FORCE KEY ...............(KRCTN) =  1
STRESS COMPUTATION KEY ........(KSTR) =  1
STRAIN COMPUTATION KEY ......(KSTN) =  0
ELEMENT STRESS/STRAIN OUTPUT OPTIONS ..  (LQ1) = –1
NODAL STRESSES OUTPUT OPTIONS  ....  (LQ2) = 2
DISPLACEMENT OUTPUT OPTIONS ....(LQ7) = 0
STRESS FREE TEMPERATURE  .......(TSFRE) = 0.00000E+00
TOLERANCE FOR NODE POINT FORCE BALANCE (TOL ) = 0.00000E+00

1 *** E M R C  N I S A  *** ——— VERSION 91.0 (06/17/91)    LOAD CASE ID NO. 1
5:53
SPECIFIED DISPLACEMENT DATA (*SPDISP DATA GROUP)

NODE NO.      LABEL DISPLACEMENT VALUE LAST NODE INC  LABELES
   1          UX      0.00000E+00          911    26  UY        UZ
1191          UX      0.00000E+00         1215     3  UY        UZ
1217          UX      0.00000E+00         1241     3  UY        UZ
 104          UY     –4.60000E–02          936   104

1 *** E M R C  N I S A  *** ———  VERSION 91.0 (06/17/91)  LOAD CASE ID. 1
5:53

SELECTIVE PRINTOUT CONTROL PARAMETERS (*PRINTCNTL DATA GROUP)

OUTPUT TYPE  --- SET NUMBERS(NEGATIVE MEANS NONE, ZERO MEANS ALL)

LOAD VECTOR                      –1
ELEMENT INTERNAL FORCES          –1
ELEMENT STRAIN ENERGY            –1
RIGID LINK FORCES                –1
REACTIONS                         1
DISPLACEMENTS                     1
ELEMENT STRESSES                 –1
AVERAGED NODAL STRESSES          –1

1 *** E M R C  N I S A  *** ——— VERSION 91.0 (06/17/91)
5:55

PROCESS NODAL COORDINATES DATA

PROCESS ELEMENT CONNECTIVITY DATA
1 *** E M R C  N I S A  *** ——— VERSION 91.0 (06/17/91)

SUMMARY OF ELEMENT TYPES USED

NKTP          NORDR        NO. OF ELEMENTS

4               1                864
4              10                 24
```

```
TOTAL NUMBER OF ELEMENTS............  =   888
TOTAL NUMBER OF NODES ...............  =  1314
TOTAL NUMBER OF ACTIVE NODES  .... =  1314
LARGEST NODE NUMBER ................  =  1449
MAXIMUM X-COORD = 0.00000E+00   MAXIMUM X-COORD = 0.61200E+00
MAXIMUM Y-COORD = -0.20000E-01  MAXIMUM Y-COORD = 0.12600E+00
MAXIMUM Z-COORD =  0.00000E+00  MAXIMUM Z-COORD = 0.25000E+00

1 *** E M R C  N I S A  *** ——— VERSION 91.0 (06/17/91)

GEOMETRIC PROPERTIES OF THE MODEL

TOTAL VOLUME            = 1.30990E-02
TOTAL MASS              = 0.00000E+00
X COORDINATE OF C.G.  = 0.00000E+00
Y COORDINATE OF C.G.  = 0.00000E+00
Z COORDINATE OF C.G.  = 0.00000E+00

MASS MOMENT OF INERTIA WITH RESPECT TO GLOBAL AXES AT GLOBAL
ORIGIN
1XX = 0.00000E+00      1XY = 0.00000E+00
IYY = 0.00000E+00      IYZ = 0.00000E+00
IZZ = 0.00000E+00      IXZ = 0.00000E+00

MASS MOMENT OF INERTIA WITH RESPECT TO CARTESIAN AXES AT C.G.
1XX = 0.00000E+00      1XY = 0.00000E+00
IYY = 0.00000E+00      IYZ = 0.00000E+00
IZZ = 0.00000E+00      IXZ = 0.00000E+00

WAVE  FRONT  STATUS  BEFORE  MINIMIZATION

MAXIMUM WAVE FRONT ...... = 552
RMS WAVE FRONT .................. = 398
AVERAGE WAVE FRONT ........ = 378
TOTAL NO. OF DOF IN MODEL ... = 3942

WAVE FRONT STATUS AFTER MINIMIZATION (ITERATION NO. 1 )

MAXIMUM WAVE FRONT    ...... = 258
RMS WAVE FRONT  ................ = 144
AVERAGE WAVE FRONT ......... = 139
TOTAL NO. OF DOF IN MODEL ... = 3942

WAVE FRONT STATUS AFTER MINIMIZATION (ITERATION NO.3)

MAXIMUM WAVE FRONT .... = 258
RMS WAVE FRONT ............ = 144
AVERAGE WAVE FRONT ........ = 139
TOTAL NO. OF DOF IN MODEL ... = 3942
WAVE FRONT STATUS AFTER MINIMIZATION (ITERATION NO.4)

MAXIMUM WAVE FRONT ..... = 255
RMS WAVE FRONT ......... = 145
AVERAGE WAVE FRONT ....... = 140
TOTAL NO. OF DOF IN MODEL ... = 3942
```

WAVE FRONT STATUS AFTER MINIMIZATION (ITERATION NO.5)

MAXIMUM WAVE FRONT ...... = 255
RMS WAVE FRONT .... = 154
AVERRAGE WAVE FRONT ...... = 148
TOTAL NO. OF DOF IN MODEL .... = 3942

*****WAVE FRONT MINIMIZATION WAS SUCCESSFUL, ITERATION NO. 1 IS SELECTED WAVE FRONT PARAMETERS ARE:
MAXIMUM WAVE FRONT = 258
RMS WAVE FRONT = 144
AVERAGE WAVE FRONT = 139

PROCESS *SPDISP (SPECIFIED DISPLACEMENT) DATA FOR LOAD CASE ID NO. 1
TOTAL NUMBER OF VALID DOFS IN MODEL ...... = 3942
TOTAL NUMBER OF UNCONSTRAINED DOFS ..... = 3771
TOTAL NUMBER OF CONSTRAINED DOFS ...... = 171
TOTAL NUMBER OF SLAVES IN MPC EQS ...... = 0

1 *** E M R C N I S A *** ——— VERSION 91.0 (06/17/91)
LOAD CASE ID NO. 1

***WAVE FRONT SOLUTION PARAMETERS ***
MAXIMUM WAVE FRONT (MAXPA) = 249
RMS WAVE FRONT = 141
AVERAGE WAVE FRONT = 136
LARGEST ELEMENT MATRIX RANK USED (LVMAX) = 24
TOTAL NUMBER OF DEGREES OF FREEDOM = 3771
ESTIMATED NUMBER OF RECORDS ON FILE 30 = 52

1 *** E M R C N I S A *** ——— VERSION 91.0 (06/17/91)
LOAD CASE ID NO. 1
27:42

***** REACTION FORCES AND MOMENTS AT NODES *****

LOAD CASE ID NO.1
NODE FX FY FZ MX MY MZ

104 0.00000E+00 –3.76834E-01 0.00000E+00 0.00000E+00
0.00000E+00 0.00000E+00

208 0.00000E+00 –6.29630E-01 0.00000E+00 0.00000E+00
0.00000E+00 0.00000E+00

312 0.00000E+00 –4.77650E-01 0.00000E+00 0.00000E+00
0.00000E+00 0.00000E+00

416 0.00000E+00 –3.72895E-01 0.00000E+00 0.00000E+00
0.00000E+00 0.00000E+00

520 0.00000E+00 –3.37551E-01 0.00000E+00 0.00000E+00
0.00000E+00 0.00000E+00

624 0.00000E+00 –3.72895E-01 0.00000E+00 0.00000E+00
0.00000E+00 0.00000E+00

728 0.00000E+00 –4.77650E-01 0.00000E+00 0.00000E+00
0.00000E+00 0.00000E+00

832 0.00000E+00 –6.29630E-01 0.00000E+00 0.00000E+00
0.00000E+00 0.00000E+00

936 0.00000E+00 –3.76834E-01 0.00000E+00 0.00000E+00
0.00000E+00 0.00000E+00

SUMMATION OF REACTION FORCES IN GLOBAL DIRECTIONS

FX	FY	FZ
–3.240727E-12	–3.128608E-13	2.470107E-12

1 *** E M R C N I S A *** ——— VERSION 91.0 (06/17/91)
LOAD CASE ID NO. 1

****** DIASPLACEMENT SOLUTION ******

LOAD CASE ID NO.1

NODE UX UY UZ ROTX ROTY ROTZ
104 5.62012E-03 –4.60000E-02 –6.93937E-06 0.00000E+00
0.00000E+00 0.00000E+00

208 5.62012E-03 –4.60000E-02 1.62906E-05 0.00000E+00
0.00000E+00 0.00000E+00

312 5.62012E-03 –4.60000E-02 1.79085E-05 0.00000E+00
0.00000E+00 0.00000E+00

416 5.62012E-03 –4.60000E-02 1.01867E-05 0.00000E+00
0.00000E+00 0.00000E+00

520 5.62012E-03 –4.60000E-02 –3.88374E-15 0.00000E+00
0.00000E+00 0.00000E+00

624 5.62012E-03 –4.60000E-02 –1.01867E-05 0.00000E+00
0.00000E+00 0.00000E+00

728 5.62012E-03 –4.60000E-02 –1.79085E-05 0.00000E+00
0.00000E+00 0.00000E+00

832 5.62012E-03 –4.60000E-02 –1.62906E-05 0.00000E+00
0.00000E+00 0.00000E+00

936 5.62012E-03 –4.60000E-02 6.93937E-06 0.00000E+00
0.00000E+00 0.00000E+00

LARGEST MAGNITUDES OF DISPLACEMENT VECTOR =
1.20547E-02 –6.16490E-02 –7.23358E-04 0.00000E+00 0.00000E+00
0.00000E+00
AT NODE 1428 1050 917 1 1 1

1 *** E M R C N I S A *** ——— VERSION 91.0 (06/17/91) LOAD CASE ID NO. 1

**** NODAL PRINCIPAL STRESSES – LOAD CASE ID NO.1 ****

NODE SG1 SG2 SG3 EQUIV.
STRESS
MAX. SHEAR
VON MISES OCTAHEDRAL SHEAR

LARGEST MAGNITUDES OF STRESS FUNCTIONS
7.94462E+03 3.76937E+03 –7.94468E+03 4.01583E+03

7.11230E+03 3.35277E+03
AT NODES 497 1215 419 419 419 419

1 *** E M R C N I S A *** ——— VERSION 91.0 (06/17/91)
LOAD CASE ID NO. 1

OVERALL TIME LOG IN SECONDS

INPUT (READ, GENERATE) .... = 43.180
DATA STORING AND CHECKING....................... = 16.200
REORDERING OF ELEMENTS...................... = 75.320
FORM ELEMENT MATRICES........................... = 160.600
FORM GLOBAL LOAD VECTOR............................ = 0.000
MATRIX TRANSFORMATION DUE TO MPC...................= 0.000
PRE - FRONT.................................. = 3.360
SOLUTION OF SYSTEM EQUATIONS............................. = 366.420
INTERNAL FORCES AND REACTIONS............ = 1.940
STRESS CALCULATION........................ = 124.460
LOAD COMBINATION............................ = 0.000
TOTAL CPU............................ = 791.480

TOTAL ELAPSED TIME IS.................................= 1670.000

** NOTE ** – RUN IS COMPLETED SUCCESSFULLY
TOTAL NO. OF COMPLETED CASES = 1
LAST COMPLETED CASE -ID = 1
FILE 26 IS SAVED, NAME = set_disp26.dat, STATUS=NEW
FILE 27 IS SAVED, NAME = set_disp27.dat, STATUS=NEW

Appendix B: Point Force

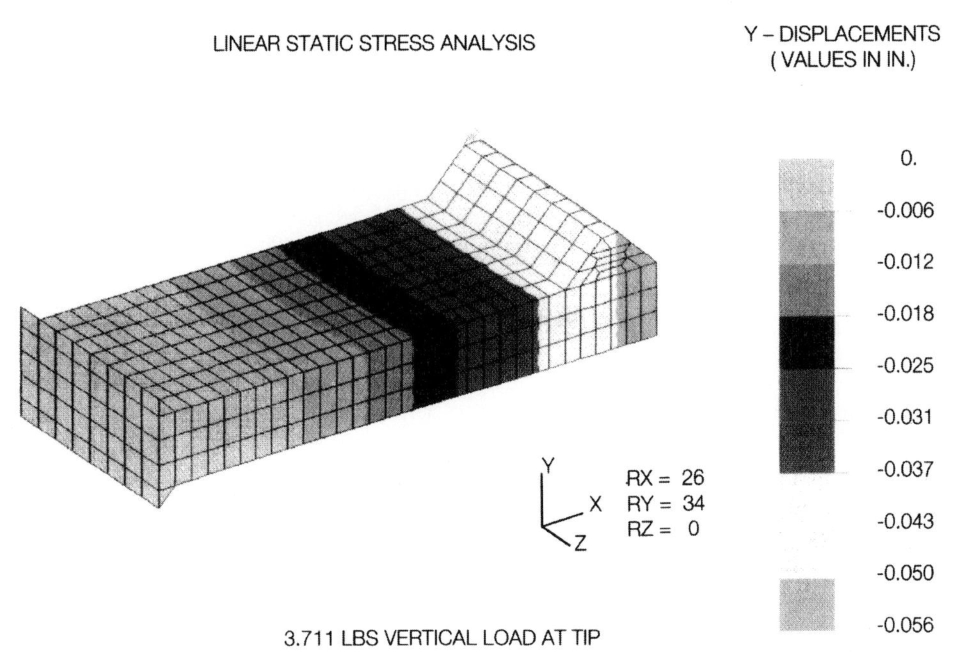

Figure B.1 Y-displacement plot, when point force loading where applied of the model, of the linear static stress analysis conducted for the one-way rectangular cross section continuous beam case history

FEA Results Output File

LISTING OF EXECUTIVE COMMANDS

```
LINE    1 **EXECUTIVE
LINE    2 ANAL = STAT
LINE    3 SAVE = 26, 27
LINE    4 FILE = SET-FORCE
LINE    5 *ELTYPE
1 *** E M R C  N I S A  ***  VERSION 91.0(06/17/91)
*NISA*          COMPUTER     PROGRAM       RELEASE        NO. 91.0
STATIC ANALYSIS
1 *** E M R C  N I S A  *** ——— VERSION 91.0 (06/17/91)
4:8
```

```
SELECTION OF ELEMENT TYPES FROM THE NISA ELEMENT LIBRARY
(*ELTYPE DATA GROUP)


NSRL  NKTP  NORDR  NODES/EL  DOF/NODE

1     4     1      8         3
2     4     10     6         3

1 *** E M R C  N I S A  *** ——— VERSION 91.0 (06/17/91)
5:52

MATERIAL PROPERTY TABLE (*MATERIAL DATA GROUP)

MATERIAL INDEX  1

EX     1  0      3.1200000E+05      0.0000000E+00  0.0000000E+00
0.0000000E+00    0.0000000E+00
NUXY  1  0       4.0000000E-01      0.0000000E+00  0.0000000E+00
0.0000000E+00    0.0000000E+00

1 *** E M R C  N I S A  *** ——— VERSION 91.0 (06/17/91)
5:53

DEFINITION OF SETS (*SETS DATA GROUP)

SET-ID LABEL ——— MEMBERS (S=SINGLE, R=RANGE, E=EXCLUDED) ———

1       R      104,    936,    104
1 *** E M R C  N I S A  *** ——— VERSION 91.0 (06/17/91)
LOAD CASE ID NO. 1
5:35
OUTPUT CONTROL FOR LOAD CASE ID NO.  1

INTERNAL FORCE AND STRAIN ENERGY KEY.... (KELFR) = 0
REACTION FORCE KEY ...............(KRCTN) = 1
STRESS COMPUTATION KEY ........(KSTR) = 1
STRAIN COMPUTATION KEY ......(KSTN) = 0
ELEMENT STRESS/STRAIN OUTPUT OPTIONS .. (LQ1) = –1
NODAL STRESSES OUTPUT OPTIONS  .... (LQ2) = 2
DISPLACEMENT OUTPUT OPTIONS ....(LQ7) = 0
STRESS FREE TEMPERATURE .......(TSFRE) = 0.00000E+00
TOLERANCE FOR NODE POINT FORCE BALANCE (TOL ) = 0.00000E+00

1 *** E M R C  N I S A  *** ——— VERSION 91.0 (06/17/91)   LOAD CASE ID NO. 1
5:53

SPECIFIED DISPLACEMENT DATA (*SPDISP DATA GROUP)

NODE NO.     LABEL DISPLACEMENT VALUE LAST NODE INC  LABELES
   1         UX     0.00000E+00           911    26  UYUZ
1191         UX     0.00000E+00          1215     3  UYUZ
1217         UX     0.00000E+00          1241     3  UYUZ
```

```
1 *** E M R C  N I S A  *** ——— VERSION 91.0 (06/17/91)
LOAD CASE ID.        1
5:53

CONCENTRATED NODAL FORCE AND MOMENT DATA (*CFORCE DATA
GROUP)

NODE NO.   LABEL FORCE VALUE LASTNODE   INC   LFN
104   FY   –2.31938E-01   104   1   0
208   FY   –4.63875E-01   208   1   0
312   FY   –4.63875E-01   312   1   0
416   FY   –4.63875E-01   416   1   0
520   FY   –4.63875E-01   520   1   0
624   FY   –4.63875E-01   624   1   0
728   FY   –4.63875E-01   728   1   0
832   FY   –4.63875E-01   832   1   0
936   FY   –2.31938E-01   936   1   0

1 *** E M R C  N I S A  *** ——— VERSION 91.0 (06/17/91)
LOAD CASE ID.        1
5:53

SELECTIVE PRINTOUT CONTROL PARAMETERS (*PRINTCNTL DATA GROUP)

OUTPUT TYPE     --- SET NUMBERS (NEGATIVE MEANS NONE, ZERO MEANS
ALL)
LOAD VECTOR                   –1
ELEMENT INTERNAL FORCES       –1
ELEMENT STRAIN ENERGY         –1
RIGID LINK FORCES             –1
REACTIONS                      0
DISPLACEMENTS
ELEMENT STRESSES              –1
AVERAGED NODAL STRESSES       –1

1 *** E M R C  N I S A  *** ——— VERSION 91.0 (06/17/91)
5:55

  PROCESS NODAL COORDINATES DATA

  PROCESS ELEMENT CONNECTIVITY DATA
1 *** E M R C  N I S A  *** ——— VERSION 91.0 (06/17/91)

SUMMARY OF ELEMENT TYPES USED

NKTP        NORDR       NO. OF ELEMENTS

4             1         864
4            10         24

TOTAL NUMBER OF ELEMENTS.............= 888
TOTAL NUMBER OF NODES ............... = 1314
TOTAL NUMBER OF ACTIVE NODES.... = 1314
LARGEST NODE NUMBER ................. = 1449
```

MAXIMUM X-COORD = 0.00000E+00 MAXIMUM X-COORD = 0.61200E+00
MAXIMUM Y-COORD = –0.20000E-01 MAXIMUM Y-COORD = 0.12600E+00
MAXIMUM Z-COORD = 0.00000E+00 MAXIMUM Z-COORD = 0.25000E+00

1 *** E M R C N I S A *** ——— VERSION 91.0 (06/17/91)

GEOMETRIC PROPERTIES OF THE MODEL
TOTAL VOLUME = 1.30990E-02
TOTAL MASS = 0.00000E+00
X COORDINATE OF C.G. = 0.00000E+00
Y COORDINATE OF C.G. = 0.00000E+00
Z COORDINATE OF C.G. = 0.00000E+00

MASS MOMENT OF INERTIA WITH RESPECT TO GLOBAL AXES AT GLOBAL ORIGIN
1XX = 0.00000E+00 1XY = 0.00000E+00
IYY = 0.00000E+00 IYZ = 0.00000E+00
IZZ = 0.00000E+00 IXZ = 0.00000E+00

MASS MOMENT OF INERTIA WITH RESPECT TO CARTEZIAN AXES AT C.G.
1XX = 0.00000E+00 1XY = 0.00000E+00
IYY = 0.00000E+00 IYZ = 0.00000E+00
IZZ = 0.00000E+00 IXZ = 0.00000E+00

WAVE FRONT STATUS BEFORE MINIMIZATION

MAXIMUM WAVE FRONT ...... = 552
RMS WAVE FRONT .... = 398
AVERAGE WAVE FRONT .... = 378
TOTAL NO. OF DOF IN MODEL ... = 3942

WAVE FRONT STATUS AFTER MINIMIZATION (ITERATION NO. 1)

MAXIMUM WAVE FRONT ...... = 258
RMS WAVE FRONT .... = 144
AVERAGE WAVE FRONT .... = 139
TOTAL NO. OF DOF IN MODEL ... = 3942

WAVE FRONT STATUS AFTER MINIMIZATION (ITERATION NO.3)

MAXIMUM WAVE FRONT .... = 258
RMS WAVE FRONT .... = 144
AVERAGE WAVE FRONT .... = 139
TOTAL NO. OF DOF IN MODEL ... = 3942

WAVE FRONT STATUS AFTER MINIMIZATION (ITERATION NO.4)

MAXIMUM WAVE FRONT ..... = 255
RMS WAVE FRONT .... = 145
AVERAGE WAVE FRONT .... = 140
TOTAL NO. OF DOF IN MODEL ... = 3942

WAVE FRONT STATUS AFTER MINIMIZATION (ITERATION NO.5)

MAXIMUM WAVE FRONT ...... = 255
RMS WAVE FRONT .... = 154

AVERRAGE WAVE FRONT ...... = 148
TOTAL NO. OF DOF IN MODEL .... = 3942

*****WAVE FRONT MINIMIZATION WAS SUCCESSFUL, ITERATION NO. 1 IS
SELECTED WAVE FRONT PARAMETERS ARE –
MAXIMUM WAVE FRONT = 258
RMS WAVE FRONT = 144
AVERAGE WAVE FRONT = 139

PROCESS *SPDISP (SPECIFIED DISPLACEMENT) DATA FOR LOAD CASE ID NO. 1
TOTAL NUMBER OF VALID DOFS IN MODEL ...... = 3942
TOTAL NUMBER OF UNCONSTRAINED DOFS ..... = 3780
TOTAL NUMBER OF CONSTRAINED DOFS ...... = 162
TOTAL NUMBER OF SLAVES IN MPC EQS ...... = 0

1 *** E M R C N I S A *** ——— VERSION 91.0 (06/17/91)
LOAD CASE ID NO. 1

***WAVE FRONT SOLUTION PARAMETERS ***
MAXIMUM WAVE FRONT (MAXPA) = 258
R.M.S. WAVE FRONT = 144
AVERAGE WAVE FRONT = 138
LARGEST ELEMENT MATRIX RANK USED (LVMAX) = 24
TOTAL NUMBER OF DEGREES OF FREEDOM = 3780
ESTIMATED NUMBER OF RECORDS ON FILE 30 = 52

1 *** E M R C N I S A *** ——— VERSION 91.0 (06/17/91)
LOAD CASE ID NO. 1
27:42

***** REACTION FORCES AND MOMENTS AT NODES *****

LOAD CASE ID NO.1

NODE FX FY FZ MX MY MZ

1 1.23015E+00 –2.97004E-01 5.79761E-01 0.00000E+00
0.00000E+00 0.00000E+00
27 6.00778E-01 –4.24355E-01 2.84386E-01 0.00000E+00
0.00000E+00 0.00000E+00
53 6.00795E-01 –4.24352E-01 –2.84388E-01 0.00000E+00
0.00000E+00 0.00000E+00
79 –1.23017E+00 –2.97002E-01 –5.79762E-01 0.00000E+00
0.00000E+00 0.00000E+00
105 1.84443E+00 –2.91119E-01 4.95901E-01 0.00000E+00
0.00000E+00 0.00000E+00
131 8.45580E-01 –4.58696E-01 2.38757E-01 0.00000E+00
0.00000E+00 0.00000E+00
157 –8.45600E-01 -4.58692E-01 –2.38746E-01 0.00000E+00
0.00000E+00 0.00000E+00
183 –1.84445E+00 –2.91116E-01 -4.95893E-01 0.00000E+00
0.00000E+00 0.00000E+00
209 2.18672E+00 –1.76931E-01 3.17524E-01 0.00000E+00
0.00000E+00 0.00000E+00

235 9.44319E-01 –2.72644E-01 1.55582E-01 0.00000E+00
0.00000E+00 0.00000E+00
261 –9.44316E-01 –2.72644E-01 –1.55568E-01 0.00000E+00
0.00000E+00 0.00000E+00
287 –2.18672E+00 –1.76931E-01 -3.17513E-01 0.00000E+00
0.00000E+00 0.00000E+00
313 2.41479E+00 –1.43099E-01 1.35066E-01 0.00000E+00
0.00000E+00 0.00000E+00
339 1.03693E+00 –2.33928E-01 6.79948E-02 0.00000E+00
0.00000E+00 0.00000E+00
365 –1.03691E+00 –2.33931E-01 –6.79863E-02 0.00000E+00
0.00000E+00 0.00000E+00
391 –2.41478E+00 –1.43102E-01 1.35059E-01 0.00000E+00
0.00000E+00 0.00000E+00
417 2.48344E+00 –1.37930E-01 1.89043E-13 0.00000E+00
0.00000E+00 0.00000E+00
443 1.06349E+00 –2.26386E-01 9.84768E-14 0.00000E+00
0.00000E+00 0.00000E+00
469 –1.06347E+00 –2.26391E-01 –9.06497E-14 0.00000E+00
0.00000E+00 0.00000E+00
495 –2.48342E+00 –1.37933E-01 –1.89349E-13 0.00000E+00
0.00000E+00 0.00000E+00
521 2.41479E+00 –1.43099E-01 –1.35066E-01 0.00000E+00
0.00000E+00 0.00000E+00
547 1.03693E+00 –2.33928E-01 –6.79948E-02 0.00000E+00
0.00000E+00 0.00000E+00
573 –1.03691E+00 –2.33931E-01 –6.79863E-02 0.00000E+00
0.00000E+00 0.00000E+00
599 –2.41478E+00 –1.43102E-01 1.35059E-01 0.00000E+00
0.00000E+00 0.00000E+00
625 2.18672E+00 –1.76931E-01 –3.17524E-01 0.00000E+00
0.00000E+00 0.00000E+00
651 9.44319E-01 –2.72644E-01 –1.55582E-01 0.00000E+00
0.00000E+00 0.00000E+00
677 –9.44316E-01 –2.72644E-01 1.55568E-01 0.00000E+00
0.00000E+00 0.00000E+00
703 –2.18672E+00 –1.76931E-01 3.17513E-01 0.00000E+00
0.00000E+00 0.00000E+00
729 1.84443E+00 –2.91119E-01 –4.95901E-01 0.00000E+00
0.00000E+00 0.00000E+00
755 8.45580E-01 –4.58696E-01 –2.38757E-01 0.00000E+00
0.00000E+00 0.00000E+00
781 –8.45600E-01 –4.58692E-01 2.38746E-01 0.00000E+00
0.00000E+00 0.00000E+00
807 –1.84445E+00 –2.91116E-01 4.95893E-01 0.00000E+00
0.00000E+00 0.00000E+00
822 1.23015E+00 –2.97004E-01 –5.79761E-01 0.00000E+00
0.00000E+00 0.00000E+00
859 6.00778E-01 –4.24355E-01 –2.84386E-01 0.00000E+00
0.00000E+00 0.00000E+00
885 –6.00795E-01 –4.24355E-01 2.84388E-01 0.00000E+00
0.00000E+00 0.00000E+00
911 –1.23017E+00 –2.97002E-01 5.79762E-01 0.00000E+00
0.00000E+00 0.00000E+00
1191 –8.80627E-02 3.90951E-01 1.67639E-01 0.00000E+00
0.00000E+00 0.00000E+00

1194 –1.24923E-01 7.03079E-01 –4.86747E-02 0.00000E+00
0.00000E+00 0.00000E+00
1197 –2.43708E-01 8.64341E-01 –4.67474E-02 0.00000E+00
0.00000E+00 0.00000E+00
1200 –2.94883E-01 9.56620E-01 –1.75823E-02 0.00000E+00
0.00000E+00 0.00000E+00
1203 –3.09217E-01 9.85387E-01 –3.69704E-14 0.00000E+00
0.00000E+00 0.00000E+00
1206 –2.94883E-01 9.56620E-01 1.75823E-02 0.00000E+00
0.00000E+00 0.00000E+00
1209 –2.43708E-01 8.64341E-01 4.67474E-02 0.00000E+00
0.00000E+00 0.00000E+00
1212 –1.24923E-01 7.03079E-01 4.86747E-02 0.00000E+00
0.00000E+00 0.00000E+00
1215 –8.80627E-02 3.90951E-01 1.67639E-01 0.00000E+00
0.00000E+00 0.00000E+00
1217 8.80614E-02 3.90947E-01 1.67638E-01 0.00000E+00
0.00000E+00 0.00000E+00
1220 1.24921E-01 7.03074E-01 4.86762E-02 0.00000E+00
0.00000E+00 0.00000E+00
1223 2.43708E-01 8.64341E-01 4.67490E-02 0.00000E+00
0.00000E+00 0.00000E+00
1226 2.94885E-01 9.56625E-01 1.75832E-02 0.00000E+00
0.00000E+00 0.00000E+00
1229 3.09219E-01 9.85394E-01 1.52101E-14 0.00000E+00
0.00000E+00 0.00000E+00
1232 2.94885E-01 9.56620E-01 –1.75832E-02 0.00000E+00
0.00000E+00 0.00000E+00
1235 2.43708E-01 8.64341E-01 –4.67490E-02 0.00000E+00
0.00000E+00 0.00000E+00
1238 1.24921E-01 7.03074E-01 –4.86762E-02 0.00000E+00
0.00000E+00 0.00000E+00
1241 8.80614E-02 3.90947E-01 –1.67638E-01 0.00000E+00
0.00000E+00 0.00000E+00

SUMMATION OF REACTION FORCES IN GLOBAL DIRECTIONS

FX FY FZ
–2.119041E-12 3.711001E+00 2.083833E-12

1 *** E M R C N I S A *** ——— VERSION 91.0 (06/17/91)
LOAD CASE ID NO. 1

****** DIASPLACEMENT SOLUTION ******

LOAD CASE ID NO.1

NODE UX UY UZ ROTX ROTY ROTZ

104 5.14670E-03 –4.20879E-02 1.21035E-05 0.00000E+00
0.00000E+00 0.00000E+00
208 5.15380E-03 –4.21168E-02 2.48283E-05 0.00000E+00
0.00000E+00 0.00000E+00
312 5.14526E-03 –4.21475E-02 2.27875E-05 0.00000E+00
0.00000E+00 0.00000E+00

416 5.13840E-03 –4.21692E-02 1.27362E-05 0.00000E+00
0.00000E+00 0.00000E+00
520 5.13587E-03 –4.21768E-02 –8.15310E-16 0.00000E+00
0.00000E+00 0.00000E+00
624 5.13840E-03 –4.21692E-02 –1.27362E-05 0.00000E+00
0.00000E+00 0.00000E+00
728 5.14526E-03 –4.21475E-02 –2.27875E-05 0.00000E+00
0.00000E+00 0.00000E+00
832 5.15380E-03 –4.21168E-02 –2.48283E-05 0.00000E+00
0.00000E+00 0.00000E+00
936 5.14670E-03 –4.20879E-02 –1.21035E-05 0.00000E+00
0.00000E+00 0.00000E+00

LARGEST MAGNITUDES OF DISPLACEMENT VECTOR =
1.10589E-02 –5.64997E-02 –6.62681E-04 0.00000E+00 0.00000E+00
0.00000E+00
AT NODE 1428 1050 917 1 1 1
1 *** E M R C N I S A *** —— VERSION 91.0 (06/17/91)
LOAD CASE ID NO. 1

**** NODAL PRINCIPAL STRESSES – LOAD CASE ID NO.1 ****

NODE SG1 SG2 SG3 EQUIV. STRESS

MAX. SHEAR
VON MISES OCTAHEDRAL SHEAR

LARGEST MAGNITUDES OF STRESS FUNCTIONS
7.27692E+03 3.45248E+03 –7.27698E+03 3.67833E+03

6.51458E+03 3.07100E+03
AT NODES 497 1215 419 419 419 419

1 *** E M R C N I S A *** —— VERSION 91.0 (06/17/91)
LOAD CASE ID NO. 1

OVERALL TIME LOG IN SECONDS

INPUT (READ, GENERATE) .... = 39.540
DATA STORING AND CHECKING........................ = 14.78016.200
REORDERING OF ELEMENTS...................... = 73.260
FORM ELEMENT MATRICES............................ = 160.600
FORM GLOBAL LOAD VECTOR............................ = 0.100
MATRIX TRANSFORMATION DUE TO MPC.................... = 0.000
PRE - FRONT.................................. = 3.240
SOLUTION OF SYSTEM EQUATIONS............................. = 379.320
INTERNAL FORCES AND REACTIONS............. = 1.820
STRESS CALCULATION........................ = 124.940
LOAD COMBINATION............................. = 0.000
TOTAL CPU............................ = 797.060

TOTAL ELAPSED TIME IS.................................. = 1618.000
** NOTE ** – RUN IS COMPLETED SUCCESSFULLY
TOTAL NO. OF COMPLETED CASES = 1
LAST COMPLETED CASE -ID = 1
FILE 26 IS SAVED, NAME=set_force26.dat, STATUS=NEW
FILE 27 IS SAVED, NAME=set_force27.dat, STATUS=NEW

Appendix C: Molding Process Data Record

To be used during prototyping, development, preproduction, and production for components made by the injection-molding process

MOLDING PROCESS DATA RECORD

Job:	Tool Description:	Machine Setup:	Instrumentation:
Operators:	Screw:		Safety Check:
Engineers:	Press No.:	Nozzle No.:	

Date / Time	Run Number	Resin	Lot Number	Temperatures (°C) / (°F)					Tool			Pressures (MPa) / (psi)				Cycle Times (sec)							Pad (mm)	RPM	Weights (g)		Observations:
				Rear	Center		Front	Nozzle	Fixed	Movable	Melt	Injection 1st stage	Injection 2nd stage	Clamp (tons)	Back	Injection	Hold	Open	Overall	Booster	Ram in Motion	Screw Retraction			Full Shot	Part Only	

Comments on molding process, start-up, etc.

Appendix D: Tool Repair & Inspection Record

To be used during prototyping, development, preproduction, and production for components made by the injection-molding process

Tool Repair & Inspection Record

Tool Name: ______________________________ Tool Number: ______________________________

Part Number(s): ______________________________ Tool Size:

Customer's Part Number(s): ______________________________ ________ × ________ × ________
Length Width Height

Fits Following Presses: ______________________________ Date Received: ______________________________

Special Setup Equipment Required: ______________________________

Check (✔) if OK for next run. Mark (**R**) if must be repaired before next run. Explain.

Tool Condition													Maintenance Request Description	Date	Maintenance Completed Description	Date
Inspection Date	Sprue Bushing	Runner	Sprue Puller	**A** Cavity & Core	**B** Cavity & Core	Ejectors	Cooling Flow	Side Action	Vents	Wiring & Heaters	Other 1	Other 2				

References

1. *Annual Book of ASTM Standards*, American Society for Testing and Materials, Philadelphia, (1992).
2. C. E. Adams, *Plastic Gearing: Selection and Application*, Marcel Dekker, New York, (1986).
3. *Advanced Materials & Processes*, ASM International, Materials Park, OH, (1985 – 1997).
4. J. Aklonis and A. Tobolsky, *Stress Relaxation and Creep Master Curves for Several Polystyrenes*, Journal of Applied Physics, **36** (11), (1965).
5. E. Atrek (editor), *New Directions in Optimum Structural Design*, John Wiley & Sons, New York, (1984).
6. E. A. Avallone and T. Baumeister III, *Mark's Standard Handbook for Mechanical Engineers*, 9th edition, McGraw-Hill, New York, (1987).
7. J. Avery, *Injection Molding Alternatives*, Hanser Publishers, Cincinnati, OH, (1998).
8. A. J. Baker and D. W. Pepper, *Finite Element 1-2-3*, McGraw-Hill, New York, (1991).
9. R. A. Banister, *Designing Hinges That Live*, Machine Design, Penton Publishing Inc. Cleveland, OH, (July 23, 1987).
10. R. D. Beck, *Plastic Product Design*, Van Nostrand Reinhold, New York, (1980).
11. A. Benatar and Z. Cheng, *Ultrasonic Welding of Thermoplastics II: Far-Field*, Edison Welding Institute Research Report, (April 1989).
12. A. F. Benson, *Assembling HP's Notebook Computer Is a Snap!*, Assembly, Vol. 36, Hitchcock Publishing Co., (July/August 1993).
13. J. Bicerano, *Prediction on Polymer Properties*, Marcel Dekker, New York, (1993).
14. J. J. Bikerman, *The Science of Adhesive Joints*, Academic Press, New York, (1968).
15. P. R. Bonenberger, Stretching the Limits of DFM, Machine Design, Penton Publishing, Cleveland, OH, (Sept. 12, 1994).
16. P. R. Bonenberger, *A New Design Methodology for Integral Attachments*, ANTEC (SPE), Brookfield, CT, (1995).
17. P. R. Bonenberger, *The First Snap Fit Handbook*, Hanser Gardner Publications, Cincinnati, OH, (2000).
18. P. R. Bonenberger, *The Role of Enhancement Features in High Quality Integral Attachments*, ANTEC (SPE), Brookfield, CT, (1995).
19. E. Bornschlegel, *Successful International Harmonization – Targeted Help for Users of Plastics by CAMPUS®*, Der Lichtbogen, **209** (11), (November 1989).

20. J. Bowman and K.E. Pawlak, Snap Fit Cap Design Using Rapid Prototyping and Taguchi Methods, ANTEC (SPE), Brookfield, CT, (1993).
21. D. Braun, *Simple Methods for Identification of Plastics*, 2nd Edition, Hanser Publishers, Munich, (1986).
22. C.A. Brebbia, J.C.F. Telles, and L.C. Wrobel, *Boundary Element Techniques, Theory and Application in Engineering*, Springer Verlag, Berlin, (1984).
23. C.A. Brebbia, *The Boundary Element Method for Engineers*, McGraw-Hill, London, (1977).
24. H. Breuer, G. Dupp, J. Schmitz, and R. Tüllmann, *A Standard Materials Databank - an Idea now Adopted*, Kunststoffe German Plastics, **80** (11), (1990).
25. H. Breuer, G. Dupp, R. Jantz, G. Wübken, M.H. Tiba, and R. Tüllmann, *CAMPUS vor Weltweiter Verbreitung Teil 2: Mit Version 3 ist der Durchbruch geschafft*, Kunststoffe, **84** (8), (1994).
26. T. Brinkmann, *CAMPUS® unter Windows*, Plastverarbeiter, (January 1996).
27. L. Brooke, *Design with a Snap*, Automotive Industries, Chilton Company, a division of Capital Cities/ABC Inc., (January 1992).
28. W. Brostow and R.D. Corneliussen (editors), *Failure of Plastics*, Hanser Publishers, New York, (1986).
29. O.S. Brueller, *On the Nonlinear Response of Polymers to Periodical Sudden Loading and Unloading*, Polymer Engineering and Science, **25** (10), (1985).
30. J.A. Brydson, *Plastic Materials*, Butterworth Scientific, London, (1982).
31. C.B. Bucknall, I.C. Drinkwater, and G.R. Smith, *Hot Plate Welding: Factors Affecting Weld Strength*, Polymer Engineering and Science, **20** (6), (1980).
32. R.L. Burden and J.D. Faires, *Numerical Analysis*, 3rd Edition, Prindle, Weber & Schmidt, New York, (1985).
33. R. Callanan, *Introduction to RF Sealing for Clamshell Blister Packages*, Journal of Packaging Technology, (1991).
34. H.S. Carslaw and J.C. Jaeger, *Conduction of Heat in Solids*, Clarendon Press, Oxford, (1959).
35. *Characteristics of Thermoplastics for Ultrasonic Assembly Applications*, Technical Bulletin, Sonics & Materials Inc., Danbury, CT, (1988).
36. W. Chow, *How to Design for Snap Fit Assembly*, Plastic Design Forum, Advanstar Communications, Duluth, MN, (March/April 1977).
37. N. Crangulescu, *Machine Design*, Penton Publishing, Cleveland, OH, (February 1997).
38. R.M. Christensen, *Theory of Viscoelasticity*, Academic Press, New York, (1982).
39. D.W. Clegg and A.A. Collyer, *Mechanical Properties of Reinforced Thermoplastics*, Elsevier, Amsterdam, (1986).
40. N. Cristescu, *Dynamic Plasticity*, North-Holland, Amsterdam, (1967).
41. J.J. Cunningham and J.R. Dixon, *Designing with Features: The Origin of Features*, ASME Computers in Engineering. San Francisco, CA, (July - August 1988).

42. *Designing Parts for Ultrasonic Welding*, Branson Ultrasonic Corporation, Danbury, CT, (1989).
43. *Designing with Plastic: The Fundamentals*, Hoechst Celanese, (1986).
44. K. Dohring, *Lost Core Molding: The Technology and Future Outlook*, 5th International Molding Conference, Conference Proceedings, New Orleans, LA, (1995).
45. *Dow Engineering Thermoplastic Basic Design Manual*, Mechanical Assembly, Midland, MI, (1987 – 1988).
46. *Dow Snap Fit Designer Software*, Version 1.0, Dow Chemical, Midland, MI, (1990 – 1991).
47. E.I. DuPont de Nemours and Co., *Design Handbook for DuPont Engineering Plastics*, Module I to IV, Wilmington, DE, (1989).
48. G. Dupp, *Vorgeschichte, Ziele und Inhalte der Kunststoffdatenbank CAMPUS® – Gemeinschaftsproduktion von vier Rohstoffherstellern*, Kunststoff Journal, **4**, (1988).
49. J.B. Dym, *Product Design with Plastics: A Practical Manual*, Industrial Press, New York, (1983).
50. *Electromagnetic Welding System for Assembling Thermoplastic Parts*, Emabond Systems, division of Ashland Chemicals, Norwood, NJ, (1987).
51. Dr. Endemann, *BASF WIS SNAPS*, Version 1.6, software for designing snap fits, BASF AG, Ludwigshafen, Germany, (1993).
52. Dr. Endemann, *BASF GRAPH1*, Version 1.6, software for graphically representing stress-strain functions, BASF AG, Ludwigshafen, Germany, (1993).
53. *FEASnap™ – Snap-Fit Design Software*, User Manual, Bayer Corporation, Pittsburgh, PA (1997).
54. *FEASnap™ – Snap-Fit Design Software*, Version 1.0, Bayer Corporation, Pittsburgh, PA (1997).
55. I. Finnie and W.R. Heller, *Creep of Engineering Materials*, McGraw-Hill, New York, (1959).
56. M. Fortin and R. Glowinski (editors), *Méthodes de Lagrangien Augmenté*, Dunod, Paris, (1981).
57. J. Frados (editor), *Plastics Engineering Handbook*, 5th Edition, edited by M.L. Berins, SPI, (1991).
58. R.H. Gallagher, *Finite Element Analysis Fundamentals*, Prentice-Hall, Englewood Cliffs, NJ, (1975).
59. G.A. Georgiou and I.A. MacDonald, *Ultrasonic and Radiographic NDT of Butt Fusion Joints in Polyethylene*, Technology Briefing (TWI), No. 465, (February 1993).
60. A.B. Glanville and E.N. Denton, *Injection-Mould Design Fundamentals*, Industrial Press, New York, (1965).
61. A.B. Glanville, *The Plastics Engineer's Data Book*, Industrial Press, New York, (1974).
62. D. Grewell, *Applications with Infrared Welding of Thermoplastics*, ANTEC (SPE), Brookfield, CT (1999).
63. R. Grimm, *Through-Transmission Infrared Welding (TTIR) of Teflon® TFE (PTFE)*, ANTEC (SPE), Brookfield, CT, (2000).

64. K. Ito, *Cold Processing of Crystalline Polymers*, Modern Plastics, No. 8, (1966).
65. R.D. Hanna, *Molded-in Hinge Polypropylene Components*, ANTEC (SPE), Brookfield, CT, (1961).
66. R.W. Hertzberg and J.A. Manson, *Fatigue of Engineering Plastics*, Academic Press, (1980).
67. C. Higdon and E. Archie (editors), *Mechanics of Materials*, 4th Edition, John Wiley & Sons, New York, (1985).
68. G.S. Holister and C. Thomas, *Fiber Reinforced Materials*, Elsevier Publishing Co., London, (1966).
69. *How Ingredients Influence Unsaturated Polyester Properties*, Amoco Chemicals Corp., Bulletin IP-70, (1980).
70. T.S. Hsu, *Stress and Strain Data Handbook*, Gulf Publishing Company, Houston, TX, (1986).
71. D. Hull, *An Introduction to Composite Materials*, Cambridge University Press, Cambridge, (1981).
72. D.O. Hummel and F. Scholl, *Atlas of Polymer and Plastic Analysis*, 2nd edition, Hanser Publishers, Munich, (1982).
73. R. Jantz, and M.M. Matsco, *The World-Wide Material Database Using Uniform Standards for Plastics Design*, ANTEC (SPE), Brookfield, CT, (1992).
74. C.T. Johnk, *Engineering Electromagnetic Fields and Waves*, John Wiley & Sons, New York, (1988).
75. I. Jones and N. Taylor, *Use of Infrared Dies for Transmission Welding of Plastics*, ANTEC (SPE), Brookfield, CT, (2000).
76. R.M. Jones, *Mechanics of Composite Materials*, McGraw-Hill, New York, (1975).
77. K. Jost, *Chrysler's New V6 Engine Family*, Automotive Engineering, (January 1997).
78. D.H. Kaelble, *Computer-Aided Design of Polymers and Composites*, Marcel Dekker, New York, (1985).
79. V.A. Kagan, *Innovations in Laser Welding Technology*, SAE World Congress, Detroit, MI, (2002).
80. B.S. Kasatkin, A.B. Kudrin, and L.M. Lobanov, *Experimental Methods of Stresses and Deformation Investigations*, Manual Naukova Dumba, Kiev, Ukraine, (1981).
81. H.S. Kaufman and J.J. Falcetta (editors), *Transitions and Relaxations in Polymers: Introduction to Polymer Science and Technology*, John Wiley & Sons, New York, (1977).
82. B. Kenneth, *Package Design Engineering*, John Wiley & Sons, New York, (1959).
83. A.J. Kinloch, *Adhesion and Adhesives: Science and Technology*, Chapman and Hall, London, (1987).
84. W.M. Kolb, *Curve Fitting for Programmable Calculators*, 3rd Edition, Syntec, (1984).
85. J.L. Lamprecht, *ISO-9000: Preparing for Registration*, Marcel Dekker, New York, (1992).
86. G.C. Larsen and R.F. Larsen, *Parametric Finite-Element Analysis of U-Shaped Snap-Fits*, ANTEC (SPE), Brookfield, CT, (1994).

87. R.F. Larsen and G.C. Larsen, *The Next Generation of PC Software for Traditional and FEA Snap-Fit Design*, ANTEC (SPE), Brookfield, CT, (1994).
88. A.F. Leatherman, *Induction Bonding*, Modern Plastics Encyclopedia, McGraw-Hill, New York, (1988).
89. J. Leighton, T. Brantley, and E. Szabo, *RF Welding of PVC and Other Thermoplastic Compounds*, ANTEC (SPE), Brookfield, CT, (1992).
90. W. Leventon, *New Software Simplifies Snap-Fit Design*, Design News, (2/10/92).
91. W. Lin, O. Buneman, and A.K. Miller, *Induction Heating Model for Graphite Fiber/Thermoplastic Matrix Composite Materials*, SAMPE Journal, No. 27, (1991).
92. B. Lincoln, K.J. Gomes, and J.F. Braden, *Mechanical Fastening of Plastics - An Engineering Handbook*, Marcel Dekker, New York, (1984).
93. D. Lobdell, *Snap-Fit Corrections*, Plastics World, (August 1992).
94. A.F. Luscher, G.A. Gabriele, P.R. Bonenberger, and R.W. Messler, *A Clasification Scheme for Integral Attachment Features*, ANTEC (SPE), Brookfield, CT, (1995).
95. M. Maniscalco, *Snap Fit Software Closes the Loop*, Injection Molding Magazine, (January 1997).
96. N.R. Mann, R.E. Schaffer, and N.D. Singpurwall, *Methods for Statistical Analysis of Reliability and Life Data*, John Wiley & Sons, New York, (1974).
97. K. Masubuchi, *Analysis of Welded Structures*, Pergamon Press, New York, (1980).
98. *Machine Design*, Penton Publishing Inc. Cleveland, OH, (1970 - 1997).
99. R.R. Mayer and G.A. Gabriele, *Systematic Cataloging of Integral Attachment Strategy Case Studies*, ANTEC (SPE), Brookfield, CT, (1995).
100. L.H. McCarty, *Radio Frequency Welds in Miniature*, Design News, (June 5, 1989).
101. *Modern Plastics Encyclopedia*, McGraw-Hill, New York, (1992).
102. N.I. Muskhelishvili, *Some Basic Problems of the Mathematical Theory of Elasticity*, P. Noordhoff, Gröningen, The Netherlands, (1953).
103. M.H. Naitove, *Resin Suppliers Push for Uniform Global Test Standards*, Plastics Technology, (July 1996).
104. *NASA Tech Briefs*, Associated Business Publications Co., Ltd., New York, (1985 - 1997).
105. J.A. Newman and F.J. Backhoff, *Welding of Plastics*, Reinhold Publishing Corp., New York, (1959).
106. L.E. Nielson, *Mechanical Properties of Polymers and Composites*, Marcel Dekker, New York, (1974).
107. K. Oberbach, *Fundamental Datatables and Database (CAMPUS®) - a Challenge and Opportunity*, Kunststoffe German Plastics, **79** (8), (1989).
108. K. Oberbach, and L. Rupprecht, *Plastics Properties for Databank and Design*, Kunststoffe German Plastics, **77** (8), (1987).
109. J.T. Oden, *Finite Elements of Nonlinear Continua*, McGraw Hill, New York, (1972).
110. R.M. Ogorkiewicz, *Engineering Properties of Thermoplastics*, John Wiley & Sons, New York, (1970).

111. M.R. Olds, *Hot Tack: Key to Better Seals*, Package Engineering, (November 1976).
112. C.R. Oswin, *Plastic Films and Packaging*, John Wiley & Sons, New York, (1975).
113. *Plastic Design Forum*, Advanstar Communications, Duluth, MN, (1980 - 1993).
114. *Plastic Engineering*, Society of Plastic Engineers Inc., Brookfield, CT, (1980 - 1997).
115. *Plastic Snap-Fit Design Interlocks in Unique and Useful Ways*, Product Engineering, (May 1977).
116. H. Potente, *Analysis of the Heated Plate Welding of Pipes Made of Semi-Crystalline Thermoplastics*, Kunststoffschweissen und Kleben; Vorträge der Internationalen Tagung, Düsseldorf, (1983).
117. H. Potente, P. Michel, and B. Ruthmann, *Eine Analyse des Vibrationsschweissens*, Kunststoffe, **77**, Hanser Publishers, Munich, (1987).
118. H. Potente and M. Uebbing, *Computer-Aided Layout of the Vibration Welding Process*, ANTEC (SPE), Brookfield, CT, (1992).
119. H. Potente and F. Becker, *Weld Strength Behavior of Laser Butt Welds*, ANTEC (SPE), Brookfield, CT (1999).
120. N.S. Rao, *Design Formulas for Plastics Engineers*, Hanser Publishers, New York, (1991).
121. D. Reiff, *Integral Fastener Design*, Plastic Design Forum, (September/October 1991).
122. R.J. Roark, *Formulas for Stress and Strain*, 5th Edition, McGraw-Hill, New York, (1975).
123. J. Rotheiser, *Joining of Plastics*, Hanser Publishers, Cincinnati, OH, (1999)
124. S. Roy and J.N. Reddy, *A Finite Analysis of Adhesively Bonded Composite Joints with Moisture Diffusion and Delayed Failure*, Computers and Structures, **24**, (6), (1988).
125. D. Satas, Plastics Finishing and Decorating, Van Nostrand Reinhold, New York, (1986).
126. A.K. Schlarb and G.W. Ehrenstein, *Vibration Welding. A Materials Technology View of Mass Production Methods*, Kunststoffe, **78**, Hanser Publishers, Munich, (1988).
127. J. Schmitz, E. Bornschlegel, G. Dupp, and G. Erhard, *Kunststoff-Datenbank CAMPUS®, Einheitliche Software der großen Vier*, Plastverarbeiter, 4, (1988).
128. J. Schmitz and K. Oberbach, *Material Properties for Database - An offer of the Raw Material Suppliers*, ANTEC (SPE), Brookfield, CT, (1988).
129. R. Shastri, K.S. Mehta, M.H. Tiba, W.F. Müller, H. Breuer, E. Baur, R.A. Latham, J.A. Grates, P.M. Sarnacke, J.S. Kennedy, and G.P. Diehl, *CAMPUS® - Presentation of Comparable data on Plastics Based on Uniform International Standards*, NPE, Conf. Proceedings, Vol. II, (1994).
130. R. Shastri, K.S. Mehta, M.H. Tiba, W.B. Hoven-Nieveistein, H. Breuer, E. Baur, R.A. Latham, J.A. Grates, P.M. Sarnacke, J.S. Kennedy, and G.P. Diehl, *CAMPUS® - Standardized Presentation of Data on Plastics*, ANTEC (SPE), Brookfield, CT, (1996).
131. S. Shillitoe, A.J. Day, and H. Benkreira, *A Finite Element Approach to Butt Fusion Welding Analysis*, Proceedings of the Institute of Mechanical Engineers, Part E, Vol. 204, (1990).
132. S. Shuzeng and W. Yousheng, *Prediction of Long Term Behavior of Fiber Reinforced Plastics*, Proceedings of the 7th International Conference on Composite Materials (ICCM), China, (1989).

133. H.R. Simonds, A.J. Weith, M.H. Bigelow, *Handbook of Plastics*, Van Nostrand Reinhold, New York, (1977).
134. I. Skeist (editor), *Handbook of Adhesives*, Van Nostrand Reinhold, New York, (1977).
135. *Snap Design User's Guide*, Version 2.0, Closed Loop Solutions, Inc., Troy, MI, (1997).
136. *Snap-Fit Design Guide*, Modulus Design Group of Allied Signal Engineered Plastics, Morristown, NJ, (1987).
137. *Snap-Fit Joints in Plastics, A Design Manual*, Miles Corporation, Troy, MI, (1990).
138. R.C. Snodgren, *Handbook of Surface Preparation*, Palmerston, New York, (1972).
139. L. Sors, L. Bardocz, and I. Radnoti, *Plastic Molds and Dies*, Van Nostrand Reinhold, New York, (1981).
140. S. Stevens, *Structure Evaluation of Polyethylene and Polypropylene Hot Plate Welds*, Technology Briefing (TWI), No. 466, (1993).
141. K. Stoeckhert (editor), *Mold-Making Handbook for the Plastic Engineer*, Hanser Publishers, Munich, (1983).
142. V.K. Stokes, *Cross-Thickness Vibration Welding of Thermoplastics*, ANTEC (SPE), Brookfield, CT, (1992).
143. V.K. Stokes, *Vibration Welding of Thermoplastics Part 1: Phenomenology of the Welding Process*, Polymer Engineering and Science, **28**, (11), (1988).
144. V.K. Stokes, *Vibration Welding of Thermoplastics Part 2: Analysis of the Welding Process*, Polymer Engineering and Science, **28**, (12), (1988).
145. L.C.E. Strik, *Physical Aging of Amorphous Polymers and Other Materials*, Elsevier, Amsterdam, (1978).
146. M. Sturdevant, *The Long-term Effects of Ethylene Oxide and Gamma Radiation on the Properties of Rigid Thermoplastic Materials*, ANTEC (SPE), Brookfield, CT, (1993).
147. N.P. Suh, *The Principles of Design*, Oxford University Press, Oxford, (1990).
148. G. Trantina and M. Minnichelli, *Automated Program for Designing Snap Fits*, Plastics Engineering, (August 1987).
149. P.E. Teague, *Fasteners Take a Custom Twist*, Design News – Cahners Publishing Co., Vol. 47, No. 18, (1991).
150. P.E. Teague, *Fasteners that Mate with New Materials*, Design News – Cahners Publishing Co., Vol. 48, No. 18, (1992).
151. S. Timoshenko and G.H. MacCullough, *Elements of Strength of Materials*, 3rd Edition, D. Van Nostrand Company Inc., New York, (1957).
152. S. Timoshenko and S. Woinowsky-Krieger, *Theory of Plates and Shells*, McGraw Hill, New York, (1959).
153. D.S. Tres, *Snap Fit Design Software for Engineering Plastics*, RETEC (SPE) conference, Rochester, NY, (1993).
154. P.A. Tres, *Blow Molding: Process and Part Design Fundamentals*, Manual, SPE, Chicago, (1995).

155. P. A. Tres, *Bright Future for Plastics*, Plastics Insights, Vol. 7, Issue 7, Hanser Gardner Publications, Cincinnati, OH (2002).
156. P. A. Tres, *Designing Injection Molded Parts for Assembly: Understanding Safety Factors*, International Plastics Design & Processing Conference at Kunststoff '95, SME conference proceedings, Krefeld, Germany, (1995).
157. P. A. Tres, *Designing Plastic Parts for Assembly*, Manual, University of Wisconsin, College of Engineering, Madison, WI, (1992).
158. P. A. Tres, *Fundamentals of Automotive Plastic Parts Design*, Manual, SAE International Congress, Detroit, MI, (1996).
159. P. A. Tres, *Hinge Design System Software*, Version 1.0, DuPont Automotive, Troy, MI, (1988).
160. P. A. Tres, *Hinge Design System Software*, Version 2.0, DuPont Automotive, Troy, MI, (1989).
161. P. A. Tres, *Lost Core Injection Molding Technology*, SME Automotive Plastics '94, Troy, MI, (1994).
162. P. A. Tres, *Plastic Part Design for Assembly*, proceedings of the National Manufacturing Week Conference, Reed Exhibitions, Chicago, IL (1998).
163. P. A. Tres, *Robust Design: '98MY DaimlerChrysler Upper Intake Manifold*, proceedings of the 5th International Manifold Forum, Spitzingsee, Germany (1998).
164. P. A. Tres, K. McDonald, D. S. Tres, and C. Jenings, *Snap-Fit Design System Software and Manual*, Version 3.1, DuPont Automotive, Troy, MI, (1991).
165. P. A. Tres, K. McDonald, D. S. Tres, and C. Jenings, *Snap-Fit Data Entry Software and Manual*, Version 3.1, DuPont Automotive, Troy, MI, (1991).
166. P. A. Tres, *Snap-Fit Snapshot*, Assembly, Vol. 36, Hitchcock Publishing Co., (July/August 1993).
167. P. A. Tres, *Bright Future for Plastics*, Plastics Insights, Cincinnati, OH, Vol. 7, Issue 7, (December 2001/Jan. 2002).
168. P. A. Tres, *Plastics ... Trends in the Industry*, Keynote speech – National Plastics Exhibition (NPE '03), Chicago, IL, (2003).
169. P. A. Tres, *Designing Plastic Parts for High Speed Assembly*, Assembly Technology Expo, Rosemont, IL, (2004).
170. P. A. Tres, *Hollow Glass Microspheres Stronger Spheres Tackle Injection Molding*, Plastics Technology, Gardner Publications, New York, NY, (2007).
171. P. A. Tres, *Simulation – A Primary Part of the Design Process*, Modern Plastics Worldwide – Cannon Communications LLC, Los Angeles, CA, (July 2007).
172. K. Wood, *Microspheres: Fillers Filled with Possibilities*, CompositesWorld, Gardner Publications, Wheat Ridge, CO, (2008).
173. P. A. Tres, *Designing Plastic Parts for High-Speed Assembly*, Webinar for Assembly Magazine - BNP Media, Bensenville, IL, (2008).
174. P. A. Tres, *Understanding Critical Aspects of Plastic Part Design and Manufacturing*, Webinar in conjunction with Geometric Ltd. – Geometric Limited, Mumbai, India, (2011).

175. P. A. Tres, *Metals Meet Plastics' Steelier Side*, Plastics Technology – Society of Plastics Engineers/John Wiley & Sons, Newtown, CT, (2013).

176. P. A. Tres, *Snap Fits Enable Plastic Parts Assembly*, Assembly Magazine, BMP Media, Bensenville, IL, (2014).

177. P. A. Tres, *Snap Fits for Plastic Assembly*, The Assembly Show – BMP Media, Bensenville, IL, (2014).

178. P. A. Tres, *Automotive Design*, Plastics in Automotive Conference – Plastics News, Detroit, MI, (2015).

179. P. A. Tres, *Avoiding Automotive Plastic Design Pitfalls*, Society of Plastics Engineers – ANTEC 2015, Fellows Forum, Orlando, FL, (2015).

180. P. A. Tres, *Investigation of the Influence of Color on Plastic Product Failure-or-Snap-Fits Which Kill*, Society of Plastics Engineers – ANTEC 2015, Product Design and Development Division, Orlando, FL, (2015).

181. P. A. Tres, *Automotive Plastic Part Design*, Manual, University of Michigan, Dearborn, MI (1999–2016)

182. *Ultrasonic Plastic Assembly*, Branson Sonic Power Company, Danbury, CT, (1979).

183. S. Utku and M. M. El-Essawi, *Error Computation in Finite Element Method*, proceedings of the 2nd International Conference on Electronic Computation, ASCE, New York, (August 1979).

184. D. W. Van Krevelen, *Properties of Polymers*, Elsevier, Amsterdam, (1976).

185. B. Walker, *Handbook of Thermoplastic Elastomers*, Van Nostrand Reinhold, New York, (1979).

186. L. Wang, G. A. Gabriele, and A. F. Luscher, *Failure Analysis of a Bayonet-Finger Snap-Fit*, ANTEC (SPE), Brookfield, CT, (1995).

187. M. N. Watson and M. G. Murch, *Recent Developments in Hot Plate Welding of Thermoplastics*, Polymer Engineering and Science, No. 29, (1989).

188. W. Weibull, *Fatigue Testing and Analysis of Results*, Pergamon Press, New York, (1961).

189. *Welding Handbook*, 8th Edition, American Welding Society, (1987).

190. J. G. Williams, *Stress Analysis of Polymers*, John Wiley & Sons, New York, (1973).

191. S. I. Wu, *Polymer Interface and Adhesion*, Marcel Decker, New York, (1982).

192. J. Yang and A. Garton, *Primers for Adhesive Bonding to Polyolefins*, Journal of Applied Polymer Science, **48**, (1993).

193. H. Yeh and R. Grimm, *Infrared Welding of Thermoplastics, Characterization of Transmission Behavior of Eleven Thermoplastics*, ANTEC (SPE), Brookfield, CT, (1998).

194. R. J. Young, *Introduction to Polymers*, Chapman and Hall Publishers, London, (1981).

195. K. I. Zaitsev, *Welding of Polymeric Materials – Handbook*, Mashinostrouenie, Moscow, Russia, (1988).

World Wide Web References Related to Plastic Part Design

Note: ► symbol represents MUST SEE World Wide Web sites.

Associations/Organizations

	Organization	Website
	ABIPLAST-Associação Brasileira da Indústria do Plástico	www.abiplast.org.br
	ABIQUIM-Associação Brasileira da Indústria Química	www.abiquim.org.br
	AVK – Industrievereinigung Verstärkte Kunststoffe e.V.	www.avk-tv.de
	All India Plastics Manufacturers' Association (AIPMA)	www.aipma.net
	American Chemical Society	www.acs.org
	American Chemistry Council	www.americanchemistry.com
	ASM International	www.asminternational.org
	Asociación Nacional de la Industria Química	www.aniq.org.mx
	Association of Plastics Manufacturers Europe	www.plasticseurope.org
	ASSOCOMAPLAST	www.assocomaplast.org
	British Plastics Federation	www.bpf.co.uk
	Chemistry Australia	www.chemistryaustralia.org.au
	Canadian Plastics Industry Association	www.plastics.ca/home/index.php
	China Plastics Processing Industry Association (CPPIA)	www.cppia.com.cn
	Confederación Española de Empresarios de Plásticos	www.anaip.es
	European Plastics Converters (EuPC)	www.plasticsconverters.eu
	Fachverband der chemischen Industrie Österreichs	www.kunststoffe.fcio.at
	Fédération de la Plasturgie	www.laplasturgie.fr
	Federación Empresarial de la Industria Química Española	www.feique.org
	Federazione Gomma Plastica	www.federazionegommaplastica.it
	Food Packaging Institute	www.fpi.org

►	Gesamtverband Kunststoffverarbeitendene Industrie (GVK)	www.gkv.de
	Indian Plastics Federation	www.ipfindia.org
	Industrievereinigung Kunststoffverpackungen e.V	www.kunststoffverpackungen.de
	Instituto Tecnologico del Plastico	www.aimplas.es
	Japan Plastics Industry Federation	www.jpif.gr.jp
	JEC Composites	www.jeccomposites.com
	Kompozit Sanayicileri Derneği	www.kompozit.org.tr
	Korea Plastic Industry Cooperative	www.koreaplastic.or.kr
	Kunststoff Verband Schweiz	www.kvs.ch
	Malaysian Plastics Manufacturers Association (MPMA)	www.mpma.org.my
	Plastics Cluster	www.plastr.cz
	Plastics Federation of South Africa	www.plasticsinfo.co.za
	Plastics Industry Association (PIA; formerly SPI)	www.plasticsindustry.org
	Plastics New Zealand	www.plastics.org.nz
	Polish Union of Plastics Converters	www.pzpts.com.pl
	Portuguese Association for the Mold Industry (CEFAMOL)	www.cefamol.pt
	Russian Chemical Union	www.ruschemunion.ru
	Pôle Européen de Plasturgie	www.poleplasturgie.net
►	SAE (Society of Automotive Engineers)	www.sae.org
	SAMPE	www.sampe.org
►	SPE (Society of Plastics Engineers)	www.4spe.org
	Thai Plastic Industries Association	www.tpia.org
	The Swedish Plastics & Chemicals Federation	www.plastkemiforetagen.se
	Turkish Plastics Industry Association (PAGEV)	www.pagev.org.tr
	Verband Deutscher Maschinen- und Anlagenbau eV	www.vdma.org

Books/Magazines

	Carl Hanser Verlag	www.hanser.de
►	Chemical Week	www.chemweek.com
►	Design News	www.designnews.com
►	Injection World	www.injectionworld.com
	Hanser Publications	www.hanserpublications.com
►	KunststoffWeb	www.kunststoffweb.de
	Kunststoffe International	www.kunststoffe-international.com
	Kunststoff Magazine	www.kunststoff-magazin.de
	GAK Gummi Fasern Kunststoffe	www.gupta-verlag.com/magazines/gak-gummi-fasern-kunststoffe
►	Machine Design	www.machinedesign.com
	Plastverarbeiter	www.plastverarbeiter.de
	Dr. Gupta Verlag	www.gupta-verlag.de
	Macplast	www.macplas.it/pagine/home.asp
	Euwid Kunststoff	www.euwid-kunststoff.de
	Kunststoff Information	www.kiweb.de
	Swiss Plastics	www.swissplastics-expo.ch
	Plastics News	www.plasticsnews.com
►	Plastics Technology	www.ptonline.com
	Chemical Industry Press	www.cip.com.cn

Consulting

	ETS, Inc.	www.ets-corp.com
	Consultek Consulting Group	www.consultekusa.com
	Robert Eller Associates	www.robertellerassoc.com
	WJT Associates	www.wjtassociates.com

Forums

	Kunststoff Forum	www.kunststofforum.de
►	PolySort	www.polysort.com/forum/

Material Databases

►	CAMPUS	www.campusplastics.com
►	Prospector	www.ulprospector.com
►	Autodesk Moldflow	www.autodesk.com/products/moldflow/overview
	MatWeb LLC	www.matweb.com
	Rapra Technology	www.polymerlibrary.com

Material Suppliers

	Addiplast	www.addiplast.fr
	Aquafil	www.aquafil.com
	Arkema	www.arkema.com
	Asahi Chemical	www.asahi-kasei.co.jp/asahi/en/
	Asahi Glass Company	www.agc.com/english/index.html
	Asahimas Chemical	www.asc.co.id
	Ashland Chemical	www.ashland.com
	Astra Polimer Sanayi ve Ticaret	www.astra-polymers.com.tr
	Akzo Nobel	www.akzonobel.com
	BASF	www.basf.com
	Bhansali Engineering Polymers Limited	www.bhansaliabs.com
	Borealis	www.borealisgroup.com
	Boryszew	www.boryszew.com.pl
	Braschem	www.braskem.com.br
	Celanese	www.celanese.com
	ChangChun Group	www.ccp.com.tw
	Chevron Phillips Chemical Company	www.cpchem.com

	China National Bluestar	bluestar.chemchina.com/lanxingen/index.htm
	Covestro	www.covestro.com
	Delta Kunststoffe	www.delta-kunststoffe.de
	DingZing Chemical	www.dingzing.com
	Dow Chemical	www.dow.com
	DuPont	www.dupont.com
▶	Eastman	www.eastman.com
	Elaston Kimya	www.elastron.com
	Equate Petrochemical Company	www.equate.com
	ExxonMobil	www.exxonmobil.com
	Ferro	www.ferro.com
	FkuR Kunststoff	www.fkur.com
	Gold Baiyi Group	http://goldbaiyi.en.ecplaza.net
	Grupo Idesa	www.grupoidesa.com
	Haldia Petrochemicals	www.haldiapetrochemicals.com
	Hipol	www.hipol.com
	Interquimica	www.interquimica.com.br
	Japan Polychem Corporation	www.pochem.co.jp/english/index.html
	Kaofu Chemical Corporation	www.kaofu.com
	Karbochem	www.karbochem.co.za
	Kolon Industries	www.kolonindustries.com
	Luk Oil	www.lukoil.ru
	LyondellBasell	www.lyondellbasell.com
	Mepol Polimeri	www.mepol.it
	Mitsubishi Engineering Plastics	www.m-ep.co.jp
	Momentive	www.momentive.com
	National Petrochemical Industrial	www.natpet.com
	Ovation Polymers	www.opteminc.com
	Palram Industries	www.palram.com

	Plastika Kritis	www.plastikakritis.com
	Plastiques GyF ltée	www.plastiquesgyf.ca
	Plazit Iberica	www.gerundense.com
	Polychim Industrie	www.polychim-industrie.com
	Polyscope	www.polyscope.eu
	Radici Group	www.radicigroup.com
	Reliance Industries Ltd.	www.ril.com
	Roechling Group	www.roechling.com
	Rompetrol	www.petrochemicals.ro
	RTP Company	www.rtpcompany.com
	Sabic Innovative Plastics	www.sabic-ip.com
	Saint Gobain	www.plastics.saint-gobain.com
	Sibur Holding	www.sibur.com
	Solvay	www.solvay.com
	Thai Plastic	www.thaiplastic.co.th
	Unitika	www.unitika.co.jp
	Vegeplast	www.vegeplast.com
	Washington Penn Plastics	www.washingtonpennplastic.com
	Yuka Denshi	www.yukadenshi.co.jp
	Zhejiang Juner New Materials	www.juner.cn
	Zhongfa	www.zhongfa-china.com

Museums

	Bakelite Museum	www.thebakelitemuseum.com
▶	Cannon-Sandretto Plastics Museum	http://museo.cannon.com
▶	German Plastics Museum	www.deutsches-kunststoff-museum.de

Rapid Prototyping

►	Rapid Prototyping Home Page	www.rapidprototypinghomepage.com
	Rapid Today	www.rapidtoday.com
	Russian Academy of Sciences	www.laser.ru/rapid/indexe.html

Testing/Research

	Akron Rubber Development Laboratory	www.ardl.com
	Plastics Technology Laboratories	www.ptli.com

Tips on Design and Processing

►	Bad Human Factors Designs	www.baddesigns.com
	Guide to Thermoplastics	www.endura.com/material-selection-guide

Tooling

	D-M-E Company	www.dme.net
	Foboha	www.foboha.de
►	Harbec	www.harbec.com
	Mold Masters	www.milacron.com/our-brands/mold-masters

Universities

	Brown University	www.chem.brown.edu
►	Cornell University Materials by Design	www.mse.cornell.edu
►	Delft University	www.io.tudelft.nl
►	Institut für Kunststoffverarbeitung	www.ikv-aachen.de
	University of Leeds Polymer Science	www.leeds.ac.uk
►	University of Massachusetts-Lowell	www.uml.edu
	University of Wisconsin-Milwaukee	www.uwm.edu

About the Author Paul A. Tres

Serving the plastics and automotive industries for over 40 years, Mr. Paul A. Tres, a Senior Consultant with ETS, Inc. (www.ets-corp.com) in Bloomfield Hills, Michigan, has provided consulting services to global companies, expert witness services for attorneys, and also trained over the years in excess of 17,000 designers, engineers, and managers in the intricacies of plastic part design.

Some of the global companies served include: B/E Aerospace of Lenexa, KS; Briggs & Stratton of Wauwatosa, WI; Bombardier Recreational Products of Valcourt, QC, Canada; Continental Automotive Systems of Auburn Hills, MI and Guadalajara, Jalisco, Mexico; Dow Chemical of Midland, MI; Ford Motor Company of Dearborn, MI; General Motors of Warren, MI; Hewlett-Packard of Vancouver, WA; Honda R&D America of Marysville, OH and Honda of Canada Manufacturing of Alliston, Ontario, Canada; Johnson Controls of Holland, MI and Milwaukee, WI; Kostal North America of Troy, MI; Leggett & Platt Automotive of Lakeshore, Ontario, Canada; Mercedes-Benz US International of Tuscaloosa, AL; Meritor of Troy, MI; Nyco Minerals of Willsboro, NY, Plastics Omnium of Troy, MI and Anderson, SC; Pelco by Schneider Electric of Clovis, CA; Philips Sonicare of Bothell, WA; Siemens of Munich, Germany and Troy, MI; Solvay of Brussels, Belgium and Adrian, MI; Southco of Concordville, PA; TRW of Washington, MI; Tyco of Menlo Park, CA; Uni-Solar Ovonic of Troy, MI; Valeo of Auburn Hills, MI; Visteon of Plymouth, MI; WL Gore & Associates of Newark, DE; and Yazaki North America of Monterrey, Nuevo Leon, Mexico.

Some of the attorneys include: Bamberger, Foreman, Oswald & Hahn, LLP of Indianapolis, IN; Brobeck, Phleger & Harrison LLP of Washington, DC and Los Angeles, CA; Cash, Krugler & Fredericks LLC of Atlanta, GA; Carcione, Cattermole, Dolinski, Stucky, Markowitz & Carcione, LLP of San Mateo, CA; Conroy, Simberg, Ganon, Krevans, Abel, Lurvey, Morrow & Scheffer, PA of West Palm Beach, FL; Cozen O'Connor of Chicago, IL; Denardis & Miller, PC of Mount Clements, MI; Dickinson-Wright PLLC of Detroit, MI; Finnegan, Henderson, Farabow, Garrett & Dunner, LLP of Washington, DC; Foley, Barron & Metzger PLLC of Livonia, MI; Fulmer & Fulmer PA of Lakeland, FL; Griffin & Szipl, PC of Arlington, VA; Kreis, Enderle, Callander & Hudgins PC of

Kalamazoo, MI; Krupnick, Campbell, Malone, Buser, Slama, Hancock, Liberman & McKee of Fort Lauderdale, FL; Larkin, Axelrod, Ingrassia & Tetenbaum, LLP or Newburgh, NY; LeClair Ryan, PC of San Francisco, CA; Lynn, Jackson, Shultz & Lebrun, PC of Sioux Falls, ND; Morgan Lewis of Washington, DC; Notaro, Michalos & Zaccaria, PC of New York, NY; Nurenberg, Paris, Heller & McCarthy Co. LPA of Cleveland, OH; Ropers Majeski Kohn Bentley PC of San Jose, CA; Rymer Moore Jackson Echols, PC of Houston, TX; Sellars, Marion & Bachi, PA of West Palm Beach, FL; Slater & Zurz of Akron, OH; Tierney Law Offices of Philadelphia, PA; VanAntwerp, Monge, Jones & Edwards, LLP of Ashland, KY; Weltman, Weinberg & Reis Co., LPA of Cincinnati, OH; and Wilson Kehoa Winingham LLC of Indianapolis, IN.

Some of the companies which benefited from the plastics training provided include: Alabama Industrial Development Training of Montgomery, AL; Alcatel Lucent of Mesquite, TX; BASF of Wyandotte, MI; Bombardier of Valcourt, QC, Canada; BorgWarner of Auburn Hills, MI; Brose North America of Auburn Hills, MI; Cobasys of Springboro, OH; Daimler Trucks North America of Portland, OR; FCI Automotive of Westland, MI; Holley Automotive of Bowling Green, OH; Honda of Canada Manufacturing of Alliston, ON, Canada; Inergy Automotive Systems of Troy, MI; Invest in France Agency of Chicago, IL; Japanese Business Systems of Torrance, CA; Kostal North America of Troy, MI; Lutron Electronics of Coopersburg, PA; Maytag of Newton, IA; Minnesota Mining and Manufacturing Company of Saint Paul, MN; Pall Gelman Laboratories of Ann Arbor, MI; Purolator Filters North America of Fayetteville, NC; Sogefy Group of Rochester Hills, MI; Technimark of Asheboro, NC; The First Years of Avon, MA; TRW Restrain Systems of Washington, MI, Tyco Electronics of Menlo Park, CA; Unisys of Plymouth, MI; Valeo of Paris, France; and WL Gore and Associates of Newark, DE. The plastic training was conducted in many countries, among them: Canada, France, Germany, India, Korea, Malaysia, Mexico, and Thailand.

Paul Tres is a Fellow of the international Society of Plastics Engineers (SPE). He is also a member of the international Society of Automotive Engineers (SAE), the American Society of Mechanical Engineers (ASME), and the Plastics Academy.

Index

Symbol

50 % RH (relative humidity) 25

A

additives 366
adhesive 319
adhesive failure 319, 328
aesthetic criteria 262
air intake manifolds 197
air pockets 69
air pressure 199
alignment pin 143
aluminum oxide 100
American Society for Testing & Materials 203
amplitude 86, 90 – 92, 94, 135, 136, 142, 143
angioplasty 214
angle of deflection 83, 273, 274, 275
anisotropic 24, 81
– shrinkage 26
annealed 209
antifreeze 262
antistatic agents 13
apparent modulus 31
aramid fibers 7, 8
arteries 214
articulated joints 346
ASTM 293
asymptotic 70
axial joint 338

B

ball bearing 200
biopolymers 6
booster 85, 89, 92
BosScrew 367
boss design 72
boundary conditions 82
British Standards Association 203
built-in preload 262
burn 290
butt joint 139

C

cadmium red 262
CAMPUS 202
carbon black 160
carbon fibers 7, 8
catalytic converter 197
centroid 76
Charpy impact test 34
circumferential velocity 124
clamping pressure 134
class A 72
clay additive 8
click 309
closed loop solution 282
CLTE 36
cohesive failure 319, 328
coining fixture 250
coining head 250
coining tools 253
cold forming 252
cold working 250
Comet 1 71
computer flow diagram 236
conduction 5
constant of integration 274
constant time 96
Consumer Product Safety Commission 289
contact pressure 113, 123 – 125, 172, 173, 191, 193, 194
core casting 195
core retraction 341
core shifting 196
Corona 324
Coulomb theory 66
counter bore 359
cracking 353
creep modulus 31
critical dimension 340
crosslinks 2
crystallinity 3
curved waves 90

D

degree of freedom 309
De Havilland 71
delamination 329
denier 16
design algorithm 200
design point 271
dielectric 291
dies 162
DOF 309
dopants 5
draft angles 74
dried as molded (DOM) 25
DST 353
ductility 14, 18, 29, 51
durable hinge 249
dyne level 322

E

edge stress 295
EDM 291
elastic hinge 219, 226, 227, 235, 241, 243
electrical discharge machining 291
electroluminescence 5
electroluminescent polymer 5
electro-optic polymers 5
elongation at break 244
enclosure 153
encoder 96
energy director 102
enforced displacement 282
ergonomic studies 266
Euclidian geometry 228

F

far-field 96, 97
faster assembly 261
ferromagnetic 144
Fiat Chrysler Automobile Group 197
fiber thickness 16
fillers 366
fillet 20, 70
- radius 20
filling phase 249
fixture 91
flame treatment 325
flash 69
- trap 131
flex element 255
flow analysis 196
flow direction 26
freeze 68
frictional heat 134
friction coeficient 266
friction factor 364
fringes 293
frozen layer 249
fusible core injection molding 195

G

gain 94
gate 26, 366
- location 196, 292
geometric constant 277
glass spheres 8
glass transition temperatures 35, 335
glow 5
glycol 197
goniometer 321
graphic interpolation 20, 21, 183
grip 215
gripping fingers 130
guides 308

H

HDT 35
heating plate 128, 129
hermetic joints 136
HiLo® 352
hinge parameters 243
hold time 92, 97
hook 309
Hooke’s law 28, 60
horn 85, 89 - 91, 95, 97
- cavity 112
hygroscopic 24, 200, 209, 210

I

ICP 5
ignition cable bracket 245
imperfections, surface 292
Indeflator 215
induction coil 145
induction-heated oven 196
inertia tool 121
infrared laser radiation 158
infrared laser tool 155
inherently conductive polymers 5
inorganic flame retardants 13
instantaneous bending moment 272
instantaneous deformation 57
instantaneous stress 271
instantaneous width 270
interfacial tension 319
interference fit 169, 191, 300
intermittent cooling 136

internal damage 101
isopropyl alcohol 330
isotropic 24
Izod impact 33

J

Japanese Industrial Standards 203
joint width 331
junction 71

K

k-method 364

L

lactic acid 6
laser welding system 152
LED 6
LEP 5
Lewis parabola 41
life of the assembly 263
light emitting display 6
light emitting polymer 5
lithium 290
living hinge 47
lost core injection molding 195
lumbar 39

M

machining tolerance 200
magnetic field 145
magnitude of error 283
manifold 195
manual assemblies 308
maximum deformation 282
- force 282
maximum strain 243
mechanical interlocking 319
megapascals 27, 28, 50
meld line 366
melt flow 69
melt-out station 196
metallic powder 145
metal safe 302
microstructural analysis 41
microtome 41
Mohr's circle method 67
molding data record 142
molding process 250
molding repair 142
mold release 98
molecular orientation 250
moment of inertia 310
Monel 91
multicomponent 335
multimaterial 335
multipoint 203
multipolymer 335
muscovite 8

N

nanocomposites 8
nanofillers 8
necking behavior 229
nest 91
nesting 310
neutral axis 224
nodal principal stresses 283
nonlinear materials 79
nut factor 364

O

offset 248
oriented mirrors 162
O-ring 265
out-of-phase vibrations 113
overconstrained 310
overlap length 331
overmolding 196, 335

P

PA 6,6 211
paint 262
PAni 5
Parachute Industry Association 17
parametric 276
phlogopite 8
piezoelectric 86
pigments 162
pilot hole 352
pin-point gate 256
pivot 307
PLA 6
planet gears 41
plasma treatment 326
plastic deformation 310
plastic part design 52
platen rotation 344
playpen 16
PMMA 291
Poisson's ratio 59
polarizer 292
polar moment of inertia 54
polyaniline 5
polyfluorene 5
polylactic 6

polymerization reaction 2
polymer melt 249
polyphenylene vinylene 5
polypyrrole 5
polythiopene 5
portholes 153
post-mold shrinkage 26
PPV 5
precision snap fits 302
premiation 91
preprocessing 82
press fitting 47
pressure drop 250
pressure gradient 196
proving grounds 210
PT 355
pure bending 224

Q

quick flexes 220

R

radial force 356
radial interference 215
radial stress 357
Rankine theory 65
recyclability symbols 311
regrind 143, 264
reinforcements 366
retardation 293
RF sealing 148
rib height 73
rib width 74
robot arm 196
robust tool 303
rotation joints 338

S

safety factor 45, 185
scattering 158
seal 292
secant modulus 62, 271
self-alignment 125
self-degating tool 255
self-locking angle 279
self-tapping screws 115
serviceability 311
shank design 357
shape deformation energy 67
sharp corner 20, 71
shear joints 331
shelf life 194
shelf-time 194
single-point 203
snap fitting 47
solvent 319
Soniqtwist 106
spark plugs 239
sparks 291
spherical motion 338
spherical snap fit 261
stack molds 344
steady state 122, 135, 138
strain 59
stress concentration factor 20
stress cracking 330
stress-relaxation 33
St. Venant's theory 66
submarine gate 176
sun gear 41
surface analyst 323
surface energy 321
surface roughness 266
surface velocity 123, 124
symbols 311

T

test specimen 52
thermically conductive 5
thread angle 353
thread cutting 351
thread forming 351
three plate tool 255
throttle body 197
tolerance 36, 120, 169, 191, 217, 289, 341
torque limiter 216, 361
trailing angle 354
transducer 85, 91
transmitted torque 172
transparent 293
transverse wave 90
trapezoid 76
Tresca 66
trilobular 353
troubleshooting 326
true strain 57
true stress 57
turret 343

U

ultrasonic weld cycle 92
undercut 213, 276, 277
under-the-hood 239
Uni-screw 352
useful life 175, 311
UV stabilizers 13

V

vacuum 130
vacuum cups 130
van der Waals 1
visco-elastic 29, 62, 79, 227
voids 25, 47, 48, 69
volume deformation energy 67
von Mises' equation 68

W

wall thickness 68
warpage 68, 132
wavelength 94
weld line 171, 266
wetting 320
worm gear 40

Y

yarn 16
yield strain level 250
Young's modulus 8, 28, 52, 60, 64, 177

Z

zinc stearate 98

HANSER

A Perfect Fit

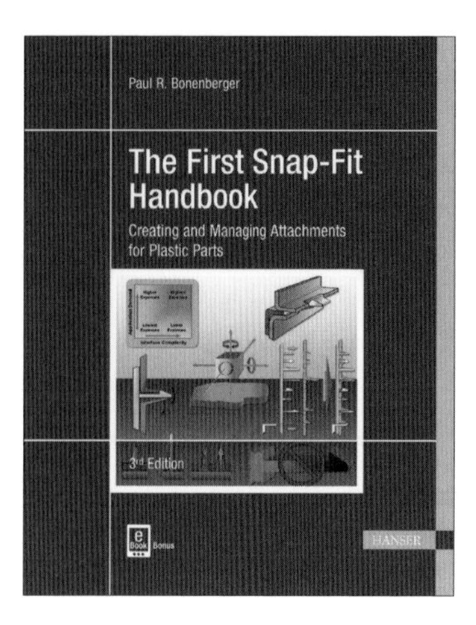

Bonenberger
The First Snap-Fit Handbook
Creating and Managing Attachments
for Plastics Parts
3rd Edition
412 pages
$ 199.99. ISBN 978-1-56990-595-1

The »system level« knowledge and design skills needed to create good snap-fit interfaces existed in the minds of self-taught snap-fit experts but was not captured in the literature.

New designers of plastic parts wishing to use snap-fit had nowhere to turn unless they were fortunate enough to have access to an experienced snap-fit designer. This book organizes and presents all design aspects of snap-fits with an emphasis on the systems level thinking required to create world-class attachments.

The third edition has been thoroughly revised to include new case histories and applications. The text has been extensively rewritten for clarity and user-friendliness and there are many new figures with expert explanations.

More information at **www.hanserpublications.com**